New Frontiers in Translation Studies

Series Editor

Defeng Li
Center for Studies of Translation, Interpreting and Cognition,
University of Macau, Macao SAR

More information about this series at http://www.springer.com/series/11894

(Hugs) Gengshen Hu

Eco-Translatology

Towards an Eco-Paradigm of Translation Studies

Springer

(Hugs) Gengshen Hu
School of International Studies
Zhengzhou University
Zhengzhou, China

ISSN 2197-8689 ISSN 2197-8697 (electronic)
New Frontiers in Translation Studies
ISBN 978-981-15-2259-8 ISBN 978-981-15-2260-4 (eBook)
https://doi.org/10.1007/978-981-15-2260-4

This Springer imprint is published by the registered company Springer Nature Singapore Pte Ltd.
The registered company address is: 152 Beach Road, #21-01/04 Gateway East, Singapore 189721, Singapore

Foreword by Dollerup

It is a great pleasure for me to write this Foreword to Professor Gengshen Hu's comprehensive introduction to Eco-translatology.

The new field of Eco-translatology came to my attention when, in 2002, Professor Gengshen Hu submitted a paper to the journal *Perspectives: Studies in Translatology* of which I was the editor-in-chief. I forwarded the article to our referees, who recommended it for publication. We all recognized that this was a novel and original approach to translation studies. The article was published and was followed by two more for *Perspectives*.

Later, I met Professor Hu during my stays as a visiting scholar at Tsinghua University (Beijing, China), and we became good and trusted friends.

The International Association for Eco-Translatology Research was founded in 2010, and I was invited to deliver a paper at the First International Symposium on Eco-Translatology in Macao in 2010. There were international visitors and approximately 60 established Chinese scholars engaged in presentations and lively discussions. There were some publications on display as a sign that the concept of Eco-translatology was catching on, not only in China but also elsewhere.

I followed developments in Eco-translatology as an adviser. I delivered a plenary presentation at the fifth symposium (2015) in Tainan, Taiwan. The number of participants in the symposium had swelled, and there were numerous young scholars and Ph.D. students. Eco-translatology was reaching out to new generations that would ensure its growth in the future. Numerous publications and theses were on display for participants to study. Eco-translatology had become a strong movement in the Chinese world, and the world outside China was increasingly paying heed to it.

It is noteworthy that Eco-translatology refers to real-life translation work, e.g., which translators (or "language workers") translate because they must earn money for food and housing. The activity is thus linked to translators' survival and careers. To survive, translators need to be able to make rational selections and adaptations in their translational eco-environments. Gengshen Hu concludes that in the same way that all the factors involved in the translation process affect translators, translators have to adapt to these factors to make decisions: "This description positions the translator in a definition of translation. The translation is a cyclic alteration of the translator's selective adaptation and adaptive selection."

The keywords in Eco-translatology are adaptation and selection in the translational eco-environment, with constant interplay between contextual factors, clients, and translators. All factors in this interplay are, in principle, of equal weight, although some aspects are more obvious than others when studied from specific angles.

Eco-translatology is explicitly the ecosystem, the text and text type, and the client select translators for specific assignments. It also stresses that translators must adapt to the environment or be eliminated according to the Darwinian principle of the survival of the fittest. This is a supplement that Professor Gengshen Hu has added to the theory, and it holds good in many contexts. The texts that are easiest to translate are easily explored and easy for the translator to adapt to the target language ecosystem, which is the ultimate goal of all translation.

It is appreciable that Eco-translatology is very complex.

I suggest that Eco-translatology has great potential. It inspires new approaches to translation studies: It can be applied to fields where it is not obvious that it is relevant. I have tried to illustrate this in a small collection, *Papers on Eco-Translatology*, in order to show its strengths, weaknesses, and possibility of providing new insights.

Eco-translatology is based on ancient Chinese notions about harmony between humans and their environment (the doctrine of *Unity of Man and Nature*). I wish to emphasize this interdependence and harmony and to underline that ancient lineage is central: the school is not prompted by any sudden changes in society but is inspired by nature, by our environments, which are greater and last longer than the individual human life. It is also permeated by the Chinese view of the collective vs. the individual (in the West). No single person can comprehend or embrace Eco-translatology in its entirety, but teams or communities of translators can. This book deals with Eco-translatology. I hope it can inspire further studies. It goes without saying that these cannot include the whole theoretic system but can cover selected parts or aspects of it.

Copenhagen, Denmark Cay Dollerup
8 June 2019

Foreword by Fang

An Emerging Paradigm of Eco-Translation Studies[1]

Professor Hugs, Gengshen HU has had a career experience similar to mine: both of us worked as translators in different research institutes in the early years. In performing sci-tech translation for many years, we accumulated wide practical translation experience, absorbed new scientific knowledge, and learned various research methods. All this helped us to generate a variety of innovative thoughts and promises to provide fruitful benefits for the rest of our lives.

Shanghai Journal of Translators for Science and Technology (now named *Shanghai Journal of Translators*), where I worked as a founder, was first published in 1986. In Serial No. 3 of that year, Professor HU published an article entitled "*Abstract Interpreting: An Approach to Scientific Interpretation*." In the following several years, he published a series of articles in the journal, taking "supplementary translation," "omission," and "variant translation" as his main topics. During the same period, he also published several consecutive articles on interpretation in *Chinese Translators Journal* and particularly in *Babel*, an international journal of translation studies. In the second half of the 1980s, there were some signs of improvement in domestic translation studies. Though many commentary and sentimental articles and superficial introductory papers about the translation experience appeared, thematic and systematic studies, especially on interpreting, similar to Professor HU's work, were still rather scarce. To my knowledge, throughout the 1980s, combined with his work experience, Professor HU made a series of explorations on sci-tech interpretation, academic interpretation, and foreign affairs

---

[1]This Foreword is written by Fang Mengzhi, a Senior Professor at Shanghai University; a Distinguished Translator in the Chinese translation field; and the founding editor of *Shanghai Journal of Translators*, a core journal of foreign languages and translation in China (1986-). The former President of the Shanghai Sci-Tech Translation Society, Professor FANG now serves as an Editorial Consultant to the *International Journal of Eco-Translatology*.

interpretation, which laid a solid foundation for his subsequent studies on intercultural communication pragmatics.

In the second half of the twentieth century, the scope of translation studies was expanded, the research objects were diversified, the theoretical background was broadened, and the foreign research approaches were frequently redirected, from the linguistic approach to the multidisciplinary approach, the cultural approach, and then the post-linguistic approach, leading to changes in the translation research foci in China. As one approach leaves the stage, another takes its place. After Western scholars had played a dominant role in the construction of translation theory for more than half a century, they began to stagnate at the turn of the century. At the moment when people were talking about the "next turn" of translation studies, Eco-translatology entered the international translation world as a new approach. In October 2001, Professor HU gave a lecture entitled *From the Darwinian Principle of Adaptation and Selection to Translation Studies* at Hong Kong Baptist University. In December the same year, he presented a paper titled "An Initial Exploration into an Approach to Translation as Adaptation and Selection," and Eco-translatology, which had been independently initiated by a Chinese scholar, became known worldwide. In this sense, Eco-translatology kept pace with the new century.

With the expansion of the macroperspective of translation as adaptation and selection, Professor HU presented a research paper entitled "Understanding Eco-Translatology" at the International Conference on Translating Global Culture: Towards an Interdisciplinary (Re)Construction in August 2006. It became increasingly clear that Eco-translatology was a newly emerging eco-paradigm of translation studies from an ecological perspective, with translation as adaptation and selection and the similarity and isomorphism between translation ecology and natural ecology as its theoretical foundation, taking a holistic view of ecology as its methodology, ecological wisdom in Chinese traditional culture as its fulcrum, and the interrelationship between translators and translational eco-environments as its research object. Thus, Eco-translatology is also the excavation, expansion, and promotion of the essence of translation as adaptation and selection.

Translatology is a comprehensive and interdisciplinary study. Systematology regards the system as an organic whole with specific functions owing to the integration of several interacting and interdependent components. From the perspective of systematology, not all disciplines associated with translatology are interdisciplinary. A comprehensive discipline is formed by taking an abstract object found in several fields as the research object, and the abstract object is also the connection point of the relevant interdisciplinary subjects. Translation is a linguistic, cultural, and thinking activity that conveys information between symbols. The connection point between linguistics and translatology is the semantic and structural transformation of linguistic signs; the connection point between culturology and translatology lies in the restriction and transmission of heterogeneous culture in the process of translation; and the connection point between cognitive science and translatology consists of the formation and the law of translation thinking. Then, we must consider the following question: Where is the connection point between ecology and translatology?

Professor HU points out that by taking ecological holism as its tenet, Oriental ecological wisdom as its basis, and "adaptation/selection" theory as its stepping stone, the eco-paradigm of translation studies systematically probes translational eco-environments, translation textual ecologies, and "translator community" ecologies as well as their interrelationships and interplay and is committed to describing and interpreting a holistic translation ecology and translation theory from the perspective of ecology. Therefore, the connection point of ecology and translatology lies in the ecological environment and translation context. The translation context (referred to as the "translation-environment ecology" by Professor HU in this book) is the sum of the subjective and objective elements related to two kinds of languages and cultures constructed by the translator in the process of translation, with four key elements: translator, original context, target language context, and interaction. The four major elements of Eco-translatology (translation-environment ecology, translation textual ecology, "translator community" ecology, and their interrelationships and interplay) are in close agreement with the four elements of the translation context. Eco-translatology is significant because it takes a holistic view of translation from the perspective of eco-reason, which compensates for the lack of reason in previous translation studies.

Eco-translatology is a self-contained translation theory system. Under the guidance of eco-holism, it emphasizes its metaphorical reference to the basic principle of "survival of the fittest," which is generally accepted by human beings. It not only shows the characteristics of a discipline at the abovementioned connection point but also constructs a holistic translation ecosystem, reveals the eco-reason of translation, and proposes Eco-translation ethics based on the classical precepts of ancient Chinese philosophy, such as "man and nature as one," "follow nature's course," "people-oriented," and "moderately harmonious." Constructed on the abovementioned basis, Eco-translatology answers the fundamental questions in translation studies: "What is translation?" "Who translates?" "How to translate?" and "Why to translate?" What is translation (what) refers to translation as an activity of the translator's selection and adaptation to the translational eco-environment and textual life transplantation; who translates (who) refers to "translator-centeredness" and "translator's responsibility" in the ethics of translation; how to translate (how) refers to the translator's "survival of the fittest," "selective adaptation," and "adaptive selection"; and why to translate(why) refers to the core tenets of translation studies, such as "doing things with translations." Meanwhile, Eco-translatology provides a new overview and exposition of translation principles, translation processes, translation criteria, translation strategies, translation methods, and other translation phenomena as well as the holistic translation ecosystem from the perspective of eco-reason.

Eco-translatology is a self-contained and independent theoretical system. It has a set of terms to describe the basic concepts of its research objects, purposes, contents, methods, laws, and theorems. Generally, the term is one of the important signs that differentiates a particular theoretic system from others. The scientification and systematization of terminology not only represent the level of the development of a certain theoretic system but also have become an important condition for ensuring the steady advancement of translatology. Eco-translatology has a distinctive terminology, such as "the Sequence Chain," eco-paradigm, eco-reason, translational eco-environment, translational ecosystem, Eco-translation ethics, textual ecology, textual transplantation, balance and harmony, multiple symbiosis, multidimensional integration, translator's responsibility, selective adaptation and adaptive selection, "three-dimensional" transformations, doing things with translations, post-event penalty, degree of holistic adaptation and selection, approach to translation as adaptation and selection, and translation as eco-balance. Some of the above terms have been included in *A Dictionary of Translation Studies in China*, edited by me, merging into the mainstream of the translation system.

This book positions Eco-translatology as an "eco-paradigm" of translation studies and points out (Chap. 7) that since ecology is a "meta-science," and the ecological approach is a transdisciplinary and holistic approach, ecology, as a methodology that dominates the universal social thinking mode, is a discipline that not only has an extensive impact on society but also is of ultimate significance to human existence and development. Therefore, translation studies from an ecological perspective are studies based on eco-reason or, in other words, studies using ecological philosophy (such as the principle of overall balance, the principle of symbiosis and coexistence, or the law of multiple unity), which are philosophical and methodological. As a general theory that holistically views translation from an ecological perspective, Eco-translatology is an integrated study involving "translation studies," "translation theory," and "translated text," which makes its universality increasingly obvious.

The efficiency of a systematic theory depends on the coverage and universality of its research field. Eco-translatology boasts a solid philosophical foundation while complying with logical rules and natural mechanisms. As soon as this theory emerged, a constant stream of research activities and a crowd of researches responded to it. Statistics show that over the past decade, there have been more than 300 master's and doctoral theses based on the fundamental theory of Eco-translatology, dozens of academic journal papers using Eco-translatology in their titles or keywords domestically and abroad, and more than a dozen related papers by foreign scholars presented and published in the series of International Symposiums on Eco-translatology. Among these, from 2004 to 2011 alone, as many as fifteen papers were edited and published by me in *Shanghai Journal of Translators*.

Some of the fifteen papers expound the core tenets, such as *Theoretical Props of the Approach to Translation as Adaptation and Selection* (Hu Gengshen, 2004); *Eco-Translatology: Interdisciplinary Integration of Translation Studies* (Hu Gengshen, 2009); *Mechanism of the Whole Ecological Translation Field Schema Theory* (Zhu Guicheng, 2010); *Translational Ecology vs Natural Ecology: Relevance, Similarity and Isomorphism* (Hu Gengshen, 2010); and *On the Ecology and Environment of Translation* (Fang Mengzhi, 2011).

Some papers apply the fundamental tenets and methods of Eco-translatology to interpret various translation phenomena, including the translation thoughts of translators, case studies of literary translations, and translation studies of various pragmatic translation texts, such as *Zhang Guruo and His "Adaptation" and "Selection"* (Sun Yingchun, 2009); *Adaptation and Selection: Tracing the Translation Principles of Yan Fu* (Huang Zhonglian, 2009); *Yan Fu's Translation of Evolution and Ethics Revisited: A Translatological Perspective* (Jiao Weihong, 2009); *Translator's Selective Adaptation and Adaptive Selection: A Review of Three English Versions of The Peony Pavilion* (Jiang Xiaohua, 2009); *On the Translation of Metaphorical Idioms in News Reports from the Perspective of Translation as Adaptation and Selection* (Liu Yanfang, 2009); and *Translation of Public Signs from the Perspective of Eco-Translatology—A Survey on the Translation of the Theme of the Shanghai World Expo* (Song Huijuan, 2010).

Eco-translatology has been developing for over 10 years. Certainly, there are ideas that cannot be easily comprehended and those who hold different views. Such issues are inevitable during the development of a newly born theory and should be discussed and evaluated. The editorial board of *Shanghai Journal of Translators* organized a discussion and published articles on one of the core questions—"translator-centeredness" in the translation process—such as *Is the Translator the Center in Eco-Translation Theory? A Discussion with Professor HU Geng-shen* (Chen Yuhong, 2011); *A Study and Argument of the Core Concepts of Eco-Translatology* (Wang Hong, 2011); and *A Response to the Question of "Translator Centeredness*" (Hu Gengshen, 2011).

In addition, in a literature review of 10 years of research, we can see the emergence, development, and perfection processes of Eco-translatology over the years: *Ten Years of Research, Ten Major Progresses—A Report on the Research and Development of Eco-Translatology* (Strong Hugs, 2011).

In recent years, one after another research paper on Eco-translatology, taking only those received by the editorial board of *Shanghai Journal of Translators* into account, has appeared, and the number has far exceeded those mentioned above. Most of these papers are applied interpretations of Eco-translatology, especially the analysis of the characteristics and gains and losses of the translated texts in the linguistic dimension, cultural dimension, and communicative dimension. However, they have seldom argued and explored translation theories or further expounded the ontology, value, epistemology, and methodology of Eco-translatology. Generally, there is still room for further advancement in three respects: extension of the research scope, improvement of the research methods, and diversification of research topics, on which topics Professor HU has commented in this book.

At the beginning of the new century, Eco-translatology emerged in response to the requirements of the times, and it has been constantly evolving. Undoubtedly, the publication of *Eco-Translatology: Towards a New Paradigm of Translation Studies*, more than a decade after the emergence of the theory, serves as a milestone, indicating that the development of Eco-translatology has entered a new stage; at the same time, it can be regarded as the declaration of a new paradigm of Eco-translatology. In my opinion, this book provides a panoramic description and interpretation of the theoretical discourse system of Eco-translatology, further improves the construction of the theoretical category, expands the methodology, and develops the translation studies eco-paradigm. It will not only facilitate the understanding of the research and development of Eco-translatology but also help unscramble and clarify theoretical viewpoints, answer questions, and resolve doubts. In addition, it will draw more attention in the academic world and draw more participants to the research and exchange of Eco-translatology. As a new benchmark for Eco-translatology, this book is a pearl, similar to the publication of *An Approach to Translation as Adaptation and Selection* 15 years ago, in Eco-translatology and research in Chinese translation theories. Doubtless, this book in English will make new contributions to the construction of translation studies as a discipline, not only in China but also worldwide.

Shanghai, China
10 June 2019

Mengzhi Fang

Preface

The last half of the twentieth century has witnessed the transition from industrial civilization to eco-civilization in human society, and globalizing eco-thoughts are essentially shaping the developmental route of world academic research. Ecological notions have been incorporated into the fields of humanities, including linguistics, comparative literature, liberal arts and cultural studies, and media and communication studies, and accordingly, "interdisciplinary" or "multidisciplinary" studies have vigorously expanded. On the one hand, however, systematic investigations in the "ecological dimension" have long passed beyond the academic concern of translation studies, though it is a discipline of striking "multi-disciplinarity." On the other hand, large-scale shortcomings and insufficiency exist along with a generalization of domestic and/or foreign attempts to approach translational issues—owing not only to the merely superficial use of ecological terms but also to efforts labeled "scattered," "parochial," "straggling," or "unsystematic" that have debased these discussions. Against such a social and academic background, the author, almost 10 years ago, embarked on a systematized route toward Eco-translatology with a view to identifying the missing link of a systematic "ecological dimension" in translation studies and unmasking the paucity of previous studies in the discipline.

If "An Initial Exploration into an Approach to Translation as Adaptation and Selection" at *The 3rd FIT Asian Translators' Forum* in 2001 was the first step toward an ecological approach to translation studies, then "Understanding Eco-Translatology," presented at the international conference titled *Translating Global Cultures: Towards Interdisciplinary (Re)Constructions* in 2006, signaled the evolution and integration of academic attempts to explore the ecological approach to translation studies. In addition, if support by the Chinese National Social Science Foundation for the project "Eco-Translatology: Ecological Perspective of Translation Studies" in 2008 and the publication of "Eco-Translatology: A Primer" in *The Chinese Translators Journal* in 2008 somewhat represented the acceptance and recognition of Eco-translatology among the Southeast Asian academic sphere of translation studies, then the establishment of the "International Association for Eco-Translatology Research" in 2010, the success of "International Symposiums on Eco-Translatology," and above all the launch of the *Journal of Eco-Translatology* show that "Eco-translatology," a theory initiated in China, has traveled far beyond the national boundaries and entered world academia.

The author thereby writes and publishes this *Eco-Translatology: Towards a New Paradigm of Translation Studies* in English as a gift for the second decade of the twenty-first century. This book, a retrospective examination of Eco-translatological studies throughout the first decade of the current century, meanwhile assumes the mission of stimulating and promoting Eco-translatology in the years to follow.

In the process of preparing this manuscript, the author adheres to the following principles: (1) the principle of the *profession*, i.e., maintaining the professional principles of translation scholarship; (2) the principle of *localization*, i.e., being rooted in Eastern culture; (3) the principle of *internationalization*, i.e., reconciling the approach with Western methods; and (4) the principle of ecology, i.e., seeking holistic eco-balance in general.

The birth and growth of Eco-translatology have benefited from various opportunities, conditions, and contributions. Eco-translatology germinated in Beijing, the Chinese capital, where Tsinghua University is located; it emerged in Hong Kong, China, where East and West converge; and it expanded in Macao, where Chinese and foreigners rub shoulders. The elite school, fresh opportunities, favorable conditions, and collaborative efforts from like-minded people worldwide all contributed to the present status of Eco-translatology. Over the past decade, the author has toiled and sought to construct an Eco-translatological discourse system. Over this decade, the author has found consolation in the fact that Eco-translatology, by explicating its position at the crux of studies on textual ecology, translation ecology, and "translator community" ecology, has given a new shape to the field of translation studies, translation theories, and translation performances. What he values most is that more than six hundred theses and doctoral dissertations and research articles on Eco-translatology have been accomplished or published; the most representative developments include the timely establishment of the International Association for Eco-translatology Research, the consecutive holding of International Symposiums of Eco-translatology, and the regular publication of *Journal of Eco-Translatology*.

On the eve of publishing *Eco-Translatology: Towards an Eco-Paradigm of Translation Studies*, the author wishes to express his sincere gratitude to the scholars Prof. ZHOU Zhaoxiang, Prof. TAN Zaixi, and Prof. ZHANG Peiyao for their help during his doctoral study; Prof. GU Zhengkun, Prof. MAO Sihui, and Prof. WANG Ning for their suggestions on his research project; Prof. YANG Zijian, Prof. FANG Mengzhi, and Prof. XU Jun for their assistance in academic growth; and Prof. Cay Dollerup (Denmark), Prof. Edwin Gentzler (USA), and Prof. Kirsten Malmkjear (UK) for their friendship and encouragement across international boundaries. In regard to the spread and promotion of Eco-translatology, the author wishes to express his thanks to Prof. JIANG Xiaohua, Prof. SONG Zhiping, Prof. MENG Fanjun, Prof. LIU Yafeng, Prof. LIU Aihua, Prof. BIAN Lihong, Prof. LI Li, Prof. HU Gongze (Taiwan), Prof. TAO Youlan, Prof. LUO Dijiang,

Prof. Marion Boers (South Africa), Prof. Douglas Robinson (USA/Hong Kong, China), Prof. Mary Snell-Hornby (Austria), Prof. Michael Cronin (Ireland), Prof. Christiane Nord (Germany), Dr. Rindon Kundu (India), Prof. Joanna Radwanska-Williams (USA/Poland), Prof. Kristina Taivalkoshi (Finland), Prof. Radegundis Stolze (Germany), Prof. Zoya G. Proshina (Russia), Prof. Yvon Gousry (France), Prof. Roberto A. Valdeón (Spain), Dr. Douglas McNaught (UK), Prof. Juliane House (Germany), Dr. Sedat Mulamyim and Prof. Anthony Pym (Australia), Dr. Anna Ponomareva (UK), Dr. Shelly Ochs (USA), Dr. Yala Ismail (Egypt), and Prof. Youngmin Kim (Korea)as well as those interested in and attracted by Eco-translatology and in particular those challenging and criticizing Eco-translatology. On the accomplishment of the draft, the author has been struck by the words that are always beyond one's words and that the higher one's aim, the harder one labors. The author has also come to realize that what one knows is proportional to what is unknown.

Eco-translatology is an emerging Eco-translation paradigm of translation studies. With metaphorical analogies between the translational ecosystem and the natural ecosystem and conceptual borrowings as its methodology, Eco-translatology probes the translational eco-environments, textual ecologies, and "translator community" ecologies as well as their interrelationships and interplay. Regarding translation as a holistic ecosystem, it describes and interprets translation activities in terms of the ecological principles of eco-holism, Oriental eco-wisdom, and *translation as adaptation and selection*. Within the Eco-translation paradigm, "translation as eco-balance," "translation as textual transplantation," and "translation as adaptation and selection" are the core concepts.

This book offers a panoramic view of Eco-translatology. Included in this book are the inception and initial development of Eco-translatology, the research foci and theoretical tenets, the main studies at the macro-, meso-, and micro-levels, empirical studies and theoretical applications, the implications for translation studies, and a roadmap for further development. The overall formation, interpretation, and illustration of the interrelationships and interplay between translational eco-environments, textual transplantation, and the translator's adaptation and selection of the "translator community" will eventually lead to the construction of an entire theoretical discourse system of Eco-translatology.

This book mainly targets translation studies theoreticians and practitioners across the globe, the faculty and undergraduate and postgraduate students of translation studies, doctoral scholars, and those who are interested in the domain of translation studies. It will also be a valuable reference for scholars and graduates from the adjacent fields of linguistics, cultural studies, cross-cultural communication studies, and related disciplines.

The author dedicates this book:

- To the adulthood (2001–2019) of research and development in Eco-translatology;

- To the 10th anniversary of the founding of the International Association for Eco-Translatology; and
- To the establishment of the Zhengzhou University Academy of Eco-Translatology, China, in May 2019.

Zhengzhou, China
Summer 2019

(Hugs) Gengshen Hu

Synopsis

Eco-translatology is an emerging paradigm of translation studies from an ecological perspective. With metaphorical analogies between the translational ecosystem and the natural ecosystem along with the conceptual borrowings as its methodology, Eco-translatology probes translational eco-environments, textual ecologies, and "translator community" ecologies as well as their interrelationships and interplay. Regarding the scene of translation as a holistic ecosystem, it describes and interprets translation activities in terms of the ecological principles of eco-holism, Oriental eco-wisdom, and translation as adaptation and selection. Within this eco-translation paradigm, "translation as eco-balance," "translation as textual transplantation," and "translation as adaptation and selection" are the core concepts.

This book offers a panoramic view of Eco-translatology that explores and expounds ecological ideas and ecological connotations in texts to be translated and in translation activities to make eco-reason, eco-principles, the ecologicalized world outlook, and methodology the general guidelines for translation studies. Included in this book are the inception and initial development of Eco-translatology, the research foci and theoretical tenets, the main studies at the macro-, meso-, and micro-levels, empirical studies and theoretical applications, the implications for translation studies, and a roadmap for further development. The overall formation, interpretation, and illustration of the interrelationships and interplay between translational eco-environments, textual transplantation, and the translator's adaptation and selection of the "translator community" will eventually lead to the construction of an entire theoretical discourse system of Eco-translatology.

This book primarily targets translation studies theoreticians and practitioners, academicians, and undergraduate and postgraduate students as well as laypersons who are interested in the discipline of translation studies. It will also be a valuable reference for scholars and graduates from the adjacent fields of pure and applied linguistics, comparative literature, cultural studies, cross-cultural communication, media studies, and other related disciplines.

Contents

Chapter 1
Preliminaries: The Name and Nature of Eco-Translatology

1.1 Introduction

Although explorations of the ecological perspective of translation studies germinated in 2001 (see details in Sect. 2.4.1), a systematic argumentation and international exposition of "Eco-translatology" did not appear until August 2006, when a conference article titled *Understanding Eco-translatology* was presented at the international conference *Translating Global Cultures: Towards Interdisciplinary (Re)Constructions*. It was published later under the title "Eco-translatology: A Primer" in *Chinese Translator's Journal* (see Hu, 2008, pp. 11–15).

Eco-translatology is an entirely novel and "formerly nonexistent"[1] perspective of translation studies proposed by a Chinese scholar; it is a systematic approach with newly coined terms and concepts to "strike out a new line"[2] that is utterly different from previous perspectives. Curious, questioning and suspicious responses to this perspective are therefore natural, and questions such as "What is Eco-translatology?" and "What are the definition and essence of Eco-translatology?" emerge naturally within and outside the academic translation field.

One dominant purpose of this book, *Eco-translatology: Towards a New Paradigm for Translation Studies* (hereafter abbreviated as *Eco-paradigm*), apart from taking a panoramic view of Eco-translatology, is to provide answers to these questions. The origin and development of Eco-translatology will be elaborated, the overall theoretical framework and scopes of different levels of Eco-translatology will be discussed and the developmental horizon and "road map" of Eco-translatology will

[1] See the relevant statement of Qian (2000, pp. 10–14) in his article "Dedicating Something Original to the World: Creativity in Foreign Language Research" in *Foreign Languages and Their Teaching*.

[2] See the relevant statement of Yang (2009, pp. xii–xiii) in his article "On Creating a Benign Eco-environment for Literary Translation" in Eastern Translation.

(Hugs) G. S. Hu, *Eco-Translatology*, New Frontiers in Translation Studies,
https://doi.org/10.1007/978-981-15-2260-4_1

be explicated in addition to, naturally, its applications and expansion, implications and significance for the disciplinary development of translation studies as a whole.

Eco-translatology covers, literally, at least "translatology" and "ecology"; hence, the ensuing section will visit both fields before elaborating on the definition and essence of Eco-translatology.

1.2 Translatology

It is natural for translatology to take priority in *Construction and Interpretation* since this monograph investigates the translatological framework, theoretical systems of translatology, and translational text operations. This prioritization is reinforced by the fact that Eco-translatology is by nature a synthetic research approach to translatology from an ecological perspective; i.e., it is rooted in and resides in translatology.

A remarkably long list of investigations and discussions on translatology, particularly its disciplinary framework and theoretical system, can be identified, with contributors inside and outside China. As early as the 1950s, Western scholars, illuminated by linguistics and owing to the newest findings in that field, explored the theoretical and practical aspects of translation from the linguistic perspective. Translation studies thus gained scientific and systemic associations, paving a scientific path for the further development of translation studies proper. In 1972, James Holmes, a Dutch-American, presented his paper "The Name and Nature of Translation Studies" at a conference on applied linguistics to announce his speculations regarding establishing translation studies as an independent discipline and sketched an overall structure of translation studies to describe the scope of the field; this structure was later represented by Gideon Toury in a graphical form. In line with "Holmes's map of translation studies," two branches of translation studies are "pure" and "applied," with the former being subdivided into theoretical translation studies (ThTS) and descriptive translation studies (DTS) and theoretical translation studies being further divided into general translation theory and partial translation theory (Munday, 2000, pp. 10–11). Through the joint efforts of scholars, "the growth of Translation Studies as a separate discipline is a success story of the 1980s" (Bassnett & Lefevere, 1990, p. ix).

More than a few scholars in China have successively investigated translation studies issues and expounded them. A multiplicity of speculations and descriptions can be found when we examine publications in recent decades. For instance, Huang (1988) deemed "translation studies a borderline science studying instructive translation theories together with practical translation laws and their applications with a complete system and structure"; Tan (1988) subdivided translation into general, specific, and applicative branches; Liu (1989) described the exterior and interior systems in the disciplinary framework of translation studies; and Yang (1989) considered "translation studies a science investigating the thinking laws and methods of translation" and presented the research objectives, nature, and three-level disciplinary structure and constructive strategic task of translation science. Wang (2001, p. 94) discussed translation studies as a science of studying the laws and arts of translation, with the principle objective of displaying the history, theories, and schools of thought domestically and abroad; explored the nature, functions, criteria, principles, and methods of translation; described the real-life translation process and stated the specialties and requirements of different types of translation; probed the scientific and artistic nature of language transformation; determined the qualities and competence required on the part of the translator and ways to cultivate and improve his or her competence, and predicted the trends of translation courses.

As mentioned above, Yang Zijian has frequently revisited the definition of translation studies: it is an independent, open, and synthetic branch of humanities that takes as its research objects the overall process of translation (including objects, subjects, processes, products, and effects), translation history, translation theory history, and methodology with the underlying notion of promoting and improving communication and reducing conflicts and harm. The key concerns are not only the general theoretical study but also the theoretical applicability (including translation practice, translation criticism, and translator training). The discipline of translation studies has two systems—a narrow one and a broad one: the former, the regimen system of the discipline, covers the underlying notions, norms, and administrative and legal systems of the discipline, and the latter, the norm system of the discipline, falls into five subsystems according to distance from practice and degree of abstraction: metatheory, translation philosophy, translation theories, translation strategies, and translation methods and tactics (Yang, 2007; Pan & Yang, 2008, pp. viii–ix).

Generally, ongoing research by scholars worldwide on the connotations, research objects, research methods, etc., has led to the continuous development of translation theories since the late 1970s. In the wake of accelerating stress over the "interdisciplinary nature" of translation studies (see Sect. 3.5.1) in particular, successive studies of translation have emerged from different disciplines, approaches, and schools, such as linguistics, comparative literature, cultural studies, communication studies, and philosophy, enabling theoretical studies of translation to prosper.

Nevertheless, the latter half of the 1990s, especially around the turn of the century, saw a relative slackening of the translation studies movement or a depression in theoretical studies, and translation exploration seemingly lapsed into a kind of late postmodern quiescence[3] Translation scholars could not avoid such baffling questions as what was the next "turn" of translation studies after its "cultural turn" ran aground? What would be the next "invigoration point" or "emergent point" of theoretical studies of translation? The prosperity and progress of translation studies brought excitement, whereas these puzzling issues of theoretical studies of translation urged reflection and exploration.

[3] As stated by some scholars, in this era of globalization, the "Western centrism" mode of thought has been challenged, and culture itself has encountered unavoidable crisis (Wang, 2004, p. 24).

1.3 Ecology

An acquaintance with ecology and ecosystems is a prerequisite because Eco-translatology is a synthesis of translation studies and the ecological perspective, and the environmental rationality, disciplinary features, research findings, and research methodology of ecology definitely shed light on Eco-translatology studies.

As we know, the term "ecology" was used for the first time in 1866 by Ernst Haeckel, a German zoologist, to denote the science of "studying the interrelations between organisms and their environments" (Wang & Zhou, 2004, p. 3). In other words, "ecology" was considered the study of the interrelation between organisms and their environments,[4] and in the early years, ecology was defined as "a sub-branch of biology to study the interrelation between organisms and between organisms and their environments."[5] Along with the deepening of recognition of the logical relation between organic bodies and their surrounding environment, modern ecology covers investigations of the logical relation between humans and nature. The rising mass environmental protection movement of the late 1960s and 1970s has pushed ecology further away from biology proper and towards the study of the intrinsic relation between humans and nature. In the territory of the humanities and social sciences in the past few years, the word "ecosystem" has been semantically expanded to aggregate natural health, maintaining balance, and "harmonious coexistence."

Ecology has clearly moved out of the ivory tower that formerly made it exclusive to biologists and ecologists and has currently become both a heated issue (see also Sect. 2.2.1) and a kind of behavioral norm. Both material and spiritual elements of the universe are endowed with ecological connotations, and all living things have been tinted green. Scholars of the former Soviet Union, such as B. A. Rosy, Jim Shen, and H. B. Norwich, stated that ecology is "a kind of world value, methodology, and scientific thinking mode." They declared that "one of the leading trends of modern natural science is its ecologization," that "the future of science relies on synthesis of ecology," that "ecology is the science of the 21st century," and that the "existence of human beings on the earth is dependent on ecological progress" ([Russia] Jim Shen, 1993, as cited in Wang & Zhou, 2004, p. 4).

Academically speaking, ecology is a philosophy, a science, an aesthetics, and a technology; a systemic science of studying the relationship between organic bodies, including human bodies, and the environment; an ancient yet green interdiscipline of natural sciences and the humanities; a science of the nature–human relation from physical, social, and rational perspectives; a worldwide value and methodology; and a mode of scientific thinking.

To put it simply, ecology is a science of environmental relations linking humans and their peers, objects, and surroundings; a livelihood and stratagem science of

[4] C.f., *An Encyclopedia of Chinese Language*, 2007, p. 1499; *Modern Chinese Dictionary*, 2008, p. 1220.

[5] Neufeldt (1995, p. 429).

human conduct closely related to the existence and development of ordinary people; and a way for humans to survive, live, and prosper. "With self-generation, self-reflection and self-reliance, ecology will pilot us all through the challenges and finally to success" (Wang & Zhou, 2004, p. 4).

Methodologically, ecology, a science based on holism, values in its research methods the entirety of interrelation and interaction (Krohne, 2001, p. 11).

In terms of ecological features, ecology is balanced, following the principle of holistic coordination, circulative self-generation, and the indestructibility of matter and conservation of energy; ecology is progressive, advocating competitive coexistence, efficient harmony, concurrent evolution, and survival of the fittest; and ecology is integral, pursuing the synthesis of wealth, health, and civilization and seeking physiological, psychological, and ethical well-being.

Ecology is a survival and developmental mechanism of competition, coexistence, regeneration, and self-generation; a systematizing and synthesizing function of the search for temporal, spatial, quantitative, structural, and orderly sustenance and harmony; and a process of pursuing constant human evolution and perfection with the ultimate aim of sustainable development.

Recently, academic research employing ecological principles and ecological theories has become a trend of academic development, springing up in various disciplines like mushrooms after rain. The application of ecological theories to humanities research, in particular, has furnished a marvelous repertoire of solutions to many real-life problems and therefore has received great attention from scholars of social sciences and humanities. Ecological theories have not only expanded the scope of academic research and shaped interesting research ideas but also triggered the emergence of a series of brand-new interdisciplines (see also Sect. 2.2.3). For instance, scientific research in the humanities on diversified ecological issues is in full swing and has greatly promoted the vigorous development and interpenetration of the humanities and ecology to bring to social ecology a large variety of new disciplines and theories of an interdisciplinary and overlapping nature, including ecological methodology, ecological philosophy, ecological thought, ecological logics, ecological psychology, ecological pedagogy, ecological politics, ecological economics, ecological ethics, ecological science of law, ecological aesthetics, ecological sociology, ecological history, and ecological culturology.

Evidently, the conceptual expansion, disciplinary features, research contents, and research methods of ecology and the "ecological properties" thereof mentioned above, especially in terms of the significance of ecology for human behavior, world values, and methodology as well as its impact on investigations into the social and human sciences, are highly inspirational and instructive for translation practitioners, theoreticians, and scholars of translation studies, "translation being a kind of behavior by nature, or in other words, a kind of 'cross-cultural behavior'" (Hans Vermeer, as quoted from Liao, 2001, p. 364). Similarly, "translation is increasingly seen as a process, a form of human behavior. A theory, therefore, should seek to establish the laws of this behavior" (Chesterman, 1993, p. 2). That is why the ecological perspective of translation studies has begun and developed (see Sect. 2.4.1), and eco-reason has become the theoretical guidance for the macro-concept studies (see Chap. 4), the meso-theories (see Chap. 5), and the micro-performances (see Chap. 6) of Eco-translatological studies.

1.4 Eco-Translatology: An Eco-Paradigm of Translation Studies

Translatology, one of the two aforementioned disciplines, takes as its "task" the transmission of linguistic messages to promote the cross-cultural communication of human beings and falls within the scope of humanities, while ecology, aiming for the construction of a holistic and balanced, dynamic and harmonious environment, focuses on the relation between human beings and their environments and falls within the scope of natural sciences.

Modern science and philosophy divide the unified world into human society and nature, resulting in the separation and opposition of natural sciences and human sciences, science and arts, all progressing in their respective ways. However, concrete queries have recently emerged: Can translatology and ecology be scientifically and seamlessly "integrated" or "married"? Where do their possible "intersection sets" and "fusions" lie, and to what degree can they intersect or be fused? What "results" or "effects" will emerge from the synthesis of translation studies and an ecological perspective? How will these "results" or "effects" act on translational acts or be applied to the explication of translation phenomena? All those questions inspire reflections and explorations.

Rooted in the developments of translation studies, ecology, and the theory of translation as adaptation and selection in particular, the birth of "Eco-translatology" has been the catalyst of the growth and establishment of terminological series such as "translational ecosystem," "translational eco-environment," "ecosystem of translation," "textual transplants," "translator's adaptations," "translator's selections," "survival through selection," "coexistence and interaction" and "emphasis on existence and harmony," and through the ongoing theoretical research in this field, positive attempts have been made to form a different perspective of the translational ecosystem and to create an ecological basis to metaphorically discuss translation as a whole based on translational practice.

Founded on all these discussions, the subdisciplinary traits of Eco-translatology are thus fully discernible:

1. a unique research perspective (a macroscopic eco-reason perspective);
2. specific research foci (textual ecology, translator-community ecology, and translation-environment ecology);
3. systemic research subjects (interaction and interrelation between translational text, translator, and translational eco-environment; the research trinity of the macroscopic framework of translation studies, mesoscopic system of translation theory studies, and microscopic textual production, etc.);

4. supporting terminology (e.g., ecology, life, survival, selective adaptation, adaptive selection, translational eco-environment, translational ecosystem, Eco-translation ethics, translator responsibility, "doing something with Eco-translations," "post-event penalty," "survival through selection," multi-symbiosis, multiple eco-integration); and
5. nascent research methodology (interdisciplinary intersections, metaphorical analogies, conceptual borrowings, and particularly the systematic synthesis of eco-holism).

In this sense, it is rational to label Eco-translatology as "cross-disciplinary"[6] and to consider it a subdiscipline of general translation studies.

Furthermore, if Eco-translatology is considered a subdiscipline of general translation studies, its theoretical basis can also be generalized as eco-holism and translational ontology.[7]

In regard to the explorations and development at present, however, although we accept ecological notions and adopt ecological principles to develop investigations, the research subject of Eco-translatology remains within translation studies. It is separate from ecology, and it is neither intended to align translation studies with ecology nor to direct its attention to the study of ecological issues. Rather, Eco-translatology is rooted in the isomorphic metaphor and conceptual analogy between the translation ecosystem and natural ecosystem, resting its major weight, to be precise, on "metaphorical reference" and revealing the illumination of metaphorical analogy through translation studies. *Green translation studies*, though accommodated in this new perspective, represent a new direction beyond the former boundaries. This leading idea has been embodied nominally, namely, "eco" has been assigned as the attributive in Eco-translatology (i.e., ecological or ecological perspective) to modify the central word, "translatology." In other words, Eco-translatology is assigned to be a perspective of translation studies in light of eco-reason.

Eco-translatology, with which we are concerned here, is thus a paradigm of ecological translation studies rooted in eco-reason and synthesized from an ecological perspective. It originated and was displayed in an upsurge of worldwide ecological theories to synthesize and describe the entire ecology of translating and translation theories proper (including the essence, processes, criteria, principles, methods, and factual cases of translation) from the ecological perspective, drawing support from the isomorphic metaphor between the translation ecosystem and the natural

[6] As has been stated by Simon (1986, pp. 234–235), an American scholar on politics and social sciences, "If the studies in an interdisciplinary research area can be sustainable and prosperous in its development, it will, then, sooner or later, become a new branch of learning," and "As long as new subjects are studied continuously and progress can be made day by day, a new discipline will be turned out from these studies."

[7] Grammatically speaking, "eco-" is just a modifier, and "translatology" should be taken as the core word in "Eco-translatology," which can be rendered from an ecological perspective of translation phenomena and therefore considered a sub-branch of translatology or translation studies from the linguistic, cultural, cognitive, and social perspectives (Zhang & Jin, 2011, p. 262).

ecosystem; conceived in ecological holism; subordinated to Eastern ecological wisdom; probing into textual ecology, translator-community ecology, translation-environment ecology, and their interrelationship; and narrated in ecological terms and an ecological manner.

The terms and concepts of Eco-translatology originate from an understanding of the natural ecosystem and are drawn from the "theory of translation as adaptation and selection." Eco-translatology analogizes natural ecosystems and translational ecosystems, transplants the theory of "survival of the fittest" from the former to the latter, and relates the theory to the inherent sense of life and the existence of the "translator-community."[8] In other words, fusing translation practice reflections on the existential system and the existential aspect of the reflective system[9] by leveraging the theory of "survival of the fittest" is where the intrinsic and timeless motive of translational success lies, and this fusion is the primitive target of the investigation of textual ecology, translator-community ecology, and translation-environment ecology of translation activities.

The present book will restructure the ambit of the scholarly view of translation studies, shifting from the age-old linguistics point of view to the current ecological reference point. The ecological rules of natural ecosystems penetrate and prevail in the intrinsic structure and interaction within a benign translational ecosystem. For instance, the "system" principle in ecology is embodied by the fact that translation studies proper is not a single entity but a complex system; the "diversity" principle by the fact that in the translation process, diversified language uses exist to overtone diversified adaptations and selections; the "dynamic" principle in the fact that changes never stop, no text can divide itself from its epochal features, and readers' perceptions of a translated text will vary over time no matter how qualified the readers are; the "response" principle in the fact that the translator will co-evolve with the text in an acting and counteracting manner; and the "yardstick" principle in the fact that reference to the whole is demanded when examining and criticizing a translated text, while scrutiny of words, phrases, and even such minute elements such as punctuation is strongly desired when discussing the syntactic structures of the translated text.

Eco-translatology, a new translation paradigm in the postmodernism context, has emerged as a cross-disciplinary or multidisciplinary progression, expansion, and transformation of contemporary translation studies, representing the trend of translation studies towards the modern, synthetically integrative perspective instead of the traditional mono-disciplinary one.

[8] 11 Including translators, readers, publishers, patrons, commentators, and other human agents involved in translation activity.

[9] For instance, Yang (1989, pp. 17–19) stated in his article "Reflections on Constructing Translation Studies" in *Chinese Translators Journal* that "translation studies is a science studying the thinking laws and methods of translation," and Feng (2008, pp. 82–85) stated in his article "An Existential Study of Translation and Its Embedded Value" in *Foreign Language Education* that the "translator should be taken as the ultimate basis."

Undeniably, in almost every epoch, a leading concept from the realm of science can be nominated as a new methodology to dominate the general societal mode of thought. "Ecological science" has recently become this dominating and shaping force in contemporary society and has been revealed as a discipline of ultimate significance to human existence and development. In this sense, Eco-translatological explorations are rooted in a sound philosophical base that promises a vigorous expansion.

In a word, "Eco-translatology" approaches translation synthetically from an ecological perspective by leaning on eco-reason traits. It is essentially a neat discourse system of translation. Indiscriminate and abstract as some of the above Eco-translatological expositions may seem, they are a kind of "preparation" for the systematic elaboration in the ensuing chapters, which I sincerely hope will succeed in expanding the notion of translation studies beyond the conventional translation theories.

1.5 Understanding Eco-Translatology Through Terminology

A specific terminology is indispensable for each theory, with terms constituting the basic elements and components for the formation of this theory. Though a systematic elaboration will follow, a preliminary expositive list of some chief terms will be conducive to first, sketching an Eco-translatological profile and second, enabling readers to grasp the gist of the book.

A list of some ecological terms[10] is compulsory here because a considerable number of Eco-translatological terms come from ecology via "borrowing" or "analogy":

- *Biosphere*—the existential and developmental state of an organism (including animals, plants, microbes, and human beings);
- *Ecological Community*—an organic aggregation set in a specified space and environment with which certain organic species interact and that shows a specific form and structure composed of morphosis and a trophic structure, denoting, in brief, the animated parts of an ecosystem;
- *Ecosystem*— a dynamic and balanced entirety; if it is a natural ecosystem, close contact and mutual action occur between ecological communities and between these communities and their environment to form, via substance exchange, energy conversion and information transmission, a dynamic and balanced entirety with a certain space, form and function. Ecological concepts, accompanied by the birth of ecology, emerge as an echo of modern biological progress. The ecosystem concept was conceived by A. G. Tansley in 1935 to accommodate

[10] Based mainly on *Terminology in Ecology* (2006), published by Mongolia Education Press and http://en.wikipedia.org/Ecology.

the natural entirety composed of organisms and environment, a simple formula being "ecosystem = ecological communities + environmental conditions."

- *Ecological Chain*—energy and substance are transmitted among the predators and the prey in an ecosystem to constitute a food chain, and the complex relationship between the predators and their prey resembles an invisible web and thus is known as the food web, which, together with the food chains, correlates all the species in one ecosystem and ensures structural and functional stability within the system.
- *Ecology*—essentially a science studying the habitats of organisms that evolved from a Greek etymological source in which *oikos* denotes dwellings and habitats and *logos* means a discipline.
  "Ecology" can be interpreted as a notion connected with managing living resources or constructing a fairyland of residence. It was defined in 1866 by Haeckel, a German biologist, to refer to a science investigating the interaction (mutual action) between organisms and their environment (covering a biotic environment and a biological environment). The noted American ecologist Odum (1956) considered it a science concerned with the structure and function of ecosystems, whereas his Chinese peer Ma Shijun considered the interaction and interrelationship between life systems and environment ecosystems the core concern of this branch of science. Within the natural world, a bion, a colony, or a community can all be regarded as a biological system, and the surrounding energy, temperature, and soil constitute the environmental system (Li, 2004, p. 2). "Ecology" is hence a science studying the relation, rules, and mechanisms of interrelationship and interaction between organisms and the elements of their environment. Organisms are in unity with their environment in ecological terms, and human beings constitute a significant member of the ecosystem and perform a leading function in the biological chain.
- *Ecological Balance*—a novel concept in modern ecological development.
  In general, when an ecosystem has an input surplus of energy and substance, its biomass will multiply, and vice versa. When input equals output over a very long period, stability consequently reigns in the structure and function of the ecosystem, and the primary stability can be restored through auto-regulation under mild interference. This is the ecosystem balance of, which therefore is called ecological balance.

In Eco-translatology, in accordance and analogy with the ecological terms and concepts stated above, these terms can be defined as follows (Hu, 2003, 2004, 2006, 2008, 2010):

- *Translator-Community*—an aggregation of "participants" (all "persons") involved in translation activities that interacts and interrelates with the generation, development, operation, results, function, and effects of translation activities, accommodating the author, target-text (TT) readership, translation critics, translation reviewers, publishers, marketers, patrons, consigners, et al., with the translator as their representative.

The translator community, ranked as important within the translational ecosystem, consists of, in ecological terms, producers, consumers, and decomposers, denoting the translators or creators of translation products, who take the central position within the translator community; the authors and TT readers, or the beneficiaries; and the researchers in the translation circle, respectively. The translation subjects—the author, the translator, and the TT readers—are mutually dependent and restrictive to constitute an open system of independent operation, namely, all "persons," an "ecosystem of translation subjects" (see Sect. 5.4.6).

- *Translation Chain*—a series of interrelated and interactive links in the "pre-translating" (i.e., preparations before translation), "during-translating" (i.e., translational acts), and "post-translating" (i.e., effects of translation) stages of translation production.
- *Essence of Translation*—Eco-translatology defines translation as "selective activities of the translator when adapting to the translational eco-environment and transplanting texts, with the translator acting the leading role, the text as its basis, and cross-cultural information transmission as its ultimate purpose," on the basis of the theory of translation as adaptation and selection.
- *Translational Ecosystem*—a system of society, communication, culture, and language that shows a certain spatial structure and temporal change and is subject to auto-regulation and openness, similar to a natural ecosystem. Illuminated by the definition of a natural ecosystem, the "translational ecosystem" can be defined as a functional unit of interaction and interdependence in translation studies constituted via consistent substance cycling and energy flow between languages and between the components and noncomponents of translation (society, communication, and culture, for instance). It can be interpreted, narrowly, as a "translational eco-environment" (see the entry below) or discussed in a broader view as an accommodation system to cover all possible activities connected with translation. Composed of a wide variety of subsystems of different ranges, the "translational ecosystem" is indefinitely divisible lengthwise and mutually connected breadthwise (see details in Sects. 4.3–4.5).
- *Translational Ecology*—the state of interrelation and interaction between translation subjects and between translation subjects and their surrounding environments, i.e., the existential and working state of translation subjects in an environment.

  The main point to stress here is the clear distinction between "translational ecology" and "ecological translation," the latter being an umbrella and comprehensive term of synthesizing and describing translation. In more specific terms, "ecological translation" accommodates both a synthetic view of translation as a whole and a metaphorical description of translational ecology via natural ecology; both the diversity maintenance of translated languages and cultures and the employment of translation to promote eco-environment protection and eco-civilization development; both the selection of texts to translate via ecological adaption and the regulation of the "translator community" via ecological ethics; and, certainly, the selection of natural texts to translate and the natural world of translation ecology via ecological notions. From the textual perspective solely, ecological translation may denote "textual transplants" (see details in Sect. 5.3.4) based on ST ecology and TT ecology.

- *Translational Eco-environment*—the worlds of the ST and the source/target language, namely, the interrelated and interactive entirety comprising the linguistic, communicative, cultural, and social aspects of translating, as well as the author, the client, and the readers (i.e., "translator community"). The translational eco-environment is essentially an aggregation of all the factors related to the translator's optimal adaptations and selections. It is divisible to the microenvironment, mesoenvironment, and macroenvironment, covering the intralingual environment and extralingual environment, material environment and spiritual environment, and, furthermore, subject environment (translator, author, readers, publisher, editors, supervisors, all "persons", et al.) and object environment (ST, TT, textual function, translation strategies and translation regularities, etc.) (see details in Sect. 3.4.1).
  In regard to translational ecology, in particular, the "entirety" of subjects' existential and working state in their environment is denoted, whereas in regard to the translational eco-environment, an "aggregation" of diversified external elements relevant to translation is nominated. These two terms, at this point, are synonymous and interchangeable but differentiated from each other in that translational ecology values the "holistic" and "integrative" state (uncountable), but the translational eco-environment gives major weight to "multitudinous" elements or a "summation" of individual environmental elements (countable).
- *Typically Important Elements*—the most important elements in the translational eco-environment.
  At the stage of translator adaptation, generally speaking, the ST is the "typically important element" in the translational eco-environment, and at the stage of translator selection, the translator himself/herself will play the role (see details in Sect. 5.5.1).
- *Preservation of the Strong and Elimination of the Weak*—a principle in the translation world distinct from that in the natural kingdom.
  The major difference lies in the fact that the adaptation of natural species (animals and plants) to the natural environment and their "elimination" under "natural selection" are absolute, signifying the "extinction," "disappearance," or "vanishing" of biological species, e.g., the extinction of the dinosaurs, South Pole wolves, and seed ferns. The translator's or TT's adaptation to the translational eco-environment in the translation world and the "elimination" under the selection of the translational eco-environment, in contrast, are relative or, in the metaphorical sense, denote "frustration," "denial," "deletion," "substitution," "misdirection," or "loss" of human behavior or feelings. In other words, the "adaption" or "mal-adaptation," "strength" or "weakness" of the translator or TT in translation activities are not absolute but metaphorically relative. Meanwhile, different TTs may have room for coexistence, as they are adaptive to different translation purposes or can satisfy different readers. "Preservation of the strong and elimination of the weak" and "coexistence" in this sense are consistent with

the basic principles of ecology (see details in Sect. 3.7.3 on "symbiosis and diversity," an ethical principle of ecological translation).

- *Holistic Degree of Adaptation and Selection*—the totality of the translator's "selective adaptation" when producing a text in linguistic, cultural, and communicative dimensions and, accordingly, the "adaptive selection" degree of attending to other elements in the translational eco-environment
  Generally, the holistic degree of adaptation and selection is positively correlated with a TT's degree of "selective adaptation" and "adaptive selection." With the application of Eco-translatology, the optimal translation is thus, comparatively speaking, the version with the highest "holistic degree of adaptation and selection" (see details in Sect. 5.5.3).

The terms and concepts listed above have been clarified in terms of their relevance and specific connotations in this field, and they form the corpus of Eco-translatological concepts.[11] The disparity between the Eco-translatological discourse system and the traditional translation description may continue to reign over the readers, but the writer anticipates that these terms and concepts will facilitate the reading of the ensuing chapters and the understanding of Eco-translatological studies as a whole.

1.6 Glancing at Eco-Translatology Through a "Shortcut" of the Nine "Three-in-One" Expressions

"Three-in-one" here refers to a series of recapitulative expressions indicating the "triad" as different angles for a panoramic overview of the studies and developments of Eco-translatology. In other words, these "three-in-one" points of view can be taken as a "shortcut" that is conducive to an understanding of Eco-translatology. Nine "three-in-one" expressions are summarized as follows:

1. "Three eco-themes" pointing to the triple *research objects* of Eco-translatology, namely, textual ecology, translator-community ecology, and translation-environment ecology. Textual ecology refers to the vital signs and living conditions of texts; translator-community ecology refers to the existential quality and competence development of all persons involved in a translation activity; and translation-environment ecology includes all translational ecosystems and eco-environments.

[11] Other core terms, such as "the Sequence Chain," ecological paradigm, eco-reason, eco-system of translating, Eco-translation ethics, textual ecology, textual transplants, symbiotic, multidimensional integration, translator responsibility, selective adaptation and adaptive selection, "three-dimensional" transformation, doing things with translations, post-event penalty, holistic degree of adaptation and selection, translation as adaptation and selection, and translation as eco-balance, will be elaborated and interpreted in the coming chapters or sections.

The definition of the triple eco-themes relies heavily on translational ecology, an orientation towards textual life, and concern over translator survival and development. These are the "three eco-themes" of Eco-translatology.

2. "Three relationships-between," pointing to the relationships among the trio of subjects that are defined as the *research objects* of Eco-translatology, namely, the relationship between the translational eco-environment and the "translator-community" (with the translator as its representative), the relationship between the source/target texts and the translational eco-environment, and relationship between "translator-community" and source/target texts. In other words, Eco-translatology endeavors to elaborate on the interaction and interrelation between the translational eco-environment and "translator-community," source/target texts and translational eco-environment, and "translator-community" and source/target texts. That is, what we mean by the three relationships—between a trio of subjects.
3. "Three translation-as," pointing to the *three core concepts*: "translation as textual transplants," "translation as adaptation and selection," and "translation as eco-balance." The first relates to the textual transformation, the second to the translator's translational acts, and the third to the translational eco-environment. "Three translation-as" is the organic entirety of the theoretical system of Eco-translatology, existing in a correlative, interactive, and mutually causal relationship with no one element separable from another and in an intrinsic and corresponding logical relationship with the "three eco-themes" and "three relationships-between" described above.
4. "Three-level studies," pointing to three *research levels* of Eco-translatology, namely, "translation studies", "translation theories," and "source/target texts." A developmental pattern of the *macroscopic* translatological structure (particularly emphasizing translation studies), the *mesoscopic* theoretical system (particularly emphasizing translation theories), and the *microscopic* textual operation (particularly emphasizing source/target texts) has been gradually formulated.
5. "Three-centeredness," pointing to the tripartite tension and ultimate trilateral relation of "source texts—translator—target texts."[12] Translation studies are supposed to center not only around STs and TTs but also around translators since Eco-translatology follows with interest the translator's "existence" and "development" and considers it compulsory that the "translator" should be "symbiotic and coexistent" with the other "two poles."
6. "Three-dimensional transformations," pointing to the translation methods of Eco-translatology that cover "multidimensional" transformations but are specified as "three-dimensional" transformations, i.e., translating is relatively conducted from linguistic, cultural, and communicative dimensions under the

[12] As stated by Li and Huang (2005, p. 96) in their collaborative article "A Brand New Theoretical Construction" in Foreign Language Education, "Translator-centeredness (initiated by Hu Gengshen) has set up a tripartite state with ST-centeredness and reader-centeredness, to balance, in the translating process, the power in theoretical explorations of translation."

translation principle of "multidimensional selective adaptation and adaptive selection."

7. "Three prerequisites," referring to the preconditions generalized as fundamental to the research and development of Eco-translatology: (1) "the Sequence Chain" displaying the human cognitive pathway; (2) the homo-analogy between translation ecology and natural ecology; and (3) the systematic investigation into translation as adaptation and selection explicated as "translation is the translator's selective activity of adapting to the translational eco-environment and transplanting the texts."

"Three pursuits," namely, the "fusion of Eastern and Western cultures," "junction of ancient and contemporary civilizations," and "channeling of arts and science," which are the academic pursuits targeted by Eco-translatology in the process of constructing the entire theoretical system of discourse, as is the goal of every academic field (see details in Sect. 8.2.2).

"Three paradigmatic traits," namely, the paradigm characteristics of practicalness, openness, and universality explicated in the investigations and developments of Eco-translatology, a synthesizing and holistic research paradigm of translation studies from an ecological perspective (see details in Sect. 8.2.3).

A "shortcut" to be acquainted with in reference to Eco-translatology, these nine "three-in-one" expressions can serve as a profile of Eco-translatological investigations and developments. Those mentioned above are merely recapitulative; the implication of theoretical conceptualization, and the "prediction," and the detailed elaboration will follow.

1.7 Eco-Translatology vs. Translation Studies and Other Prevailing Translation Theories

There is still a necessity to describe the relation between Eco-translatology and translation studies and the connection and segmentation between Eco-translatological investigations and other translation theories as an introduction to Eco-translatology.

Eco-translatology, as mentioned in Sect. 1.4 above, can be interpreted as an ecological perspective of translation studies. As a holistic study of investigating the interaction and interrelation between textual ecology, "translator-community" ecology and translation-environment ecology and a cross-discipline aimed at a synthetic description of translation activities from an ecological perspective, Eco-translatology can be labeled a sub-branch of general translation studies to investigate translation phenomena, to enrich translation theory studies, and eventually to promote the development of translation studies in general. The relation between Eco-translatology and translation studies is thus clear.

In regard to the connections and segmentation between Eco-translatological investigations and other translation theories, the author conceives, splendid and rational factors of some theoretical aspects of various Western translation schools represent novel theoretical creations and fruits, whereas weakness and inapplicability exist for temporal, regional, cultural, or horizontal restraints (Hu, 2002, pp. 80–104). When constructing and interpreting an Eco-translatological framework, therefore, basic principles of assimilating or discarding, synthesizing or surpassing, inheriting or developing should be closely followed to guide the construction of an Eco-translatological discourse system. Under this guidance, an endeavor is made to assimilate the rational core of different translation schools while discarding certain elements, to override the restriction of different schools by synthesizing common views of different translation schools, and to transmit the traditional wisdom of Eastern and Western translation studies to develop Eco-translatology by relying on translation studies. That is, the aim is to theoretically synthesize translation schools and thus to construct a "pluralistic" paradigm of Eco-translatology. The direction and positioning of Eco-translatology are thus defined.

As stated by Valdeón (2012, p. 5), the editor-in-chief of *Perspectives: Studies in Translatology*, "Eco-translatology represents an emerging paradigm with a great potential for research and study. It is also a model with clear connections with other schools in the West." Dollerup (2011, p. 34), a Danish translation theoretician, has also discussed Eco-translatology in the context of Western translation studies and pointed out that "Eco-translatology is related to these schools [three influential Western schools, namely, the principle of dynamic equivalence, the Skopos school, and descriptive translation studies]."

Eco-translatology, nevertheless, as a brand-new Eco-translation paradigm that differs from the existing ones, has its own characteristics and originality. A comparative study tells us that Eco-translatology is novel in terms of philosophical background, research perspective, core assumptions, research methodology, self-contained terminology, discourse system, translation ethics, etc. These all explain the "originality" and thus the "novelty" of Eco-translatology. (Hu, 2008, 2011; Hu & Tao, 2016)

"Eco-translatology is in keeping with—and is superior for explaining—actual translation practices even in the West," as has been stated by Dollerup (2011, p. 34)[13]

[13] For instance, "Eco-translatology, like translation studies integrating cultural factors or semiotics, covers a non-linguistic environment of wider scope and what distinguishes the former from the latter two lies in more weight on coordination and balance between factors of environment with 'eco-'. The latter two, though having integrated culture and semiotics, stick to the translator's subjective examination, integration and strategy selection among factors in translation; Eco-translatology, however, stresses the interaction between translator and other translation subjects and the impact of the holistic environment on translator and translations. It moves beyond the exploration into translational act and TT quality, rendering itself more inclusive and explaining why some translations, though thought low of by critics, can survive with a wide popularity. Eco-translatology exposes its potential when it approaches the relation between translator and translational environment, translator's strategy selection or even assessment of translation quality. Independence of one specific strategy or technique to restrict translating practice features this theory, which allows subjective judgments and rational explanations for those judgments…. A definite purpose will accompany and pilot all through the translating stages, which echoes the arguments of *skopostheorie*. Whether the anticipated effects or purposes of translations are to be attained, nevertheless, goes far beyond the research scope of *skopostheorie*, and the translational

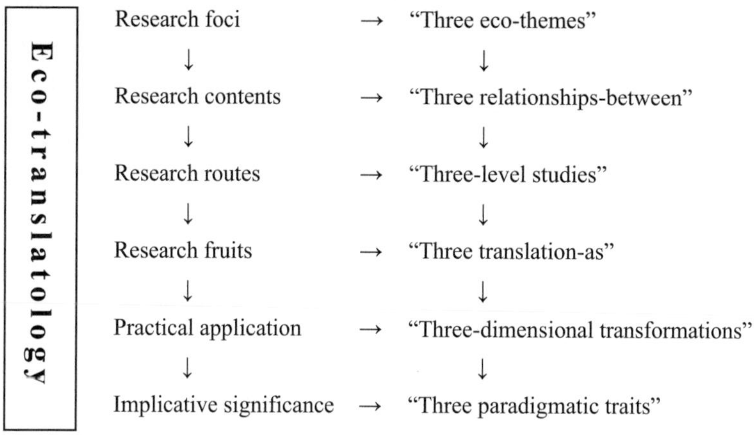

Fig. 1.1 Simplified framework of the contents of Construction and Interpretation

and echoed by his Chinese fellow scientists. "Translation studies are thus incorporated into an organic explorative system, translational eco-system, via the establishment of an Eco-translatological research paradigm, which moves translation studies beyond the comparatively 'partial' linguistic, literary or even cultural paradigms of translation studies to an unprecedented vision of magnitude via the route of inheritance, transcendence and return" (Meng, 2008, p. 73).

The theoretical discourse system of Eco-translatology, in brief, has been formulated with a systematic exploration and description from the perspectives of translation studies, translation theories, and textual production in this globalized world, based on synthesis and reflection on translation theories and the properties and rules of ecology and guided by ecological wisdom and values.

What has been stated above and what will be stated in this book show the developing line and overall logical framework of *the paradigm* and, moreover, the ontology of Eco-translatology as a flexible organism of a complex translational ecosystem composed of "textual ecology, translator-community ecology, and translation-environment ecology". The framework of Eco-translatological research logic can now be simplified as follows (Fig. 1.1).

1.8 Summary

Translation studies, a science of theoretically exploring translation issues, investigates the arts of translation. Eco-translatology, an emerging Eco-translation paradigm synthesized from the ecological perspective, is a cross-disciplinary study that

eco-environment will operate with more facility for its nature of being dynamic" (Han, 2013, pp. 122–123).

systematically probes the interaction and interrelation between textual ecology, translator-community ecology, and translation-environment ecology to synthesize and describe the entirety of translational ecology and the ontology of translation theories. It is grounded firmly in the isomorphic metaphor between translational ecology and natural ecology, taking ecological holism as its core, adhering to Eastern ecological wisdom and founded on the translation as "adaptation/selection" theory.

A sketch of Eco-translatology has been accomplished via the explication of translation studies, ecology, Eco-translatology, and the terminology and investigations of Eco-translatology. Satisfying answers to what Eco-translatology is, nevertheless, remain far beyond reach, particularly in regard to the issue of the "name" and "nature" of Eco-translatology. Above all, this book will serve as an indicator of a field of study that is still trying to find its boundaries.

An insight into Eco-translatology calls for further reading of this book, and the priorities are the background and inception, origin[14] and development of Eco-translatology in Chap. 2.

References

Bassnett, S., & Lefevere, A. (1990). Preface. In S. Bassnett & A. Lefevere (Eds.), *Translation, history and culture* (p. ix). London: Pinter Publishers.

Chesterman, A. (1993). From 'is' to 'ought': Laws, norms and strategies in translation studies. *Target, 1*, 1–20.

Dollerup, C. (2011). Eco-translatology in translation theory contexts. *Journal of Eco-Translatology, 1*, 34–40.

Feng, W. K. (2008). An existential study of translation and its embedded value. *Foreign Language Education, 4*, 82–85.

Han, W. (2013). Reinvestigation of "translational eco-environment" and "survival of the fittest". *Foreign Language Research, 1*, 122–126.

Hu, G. S. (2002). Prosperity, perplexity and exploration of translation theories. *Translation Quarterly, 25*, 80–104.

Hu, G. S. (2003). Translation as adaptation and selection. *Perspectives: Studies in Translatology, 4*, 283–291.

Hu, G. S. (2004). *An approach to translation as adaptation and selection*. Wuhan: Hubei Education Press.

Hu, G. S. (2006). *Understanding eco-translatology*. Presented at the International Conference on Translating Global Cultures: Towards Interdisciplinary (Re)Construction, Beijing, August 14–16.

Hu, G. S. (2008). Eco-translatology: A primer. *Chinese Translator's Journal, 6*, 11–15.

Hu, G. S. (2010). Eco-translatology: Backgrounds and bases for its development. *Foreign Language Research, 4*, 62–67.

[14]As stated by Ye Xiè (1627–1703), noted Chinese literary figure of the Qing Dynasty, in his *Anthology XI: essays on man and friends,* academic research should discern first its origin and then its specific developing route and hence varies to a great deal, that flexibility is the hardest to achieve and mastery is harder, even though a proposal is within reach, and that what matters ultimately is clarification of the origin and its developing route.

Hu, G. S. (2011). Eco-Translatology: Research focuses at the macro level. *Journal of Eco-Translatology, 1*, 24–33.

Hu, G. S. & Tao, Y. L. (2016). Eco-translatology: A new paradigm of eco-translation-a comparative study on approaches to translation studies. *Translation & Interpreting Review, 6*, 115–131.

Huang, L. (1988). *A course of translation arts*. Nanjing: Nanjing University Press.

Jim Shen. (1993). Views on ecology-oriented studies. As cited in Wang, R. S. & Zhou, H. (2004, p.4). *Human and Ecology*. Kunming: Yunnan People's Press.

Krohne, T. (2001). *General ecology*. Brooks: Thomson Learning.

Li, Y. S., & Huang, Z. L. (2005). A unique theoretical construction: On reading Hu Gengshen's monograph titled *an approach to translation as adaptation and selection. Foreign Language Teaching, 6*, 95–96.

Li, Z. J. (2004). *Ecology*. Beijing: Science Press.

Liao, Q. Y. (2001). *Contemporary translation studies in UK*. Wuhan: Hubei Education Press.

Liu, M. Q. (1989). On general modes of translation studies in China. *Chinese Translators Journal, 1*, 12–16.

Meng, F. (2008, August). *Turn, surpass, and return: Theoretical reflection of the theory of translation as adaptation and selection* [OL]. Retrieved from http://www.Language-international.net

Munday, J. (2000). *Introducing translation studies: Theories and applications*. London: Routledge.

Neufeldt, V. (1995). *Webster's new world college dictionary* (p. 429). New York: Macmillan.

Odum, T. (1956). Limnology and oceanography. *Limol Oceanogr, 2*, 1956.

Pan, W. G., & Yang, Z. J. (2008). A historical review on Chinese-English contrastive studies in the new era—Breakthroughs in theories and methodology. *Journal of Foreign Languages, 6*, 86–91.

Qian, G. L. (2000). Dedicating something original to the world: Creativity in foreign language research. *Foreign Languages and Their Teaching, 1*, 10–14.

Simon, H. A. (1986). Trans-disciplinary studies in science. *Bulletin of Chinese Academy of Scien ces, 3*, 223–235.

Tan, Z. X (1988). On translatology. *Foreign Language Journal, 3*, 22–28.

Valdeón, A. R. (2012). *Recycling in the translation ecosystem: The case of stable and unstable sources in news production*. Presented at the 3rd International Symposium on Eco-translatology, Chongqing, China, November 23–25, 2012.

Wang, C. L. (2001). *A grand dictionary for sci-tech editors*. Shanghai: The 2nd PLA Medical University Press.

Wang, N. (2004). *Globalization and cultural translation*. London: Marshall Cavendish International (Singapore) Private Limited.

Wang, R. S., & Zhou, H. (2004). *Humans and ecology*. Kunming: Yunan People's Press.

Yang, W. N. (2009). Constructing a favorable eco-environment for literary translation. *East Journal of Translation, 1*, 12–13.

Yang, Z. J. (1989). Reflections on constructing translation studies. *Chinese Translators Journal, 4*, 7–10.

Yang, Z. J. (2007). Sino-foreign logic comparison and academic creation. *Foreign Languages in China, 2*, 58–64.

Zhang, L. F., & Jin, W. N. (2011). On eco-translatology and its three eco-dimensions: A concurrent discussion with Professor Hu Gengshen. *Acta of Shanghai University of Science and Technology (Social Science Edition), 2*, 30–35.

Chapter 2
Eco-Translatology: Inception and Development

2.1 Introduction

Processes constitute the world. As Friedrich Engels (1820–1895) said, the world is a complex of *processes* instead of monotonous things.[1] "Process" here refers to phases or stages of development. In philosophical terms, "process," denoting temporal continuity and spatial extension, is the existential and progressive form of things and their conflicts. "Process" can also be interpreted as a means of integrating resources to generate the expected outcome. For any type of process, input and output are the principal elements, the former as bases, prerequisites, and conditions of implementing a process and the latter as the outcomes of accomplishing it. The former can be tangible or intangible products, such as theories, software, or services. Only through the process can a task be accomplished and an activity conducted.

What Eco-translatology has undergone in the first decade of the twenty-first century is a process of consolidation and progression. In the course of Eco-translatological construction, the ultimate establishment of a theoretical discourse system targeted by the present writing serves as the output of this "process," which suggests its "input," namely, the bases, prerequisites, conditions, etc., of its inception and development. Familiarity with the records of Eco-translatology, meanwhile, promises a reasonable anticipation of its course.

Given this, what will be critically studied in this chapter is the background of the inception and the bases of exploration of Eco-translatology and its continuous study and development.

---

[1] Zhang (2002). "Process" is defined in a darker light by the author of this book.

(Hugs) G. S. Hu, *Eco-Translatology*, New Frontiers in Translation Studies,
https://doi.org/10.1007/978-981-15-2260-4_2

2.2 Origins

2.2.1 *Global Ecological Trend of Thought*

Ideas are deeply rooted in a specific historical background and ethos. The inception and development of Eco-translatology follow the progressive route of contemporary society and scholarship. First, it is a translatological response to the present economic and social transformations. As is well known, the 1960s witnessed the transformation of human society from an industrial civilization to an ecological civilization. Human beings were faced with a crisis of existence and development, as declared in 1962 by Rachel Carson, an American marine biologist, when she wrote of human violence against nature in her noted *Silent Spring*. The *Declaration of Human Environment* and *Our Common Future* were announced in 1972 and 1987, respectively, by the United Nations Conference on Environment, regarding the protection of the natural environment as a problem that concerns all humankind. Eco-environmental protection has become a central concern in China since the 1970s. Later, a sustainable development strategy and scientific development strategy were formulated in succession by the Chinese government, which announced that "human society is transitioning from an industrial civilization to an ecological civilization."[2] In this context, the introduction of the neglected and overlooked "ecological" dimension of scientific research fields, including translation studies, is definitely an embodiment of conforming to the trends of our times.

Second, the ecological trend of thought is an inevitable result of cognitional transitions in modern ideology and philosophy. Since the twentieth century, these two fields have experienced a cognitional transition from subject–object dichotomy to intersubjectivity, from anthropocentrism to ecological unity. For instance, in 1967, French philosopher Jacques Derrida wrote that the "center" exists at once within and without a structure and that the center is hence not a center; in 1973, Norwegian eco-philosopher Arne Naess proposed his "deep ecology" to integrate ecology into philosophy and ethics and put forward such far-reaching eco-philosophical notions as eco-ego, eco-equality, and eco-symbiosis; and in 1995, American eco-philosopher David Griffin came up with "eco-existence," signaling the birth of eco-ontology.

The signs of progress mentioned above tell us that contemporary philosophy is turning from epistemology to ontology as well as from anthropocentrism to ecological unity. Therefore, this philosophical turn rightly broadens the horizons and mentality of translation researchers, drawing them to approach translation activities "ecologically." Eco-translatology emerged as a response.

[2] Yu (2007).

2.2.2 Ecologically Related Studies in Different Disciplines

The "return to nature" has swept the world as a global trend that increasingly drives people to seek ecological foods, ecological furnishings, ecological tourism, and a balanced eco-environment. With ecology viewed as a kind of scientific mode of thought, the word "ecological" collects profound implications and enjoys wide popularity. The concepts of ecological construction, ecological projects, ecological restoration, ecological culture, ecological architecture, ecological city, ecological forest, ecological politics, ecological movement, etc., have arisen in response. "Ecological" has become almost the most familiar and at the same time the essential keyword of news, public documents, or even street chats in recent years.

The academic community has experienced "environmentalism's overdue move beyond science, geography and social science into 'the humanities'" (Kerridge, 1998, p. 5). Scholars are engaged "in an effort to turn cultural attention back to the wider living environment" (Parham, 2002, p. 1; Westling, 2002). Following this trend, studies of ecological essence have emerged. Within language learning, which is closely related to translation studies, there are environmental linguistics, green grammar, language environmentology, eco-lexicology, eco-linguistics, language ecology, language acquisition ecology, ecology of language evolution, studies of linguistic and biological diversity, etc., in addition to eco-criticism, eco-aesthetics, eco-literature, eco-politics, eco-theology, ecological library science, eco-Marxism, eco-urbanology, eco-socioeconomics, environmental humanities, etc. (Fill, 2001; Gabbard, 2000; Mühlhaüsler, 2003).

Now that the humanities, including applied linguistics, cultural studies, comparative literature, etc., have imported ecological notions and carried out relevant "cross-disciplinary" or "interdisciplinary" investigations, is it possible for translation studies, which is essentially "interdisciplinary," to follow suit? This implicative "tension" and groping desire serve as a kind of "motive" for initiating Eco-translatological explorations.

2.2.3 Ecologically Superficial Terms Used in Translation Studies

Theories inspired by ecological notions have proliferated globally. In this context, translation studies from an ecological perspective have sprung up, and superficial ecological terms such as "ecology," "environment," "existence," "adaptation," and "selection" appear throughout such works. An increasing number of such studies or remarks have appeared in the translation field.

For instance, the first category of cultural words, "ecology," as stated by Newmark (1988, p. 95), has a distinctly ecological tint; the modification and clarification of culture, as described by Katan (1999, pp. 45–52), covers the physical environment, political environment, climate, space, environment, garments, food, smell, situation, etc.; and the "ecology of translation"[3] constitutes one section of *Translation and Globalization*, written by Cronin (2003), who calls for a "healthy balance" of translation between languages (Phillipson, 2006, p. 231).

The relationship between "universalist" translation theory and "relativist" translation theory is likened by Steiner (1975, p. 235) to that between primary human processing modes, namely, "collective environmental adaptation and local environmental adaptation" (Lance & Jacky, 1991, p. 34). Translation is stated to be "a mode of recognition and existence" by Warren (1989, p. 6), who adds that "transplanting a literary work from one language to another is just like migrating a plant or an animal from one place to another. Only by fitting into the new environments, can they change to survive there, as an individual or a nation's adaptation and growth." "Cultural environment" is frequently employed by Bassnett and Lefevere (1990, p. 4) to describe the context of translation when arguing for their "cultural turn" and using ecological analogies such as to "discover and describe" to depict the investigations of linguistic approaches to translation studies. Analysis of "textual input conditioned by environment" and manipulation of input text within the feedback processing mechanism framework have been counted as two stages of the translation process by Wilss (1996, p. 123), who writes that this manipulation, being hierarchical, must refer to various "environmental features" of target readers to formulate the final textual output in the target language. "Pivotal concern should be given to different [translational] environmental factors, such as features of the translating task, requirements of clients, the translator and his decision-making competence, etc." (Wilss, 1996, p. 184), he adds when calling for a more in-depth investigation of the decision-making process of translation.

Explorations or remarks of an ecological perspective of translation may not be as frequent in China, whereas terms such as "translational ecology" consistently appear in discussions of translation quality, translation theories, or the development of the translation industry (Cui, 2006; Ji, 2005; Zhang, 2005; etc.). A monograph titled *Translational Ecology* (Xu, 2009) was published recently by China Three Gorges Publishing, suggesting arising concern about ecological approaches from an increasing number of scholars, although its contents consist mainly of ecological theories, and its author considers "translational ecology" an offshoot of ecology.[4]

---

[3] "Ecology of translation" should not be interpreted as *Science of Translational Ecology*. In a section of Chap. 6 titled "Minority Languages in a Global Setting," which discusses language policy, the dominance of English, and its pressure on minority languages, the ecology of translation is mentioned at the end in a call for a balance of translation between languages (Phillipson, 2006, p. 231). Hence it is not a science of translation studies, since Cronin does not mention integrating ecology and translation studies to start a disciplinary construction.

[4] For example, in Chap. 3, titled Eco-environment of Translation, the existing environment of fish and birds and that of human beings, the existing and developing environments of translation, etc., and their interactions with and linking mechanisms to the surrounding environment constitute the major issues of various ecological studies (such as ichthyological ecology, ornithological ecology, human ecology, and science of translational ecology) (Xu, 2009, p. 11). Science of Translational Ecology is defined as parallel to studies in ecology, as analyzed by Jia Guanghui (http://www.baike.baidu.com/view/2102202.htm, March 20, 2010) and Liu (2010, pp. 75–78).

Earlier, the claim appears that "eco-environment as a synchronic determining factor" ranks as the first law of "the nine laws governing human cultural development" (Gu, 2004, p. 239).

Ecological terms such as "ecology," "environment," "existence," "adaptation," and even "ecology of translation" are employed, metaphorically or not, in the studies or descriptions mentioned above, moving from a unique aspect towards a broader acceptance of integrating ecological terms into translation studies. Even ecologically superficial terms pave the way for Eco-translatological studies and relevant further explorations.

2.2.4 *Ecologically Overlooked Dimensions in Translation Scholarship*

The emergence of any new concept or notion in science means, to some extent, "dissatisfaction" with certain realities. This is also true of the emergence of Eco-translatology, and, from this Eco-translatological perspective, contemporary translation studies suffer from the following limitations or weaknesses:

1. Lack of systematic investigations of the ecological perspective of translation studies:

Constant investigations employing the terms or concepts of natural ecology to describe translation activities domestically and abroad (see the previous section) indicate, on the one hand, a common understanding in the academic translation field of the rationality of translation studies based on ecological analogy and, on the other hand, pave the way for further investigations. Generally speaking, the current attempts feature deficiency and inadequacy; hence (a) unsystematic or partial comments prevail; (b) "light touch" attempts are confined to superficial ecological terms; (c) a parochial horizon hinders multidimensional or insightful interpretations; and (d) a lack of an integrative view results in obscurity. In addition, many investigations skip over the globalizing ecological trends and currents. Ecological approaches to translation studies confine themselves to one corner of general translation studies, and ecological dimensions are woefully absent from translation studies.

2. Neglect of some issues in the "cultural turn":

In regard to the development of translation studies, the "cultural turn" is a landmark change in terms of research perspective. The "cultural turn," nevertheless, has limitations and blind spots. It overstresses the cultural origins of the text, deeming culture versatile to blur the distinction between translation studies and cultural studies; it does not generate a maneuverable model, translation methods or techniques and fails to explore the translation process; and it has a weak theoretical basis and lacks attempts to infer a cycle (Zeng, 2006, p. 195). Current integrations of cultural studies and translation studies leave much to be expanded or strengthened (Lv & Hou, 2006, p. 113). For example, such issues as the "translator's plight and translator's competence development," "constituents of the translational ecosystem and their interaction," the "development of translation and evolution of human cognition," and the "holistic study of the translational ecosystem" are crucial to translation studies, whereas they are peripheral in the eyes of cultural turn scholars. Concepts ranging from the space ecosystem to the translator's living ecosystem go beyond the coverage of culture. Even the concept of the translational eco-environment cannot be correctly interpreted in terms of culture in its narrow sense.

Additionally, the cultural perspective of translation studies focuses on cultural phenomena in translation and the impacts of culture on translation. It is concrete as both a translational probe of culture and a cultural probe of translation. Comparatively speaking, ecology is a kind of "metascience," and ecological approaches are interdisciplinary in their perspective. The ecological approach to translation studies is hence a translation approach in light of ecological rationality, the application of ecological philosophy (for instance, balanced global law, symbiosis rule, diversity law, etc.) to translation studies, or a philosophically and methodologically significant study of translation. Cultural approaches to translation studies are necessary for cultural studies in translation, whereas ecological approaches to translation studies are not ecological studies in translation but a reinterpretation of translation from an ecological perspective, presenting new explanations for entire translational ecosystems and ontological translational theories and lending ecological rationality to translation studies. Therefore, as far as translation studies are concerned, cultural approaches and ecological approaches run along different lines. In this sense, characteristic of "social phenomena" and "cultural phenomena," cultural approaches are limited in terms of their philosophical and methodological significance to translation studies.

3. The indefiniteness of the "leading" discipline in cross-disciplinary integration:

Many discussions or insights[5] have recently appeared in explorations of the integrative, multidisciplinary, or cross-disciplinary traits of translation studies. Unfortunately, vague ideas; undemanding combinations; or merely typological descriptions, analysis, or reinterpretation are featured in these attempts, which omit an interdisciplinary logical affiliation and integrative basis. An overall study will be hindered if the different disciplines remain segregated (Fang, 1999, p. 19). Cross-disciplinary integration is a false concept in the absence of a "leading" discipline or converging basis. The prosperity and perplexity of translation studies (Hu, 2002) have evoked investigations of the neglected or weak link, particularly endeavors to explore the converging basis of a relevant discipline with translation studies.

[5] See *Translation Studies: An Integrated Approach* (Snell-Hornby, 1988), *Translation Studies: An Interdiscipline* (Pöchhacker & Kaindl, 1994), "Implications of System Theory to Translation Studies" (Yang, 2004), *Multidimensional Translation: A Game Plan for Audiovisual Translation in the Age of GILT* (O'Hagan, 2005), and "Translation: Conceptual Integration Process" (Wu, 2006), etc.

In sum, Eco-translatology has sprung into being against the global trend of ecological rationality, inspired by developments in relevant disciplines, and driven by incipient "ecological perspectives" within the translation studies field. It emerged as the times required, out of a "sense of mission" to fill the gap and the personal ideology of the researcher himself/herself. The developmental and background profiles described above clearly indicate that Eco-translatology is an answer to a call from society, culture, and academia as an inevitable expansion of the research horizon in translation studies.

2.3 Prerequisites for the Exploration

As mentioned in "Glancing at Eco-translatology through a 'Shortcut' of the Nine 'Three-in-One' Expressions," the "three prerequisites" of Eco-translatology are as follows:

1. "The Sequence Chain" (translation ←→ language ←→ culture ←→ human beings/society ←→ nature) displays the human cognitive pathway;
2. The homo analogy between translational ecology and natural ecology;
3. A systematic investigation into Translation as Adaptation and Selection explicated as "translation is a selective activity of the translator to fit the translational eco-environment via textual transplants."

The present section is an elaboration of these three prerequisites of Eco-translatological research and its development.

2.3.1 Epistemic Sequence of the "Sequence Chain"

It is now commonly agreed within the academic translation field that the birth and development of translation activities are bound up with the evolution of human society. Once people of different languages desire to communicate with each other, endeavors emerge that rely on translation to surmount the language obstacles to communion (Xu & Tang, 2002, p. 2). From this developmental point of view of human society, translation has massive power in society. Viewed from the praxis of social communications and cultural exchanges, culture, as we all know, emanates from human communication, and cultural exchange starts from the exchange of ideas, which, in turn, is bound up with language. "Inside or between languages, human communication equals translation" (Steiner, 1975, p. 49). From the perspective of linguistics, cultural information is carried and deposited by language, and culture and language are closely connected—culture collapses the moment it is divided from language. From the translational viewpoint, translation activities begin with the diversity of national languages; translation is a transfer of information, a

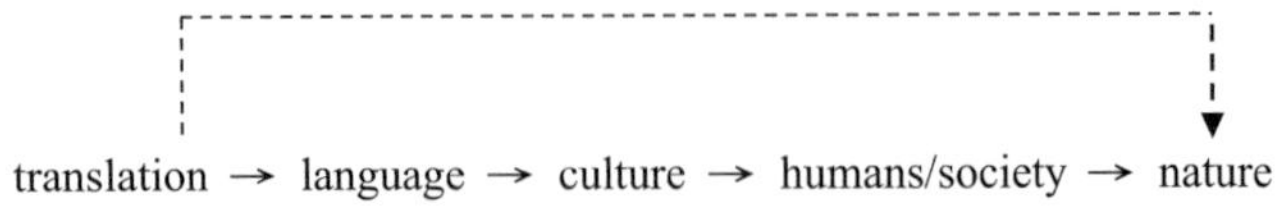

Fig. 2.1 Links from translation to nature

semiotic event, from one language to another. Human culture depends on semiotic activities, and humans are semiotic animals (Bartes, 1964).

Based on the aforementioned points of departure, in order to highlight the pivots, we move directly to the relationship between translation activities and natural ecosystems. Hu (2003, pp. 298–299), the present author of this book, has created an illustration of a chain, tentatively named "the Sequence Chain," that logically links "translation" to "nature" for an expansion of the cognitive horizon.

Because translation is the transfer of languages, and language constitutes culture, culture is a collection of human social activities, and human society constitutes nature, the following chain of interconnection makes sense:

The chain illustrated in Fig. 2.1 shows the vital links from translation to nature, and these links explicate the interconnection and interaction between translation activities and nature. Similarly, a reverse chain of interconnection can be naturally inferred as follows:

A straightforward explanation of Fig. 2.2 is as follows: humans are part of nature, their activities formulate culture carried by language, and translation is a necessity when one language fails to communicate.

The acceptance of the explanation of these "chains" of interrelation leads to the following synthetic illustration:

A synthesis of Figs. 2.1 and 2.2, Fig. 2.3 states that translation activities, such as human behavior and natural events, are on the whole interrelated, and they interact.

The above description and analyses echo what Laplace (1835), a noted French scientist, stated: "Everything is connected in the immense chain of truths, and the general law of nature pins together what seem to be irrelevant." The nature before us is a system, namely, an entirety of interacting bodies.[6] In other words, every phenomenon in nature is connected with the others, and nature is an entirety of laws mingled together (Li & Ren, 1989, p. 19). "More and more scientists claim that evolutive views are integrative concepts of ecology, even other disciplines of natural sciences (or humanities)" (Haught, 2007, p. 121). Dating back to ancient ecological holism, on the other hand, there exist such opinions as "oneness of all things," "continuity of existing things" (Wang, 2005, p. 88), which echo the four ecological "laws"[7] of American ecologist Barry Commoner. Obviously, "principles applicable in the natural world are also pertinent to studies in the humanities, including

[6] See Marx (1972), pp. 492.

[7] The four laws of ecology as described in *The Closing Circle* by Barry Commoner in 1971 are: (1) everything is connected to everything else. (2) Everything must go somewhere. (3) Nature knows best, and (4) there is no such thing as a free lunch.

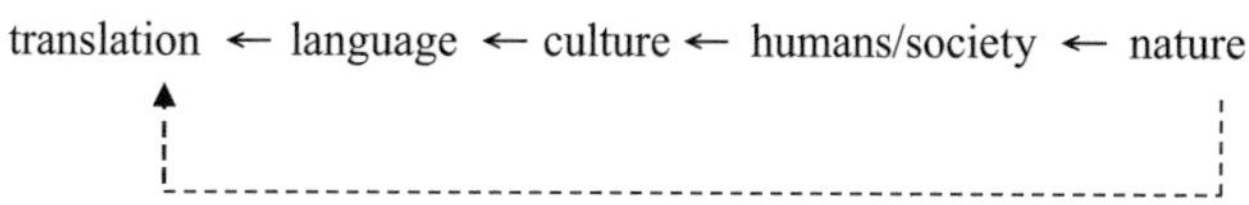

Fig. 2.2 Links from nature to translation

translation” (Hu, 2003, p. 289) as part of human communication and a correlate of

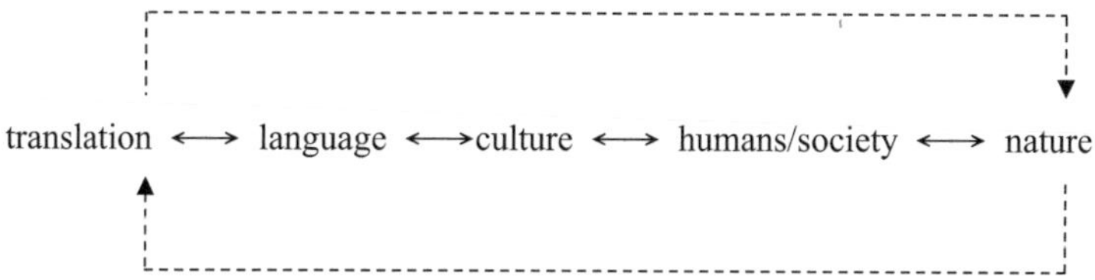

Fig. 2.3 Interrelation between translation and nature

natural events.

“The Sequence Chain” of “translation ⟷ language ⟷ culture ⟷ humans/society ⟷ nature” here profiles the route of expansion in the human cognitive horizon and progression in rational cognition. It complies with the fundamental evolutive law of human cognitive competence and the research routine of “A to B and then to C.” The chain functions, therefore, as a “deductive” or “predictive” chain to foretell the future by drawing on what has happened. As stated by a Chinese saying, “Count the past, and you will know the future Meng” (2011a). What translation studies have undergone, starting from “self-explanation” at the pre-scientific stage to the “linguistic turn” and then to the “cultural turn,”[8] displays, in its true sense, a shift in perspective and expansion of the horizon in this field. As a consequence, the progressiveness and interactivity of “the Sequence Chain” from “translation” to “nature” profiles the logic order and directive mechanism of the human cognitive horizon,[9] i.e., “translation ⟷ language ⟷ culture ⟷ humans/society ⟷ nature.”

[8] If the cognitive order stated by “the Sequence Chain” is to be followed, an anthropological turn can be expected for translation studies as a stage after the “cultural turn,” based on the fact that “translation is a human activity.” Then, an ecological turn follows, based on the fact that “translation is a natural activity.” Strictly speaking, however, the anthropological turn has difficulty distinguishing itself from the cultural turn, as anthropology fails to distinguish itself essentially from anthropological culturology. For instance, the investigations on translation subjectivity and intersubjectivity at the turn of the century were all studies of the “human” factor in the translation process. Such studies fall within the intersection of culturology and anthropology; hence the intangibility of the anthropological turn in translation studies (Meng, 2009, 2012). Furthermore, the “translation as adaptation and selection” theory identifies the “translator’s central position” and interprets the “translator’s predominance.” That, in fact, can be considered a transition from the “anthropological turn” to the “ecological turn” in the field of translation studies and hence a shift in the cognitive horizon.

[9] The interaction of translation activities and nature demonstrated by Hu Gengshen paved the way for Eco-translatology. It can be deduced that the process from translation to nature is not a linear progression. This process integrates cognitive transformation in the human brain and hence denotes

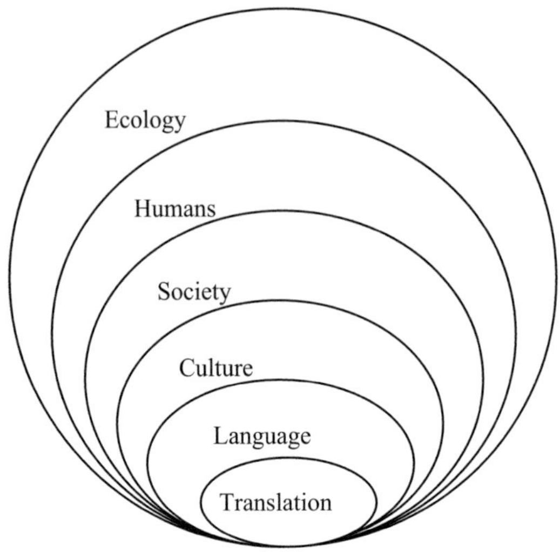

Fig. 2.4 Three-dimensional schematic diagram of "the Sequence Chain"

To demonstrate the logical order and directivity mechanism of the human cognitive horizon, to eradicate, particularly, the misunderstanding of translation to nature as a linear or unidirectional interaction,[10] and to give full expression to the level-by-level inclusion and transcendence between every two systems, the plane figure above is transformed into a three-dimensional schematic diagram (see Fig. 2.4). A detailed transcription of the diagram based on "the Sequence Chain" is as follows: translation studies are a part of language studies, which are a part of cultural studies; cultural studies are a part of sociological studies, which are a part of anthropological studies, while anthropological studies are a part of ecological studies. Intersection, interaction, and extension between elements of "the Sequence Chain" are thus demonstrated in Fig. 2.4.

It can be argued that since natural ecosystems and human social systems interact by nature, translation activities and nature, therefore, are interrelated. This interrelating, intersecting, interacting, and progressive relation denotes not only the connection between the translational ecosystem and natural ecosystem but also, above all, the existing potential of the media and similarity in between (see Sect. 2.3.2).

"The Sequence Chain" is a crucial prerequisite and basis of Eco-translatological development. To a certain extent, except for "the Sequence Chain," studies on the relation between translation activities and the natural ecosystem will be inaccessible, and except for preliminary studies on Translation as Adaptation and Selection (TAS), the path to Eco-translatology will be blocked.

multidimensionality (Zhang & Jin, 2011, p. 263).

[10] Meng Fanjun has explained this relationship with the basic principles of "the Sequence Chain" and expounded the "turns" in translation studies. See Meng (2011b).

2.3.2 Isomorphism Between Translational and Natural Ecosystems

If Eco-translatology can be interpreted as an outcome of integrating translation studies and ecology, then, relevance, affinity, and isomorphism must exist between translation and natural ecosystems, at least to a certain extent. Tracing the relevance, affinity, and isomorphism between these two ecosystems is a process of simplification and rule identification. That process will be methodologically meaningful for Eco-translatological explorations; moreover, the relevance, affinity, and isomorphism between translation and natural ecosystems constitute the primary theoretical foundation for the rationality and objectivity of Eco-translatology.

The relevance and affinity between translation and natural ecosystems have already been elaborated, and the present section focuses on the similarity and isomorphism between translation and natural ecosystems.

Similarity means likeness or rough resemblance. The characteristic of similarity denotes certain traits shared by things or relations existing between them. It expresses the impossibility of things being absolutely identical or of unconditional universality. Things may differ fundamentally, while different things may have a typological resemblance, i.e., commonality at a typological level.

In Eco-translatological terms, the similarities between translation and natural ecosystems are manifold and highly visible. The similarity and isomorphism between translation and natural ecosystems may be approached from aspects such as functional systems, basic traits, and operating rules.

First, ecology highlights the interaction or mutual influence between the eco-environment and organisms. The environment provides the material and energy needed for the growth and development of organisms, and organisms act on or remold the environment. Therefore, organisms adapt to the environment to ensure the natural growth of individuals. Interaction is denoted between elements of the ecosystem, and the change of any member will definitely bring about a change or changes in other members. The same is true of the translational ecosystem. Interested bodies within the translational ecosystem are mutually interactive and intersecting, which makes the translation ecosystem complex and enables interaction between translation ecosystems. For example, the translational management ecosystem correlates and interacts with the translational market ecosystem, the translational market ecosystem correlates and interacts with the translational education ecosystem, and the translational education ecosystem correlates and interacts with the translational ecosystem properly (Hu, 2009a).

Second, in nature, organisms and their environment interact to reach a certain eco-balance. On the one hand, the environmental conditions will determine the form, structure, physiological activities, chemical composition, genetic characteristics, and geographic distribution of organisms; on the other hand, organisms must adjust to adapt to different environmental conditions. The translation ecosystem follows suit. Translators, as representatives of the "translator-community," differ in their mode of thought, educational background, hobbies, translation views, aesthetic criteria, translation experience, etc., and texts are distinguished by text type, readers' expectations, receptor culture, circulation channel, specification environment, etc. These differences create certain adaptations and selections between the "translator-community" subjects, who have to adjust dynamically to adapt to the entire translational eco-environment (Hu, 2009a, p. 6; Hu, 2009b).

Meanwhile, translation itself is a complex ecosystem that likewise features balance. The self-adjustment capacity of the translation ecosystem relies heavily on the capacity of translators as representatives of the "translator-community" within the system. A translation ecosystem of mutual reliance and dynamic balance can be achieved through interaction between the translation subjects and objects, the translation subjects and their eco-environment.

Third, mutual benefit and symbiosis exist between individuals of different species. In other words, both parties in symbiosis can obtain what they need. It is a universal law for the natural ecosystem and for the translation ecosystem. The natural ecosystem is dynamic, as is the translation ecosystem. In the natural ecosystem, humans are conscious and purposeful, capable of improving, conditioning, and reconstructing eco-relations within the natural environment; in the translation ecosystem, the conscious and purposeful activities of translators as representatives of the "translator-community" are likewise capable of improving, conditioning, and reconstructing the translational eco-environment (for instance, strengthening the macromanagement of the translation industry, formulating the academic norms of translation studies, enriching the content of translation training, and exalting the social status of the translator-community). In brief, the fundamental laws of these two ecosystems are commensurable.

Fourth, similar principles function in these two ecosystems. For example, in the natural ecosystem, the principle of repellant competition works: the more prominent similarity two species share in the utility of resources and space, the more ecological intersection there will be and the more severe their competition will become, while in a stable environment, two or more species that are resource-conditioned and identical in their utilization pattern of resources will not coexist for long. The same is true for the translation ecosystem. For example, criticism criteria for translations of different genres vary, leaving no room for comparison, while translations of one genre are comparable, resulting in fierce competition. In addition, the increasing professionalization and commercialization of translation as a service industry ultimately intensify competition.

Fifth, similar phenomena exist within these two ecosystems, and they follow similar patterns. For instance, in the natural ecosystem, there are "the beauty and infinite complexity of the co-adaptations,"[11] and "organic beings vary under the impact of their environment." In the translation ecosystem, "adaptation," "selection," "existence," and "extinction" prevail in the translating process. The adaptive optimal selection of translators is actually a kind of conscious or unconscious

[11] For all the ensuing quotations, see Chap. 4 in Charles Darwin's *The Origin of Species* (1859).

translator action. Translation is a series of optimal selections, and a translation is the outcome of the translator's adaptation to the translational eco-environment.

In a natural ecosystem, "if any one species does not become modified and improved in a corresponding degree with its competitors, it will soon be exterminated," and "natural selection can act only by the preservation and accumulation of infinitesimally small inherited modifications, each profitable to the preserved being." In the translation ecosystem, translation is a process of the translator's constant preservation/accumulation of various optimal improvements, which give rise to a translation of the highest "holistic degree of adaptation and selection." The ultimate purpose of translation studies is to produce a translation of optimal adaptability and to facilitate the ultimate development of translation studies.

As a further example, "it may be said that natural selection is daily and hourly scrutinizing, throughout the world, every variation, even the slightest; rejecting that which is bad, preserving and adding up all that is good; silently and insensibly working, whenever and wherever opportunity offers, at the improvement of each organic being in relation to its organic and inorganic conditions of life," and "even if organic beings had adapted themselves to their living conditions in the past periods, but when conditions change, they will not be completely adaptive unless they vary accordingly." In a translation ecosystem, a translation results from the accumulation of the translator's constant rejection of "that which is bad" and preservation of "all that is good" and continuous improvement according to various specifications. "Preserve the strong and eliminate the weak" is not only natural law but also the rule that the translator follows in translation. Because the translational eco-environment is dynamic and changing, and translation criticism is subject to the dynamic translational eco-environment, "retranslation" is natural in translation activities.

Finally, and most importantly, family resemblance (Wittgenstein's idea of family resemblance; Nystrom, 2005, p. 473) knits members of a homogeneous society and those of the natural world tightly together, as stated in many studies. In particular, the "translator-community," with the translator as its representative, is simultaneously a collective of subjects in translation activities that aim ultimately for the cross-lingual and cross-cultural communication of human beings. "Humans" and "human activities" are thus involved in both ecosystems; therefore, the isomorphism between them speaks for itself.

In summary, what is noticeable is an affinity between the translation ecosystem and natural ecosystem and isomorphism between the systemic traits of these two ecosystems. This suggests analogical rules or laws between these two ecosystems; in turn, some rules or laws in the natural ecosystem may work in the translation ecosystem. Such studies and discussions are not uncommon among scholars in different fields. For example, Charles S. Pierce, a philosopher, thinker, and sociologist, announced that the cognitive process of human thinking, which is "naturalized," is applicable to natural events, and thinking itself is an activity of traits similar to natural activities (Gorlée, 1994, pp. 48–49); Keesing (2001, pp. 13–14) states that human beings, to anthropologists, are both living organisms and cultural products, with interactive and complementary cultural essence and biological behavior, if seen from the perspective of a natural system and a social system, and hence an extensive perspective is possible for human evolution; and Blackmore (1990, p. 60) comments that selection in the organic world and cultural world operates on the basis of an identical criterion, i.e., the absolute criterion of adaptability, and these two types of selection are complementary.

Similar arguments can be found among Chinese scholars. For instance, Qian (2001, pp. 1, 206), a philosophical theoretician, believes that human beings belong to nature and that every human is connected, in the form of limited manner and unlimited existence, to nature; Qian (2002, p. 133), a theoretical linguist, finds that human language and the human existential pattern are mutually containing, and holographic; and Zhan (2002, p. 9) proposed that humans and nature interact on an equal level, as far as explorations of scientific knowledge are concerned and are mutually complementary and dualistically active and passive.

Steiner (1975, p. 235), a translation theoretician, has a similar idea and suggests that translation theories fall dualistically into "universalist and relativist theories," and Lance and Jacky (1991, p. 34) conclude that these two views, generally speaking, seem to tally with two basic ways of human adaptation to the environment.

On the one hand, affinity and isomorphism do exist between traits of the translation ecosystem and natural ecosystem, as stated in the previous analysis and exposition, and these traits, in turn, constitute the basis and foundation of Eco-translatology. On the other hand, the discussions of scholars from diverse fields on the interrelation, interaction, and complementarity of the natural ecosystem and human social system, as presented above, are both reference and support for Eco-translatological explorations.

2.3.3 *Systematic Studies on Translation as Adaptation and Selection (TAS)*

Eco-translatology develops from the approach to Translation as Adaptation and Selection (TAS); hence, the latter constitutes a crucial prerequisite and basis of the former. TAS ushered into translation studies the theory of "adaptation/selection" from the Darwinian theory of evolution[12] and redefined translation as "a selection

[12] Schleicher (1863) pioneered the introduction of Darwinian theories into linguistic studies. In his article "Darwin's Theory and Linguistics," Schleicher advocated that linguistics, and Darwin's theory, should be aligned as natural sciences and that Darwin's theory works in linguistics. He stated that language, like other living forms in the world, experiences development, maturation, and death and that language contacts and conflicts impact and infiltrate mutually as a fatal struggle between living bodies and hence "survival of the fittest." Later, Sapir (1921), Jakobson (1970), Lieberman (1975), and Pawley (1983) probed the relation between language evolution and human existential conditions or interpreted phenomena of language use and changes via natural selection and adaptation (Verschueren 1999, p. 314). Verschueren (ibid) himself explicated "evolutionary epistemology" as an expansion of biological theories, the natural selection

activity of the translator's adaptation to fit the translational eco-environment, dominated by the translator to target the cross-cultural transmission of messages." Meanwhile, "context" in translation has been expanded conceptually to the "translational eco-environment" to denote "the worlds of the source text and source/target languages,"[13] an entirety of interrelated and interactive members and a collection of factors conditioning the translator's optimum adaptation and selection, "comprising the linguistic, communicative, cultural and social aspects of translating, as well as the author, the client, and the readers." The process of translating is interpreted as the "alternate cycles of translator's adaptations and selections" initiated by the translator. The translator adapts and selects, making an adaptive selection since selection is integrated into adaptation, and selectively adapts since adaptation is integrated into selection (Hu, 2001, 2003).

"Translation is adaptation and selection," as the thematic concept of TAS, explicates "what translation is"; the "translator's dominance" and "translator-centeredness" in the translation process, as the core notion of TAS,[14] explain "who translates"; "preserving the strong and eliminating the weak"/"survival through selection" and "selective adaptation"/"adaptive selection," as the mainstay translation methods of TAS, state "how to translate"; and "survival of the fittest" and in particular "doing things with translation" (i.e., the motives of a translation act), as the primitive purpose of translation, elaborate on the issue of "why to translate."

The above exploration and expounding of TAS in relation to ontological translation is crucial to the discourse construction of Eco-translatology since the former, when modified and reconstructed as the "meso-level" ontological recognition and "micro-level" Eco-translation operations of the latter, constitutes the "core" and "base" of the Eco-translatological system (see details in Chap. 5). This explains why Eco-translatology is labeled an expansion and continuation of TAS, and TAS is considered the base of Eco-translatology (regarding the connection and technical consistency between them, see Sect. 2.4.3).

---

paradigm in particular, into aspects of behavior and social culture, including language, learning, and science.

[13] In the world of academic translation, cases of comparing "diversified factors" pertinent to translation to "world" or "worlds" are not uncommon. For instance, Wittgenstein (1955, p. 31) wrote that "the *world* is the totality of facts"; Steiner (1975, p. 246) stated 40 years ago that the translator concretely reexperiences the evolution of language itself, "the relation between language and *world*, between 'language' and '*worlds*'"; Baker (1992, p. 217) described the totality of "factors" as "world," "the model of the *world* presented in the source text," and "the target culture's version of the *world*," in her terms; and Wilss (1996, p. 175) depicted all "extralinguistic" and "sociocultural" factors as "knowledge of the *world*."

[14] As pointed out by Nida and Taber (1969, p. 188), "[T]he real problems of translation are not technical, they are human," and by Hermans (2000, pp. 12–13), translation tells us more about the translator than about the translations. The focus of Eco-translatology on "organic bodies," i.e., "human" bodies, in the "translation community" is a "translator-oriented" perspective.

2.4 Continuous Studies and Developments

2.4.1 Initial Studies at the Beginning of the Twenty-First Century

Eco-translatology germinated in Hong Kong at the beginning of the twenty-first century. Hong Kong, a center of cultural contact, bridges China since its opening-up and the West. Its position as a gateway promises an explosion of translation activities and translation thoughts. Translation seminar series at the Centre for Translation of Hong Kong Baptist University have facilitated exchange and dialogue on investigations and findings in translation theories between scholars and practitioners, as well as postgraduates in this field, since 2001, with public lectures from translation scholars from China or the West.

My exploration into Eco-translatology started there in June 2001, when I was studying for my doctoral degree in the Department of English Language and Literature, and a public lecture titled "From the Darwinian Principle of Adaptation and Selection to Translation Studies"[15] was delivered in the seminar series on October 22nd to elaborate on the foundations and feasibility and to conceive of ways to shape TAS. Since then, keen interest has been expressed by over 50 seminar scholars and practitioners from Lingnan University of Hong Kong, City University of Hong Kong, Hong Kong Polytechnic University, Chinese University of Hong Kong, and Hong Kong municipal agencies.

A conference article, "An Initial Exploration into an Approach to Translation as Adaptation and Selection" (Hu, 2001),[16] was presented at *The 3rd Asian Translators' Forum* on December 6, 2001. Translation was redefined, and the translating process, principles, and methods were reinterpreted from the adaptation/selection perspective with the relation between the translator's adaptation and selection explicated through a rough sketch of the theoretical framework of Eco-translatology. The exploration of the translator's adaptation to the translational eco-environment and his/her selection of the final form of a translation paves the way for the systematic construction of TAS, thus laying the foundation of Eco-translatology. "Translation as Adaptation and Selection," my English treatise, was published in 2002 in *Perspectives: Studies in Translatology*, with a concluding statement defining this exploration as "an ecological approach to translation."[17] Essentially, TAS germinated in Hong Kong in 2001, a year witnessing this pioneering exploration of the ecological approach to translation. Such investigations had not occurred before and signaled the birth of a new approach to translation studies.

---

[15] http://www.trans.hkbu.edu.hk.

[16] Published later as "Prosperity, Perplexity and Perplex of Translation Theories" (Hu, 2002).

[17] Hu (2003, p. 278).

2.4.2 Progressive Achievements and Information Dissemination

The exchange of research findings and spread of academic thoughts are both gradable and infallible regardless of what scientific studies and development are concerned. "Publish or perish" prevails as a motto throughout worldwide academia, which correctly depicts the route of Eco-translatology. On the whole, publications, lectures, conference proceedings, etc., constitute the main channel of Eco-translatological explorations and findings. A Chinese saying is "Focus and attention together make success." Eco-translatology persists in its pursuit of "focus" with studies, treatises, columns, seminars, etc., from an Eco-translatological thematic perspective.

In addition to the two articles mentioned above, a series of treatises on Eco-translatology has been published, including "A New Perspective to Translation" (*Cambridge*, Britain), "Experiment of Translation Studies: A Questionnaire Survey of TAS" (Foreign Languages and Their Teaching, 2004, China), "From 'Translator's Subjectivity' to 'Translator-centeredness'" (*Chinese Translators Journal*, 2004, China), "Translator-centeredness" (*Perspectives: Studies in Translatology*, 2004, Denmark), "Translation Principles and Methods of Translation as Adaptation and Selection: Exemplification" (*Foreign Languages Research*, 2006, China), "Eco-translatology: A Primer" (*Chinese Translators Journal*, 2008, China), "Eco-translatology: Backgrounds and Bases for Its Development" (*Foreign Languages Research*, 2010, China), "Eco-translatology: Research Foci and Theoretical Tenets" (*Chinese Translators Journal*, 2011, China), and "Eco-features of Eco-translatology and the Implications to Translation Studies" (*Foreign Languages in China*, 2011, China).

Recently, monographs or symposia proceedings on or related to Eco-translatology have been published, for example, *An Approach to Translation as Adaptation and Selection*[18] and the *Translation and Intercultural Communication* series, with treatises such as "Rethinking Translation as Adaptation and Selection" and "An 'Ecological Turn' Signified by the 'Sequence Chain'."

Lectures and exchanges have facilitated the spread of Eco-translatological findings. Eco-translatological scholars have toured to give lectures at universities and institutes in Hong Kong, Taiwan, and Macao as well as in other parts of China, Britain, Indonesia, Belgium, etc., and have presented their findings at national and international seminars such as "Translation Globalization: International Symposium of Cross-disciplinary Theoretical Construction" (China), "The FIT Fifth Asian Translators' Forum" (Indonesia), "Tracks and Treks in Translation Studies—European Society for Translation Studies 6th EST Congress—Leuven 2010" (Belgium), etc. "Interpretation of Eco-translatology," "Translation Ecosystems: An

---

[18] Listed in "Key National Booklist of the Tenth Five-year Plan," and published in 2004 by Hubei Education Press, China.

Eco-translatological Perspective," "The Ecological Track: A Worthwhile Trek—Eco-translatology Explained," etc., are keynote reports from those conferences.

It is worth mentioning that "language, culture, society, and translational eco-environment" has been a theme at the "Cross Taiwan Straits Symposium on Translation and Intercultural Communication" series commenced in 2005. Relevant treatises presented at the third symposium held at Fu Jen University of Taiwan include "Inheritance and Innovation of Eco-translatology" by Meng Fanjun from Southwest University, "A Non-binary View on Translation Errors as Potential Choices in the Light of the Linguistic Adaptation Theory" by Song Zhiping from Northeast Normal University, "Translation Studies Depends on Its Areal Ecosphere" by Xu Jianzhong from Tianjin University of Technology, and "Development and Progress of Translation Studies from an Eco-translatological Perspective" by Hu Gengshen from Tsinghua University. Eco-translatological scholars have also voiced their views through international exchange and cooperation in addition to journal articles and conference presentations (see Sect. 2.4.5).

Publications, lectures, and exchanges do push forward the spread of Eco-translatological findings, and this approach to translation is drawing attention domestically and abroad. Applied research is rapidly increasing, whereas ontological study in the field still calls for innovative insights.

2.4.3 Incorporation of TAS into Eco-translatology

The incorporation of TAS into Eco-translatology ranks as a directional breakthrough, or a turn, of the latter. TAS, originating from the relation between organisms and the eco-environment, tints itself an ecological color, which was established at its birth in 2001 and has continued since then. Elaboration on the connection between "natural selection" and "survival of the fittest" in Darwinian principle and ecological progress, nevertheless, leave much to be desired, and exploration of the filiations between the two base disciplines, i.e., biology and ecology, has missed much. To some extent, the theoretical foundation of TAS was thus blocked from being expanded, and perplexity consequently ensued (for instance, issues regarding the relation between biology and ecology, the relation between TAS and Eco-translatology and the applicability of their theoretical foundations, and the transition between these two theories).

Clarification of perplexity is a process of ideological innovation, compensating for the shortage and furthering the investigation. Literature reviews can provide a pleasant surprise in that relevant argumentations clarifying such perplexity can be identified without any hindrance.

Darwinian "natural selection" contributed to the theory of evolution, according to Ernst Haeckel, an ecologist who named the discipline, similar to other ecologists in the early years; one grand task of ecology in modern biology has been to unveil the mystery of adaptability, as stated by biologist Henry Cowles. The adaptation of organisms to the environment is attributed to organic evolution, which, with a full elaboration in the *Origin of Species*, brings Darwin the reputation of being a great advocate (Wang & Zhou, 2004, p. 77).

The environment is diversified and changing, and so is the adaptability of organisms to the environment. The evolution of organisms results directly from this ecological relation between environment and organism. Organisms, particularly plant life, have great potential to reshape the environment, and the evolution of organisms urges, in turn, the transformation of the environment, ending with the co-evolution of both parties.

Furthermore, ecology develops from phytoecology (plant ecology), the progress of which naturally carries forward zoology (animal ecology). Plants and animals are subjects of ecology; therefore, the discipline was defined in early years as "the branch of biology that deals with the relations between living organisms and their environment" (Neufeldt, 1995, p. 189) and therefore as a science of the dialectical relationship between organisms and between organisms and the environment.

Deductions emerged from the abovementioned: on the one hand, apparent filiations exist between biology and ecology, with the former originating and developing prior to the latter and the former as an umbrella discipline covering the latter in the early years; on the other hand, it is commonly acknowledged that Darwinian theory contributed to both biological and ecological developments.

These deductions are meaningful to systematic translation studies in that, in ecological terms, "isogeny" and inheritance label the relation between TAS and Eco-translatology, which share something substantially in common.

These deductions are meaningful to Eco-translatology in that, in terms of scientific principles or laws, the transition and linkage of TAS and Eco-translatology once realized, a theoretical base and research orientation are explicated for the macro-level studies of the translation ecosystem in Eco-translatology; more importantly, a base or support is erected for the "meso-level" ontological translation theories and "micro-level" Eco-translation operations. This promises the potential of constructing an Eco-translatological discourse system characteristic of the "three-in-one" of translation studies, translation theories, and translations.

What follows here is a more specific interpretation of the above argumentation. TAS in its earlier form is positioned as a systematic translation theory, which, however, should be graded differently from investigations into translation studies, one being apart and the other being the "entirety." In other words, TAS, relative to translation studies as a whole, is part of "low- and medium-level" investigations. Here, we present macro-level translation ecosystem studies pertaining to the investigations into translation studies. A transition and connection can thus be realized between meso-level and micro-level studies and macro-level translation ecosystem studies; the linkage between "translation studies framing," "translation theory," and "translation shaping" is, therefore, a likely outcome that will integrate studies in these three areas for a spontaneous theoretical construction of Eco-translatology characteristic of "three-in-one."

Pushing through perplexity, reflection, and clarification from 2003 to 2005, I accomplished in 2006 a treatise titled "Understanding Eco-translatology"[19] (Hu, 2006) and presented it at an international conference on "Translating Global Cultures: Towards Interdisciplinary (Re) Construction" the same year to elaborate on the essence, background, current state, findings, limitations, and prospects of the approach. Deep concern and interest were the response, and "Eco-translatology: An Ecological Approach to Translation Studies" received support from China's National Social Science Fund.

2.4.4 *Theoretical Applications and Further Development*

In any theory or discipline, constant concern and applications motivate investigations into the theory. In the past decade, Eco-translatology, developed from TAS, has embodied the belief in "survival of the fittest." As stated by Yang and Xu (2011), Eco-translatology is now a fact, which shows its value. Ultimately, constant concern and application add fuel to enable steady progress.

Academic interest in and application of Eco-translatology are shown in different ways. First, book reviews and comments from scholars domestically and abroad have appeared in succession and are mounting in number as relevant monographs are turned out, together with published articles, conference proceedings, or collected keynote speeches. For instance, statistics show that *An Approach to Translation as Adaptation and Selection*, after it was first printed in late 2004, has collected about a score of commentary articles, for example, "A Translation Monograph of Exploration: On *An Approach to Translation as Adaptation and Selection*" by Liu and Xu (2004, pp. 40–43); "A Breakthrough in Theoretical Construction: On *An Approach to Translation as Adaptation and Selection*" by Li and Huang (2005, pp. 95–96); "*An Approach to Translation as Adaptation and Selection*: A Book Review" by Cai (2006, pp. 39–42); "*An Approach to Translation as Adaptation and Selection*: A Monograph of Innovation" by Song (2007, pp. 105–106); "Turn, Surpass, and Return: Theoretical Reflection of the Theory of Translation as Adaptation and Selection"[20] by Meng (2009); editor's notes of "Eco-translatology" Column in Shanghai Journal of Translators by Fang (2009, p. 2); editor's notes of "New Perspective to Translation Studies" Column on Foreign Language by Yang (2010, p. 62); and "Reflections upon the Budding Eco-translatology in China" by Liu and Xu (2011, pp. 52–56).

Meanwhile, reflections, questions, and doubts have emerged in the form of journal articles in *Journal of Sichuan International Studies University* and *Shanghai Journal of Translators*, postgraduate theses, etc., all with a good faith intention to push further in the investigations. These comments are voices of academic concern

[19] Published later as "Eco-translatology: A Primer" (Hu, 2008).

[20] See http://www.language.international.com. Retrieved on August 2008.

and, at the same time, the effects of this approach in academia that accompany the birth and growth of Eco-translatology. Positive or derogatory, they are all deemed benevolent, inspiring, and illuminative by the author.

Furthermore, Eco-translatology has been adopted as a theoretical framework of postgraduate theses or doctoral dissertations in over 100 Chinese universities or institutes,[21] including Tsinghua University, Beijing Foreign Languages University, Shanghai International Studies University, Guangdong University of Foreign Studies, Jiangsu University, Shandong University, Henan University, and Sichuan University (see Sect. 3.2.3).

2.4.5 *Development Momentum: From Streamlet to "Strong Current"*

The last decade or so has witnessed the birth and growth of Eco-translatology, which is now drawing attention from scholars in different countries (see Sect. 8.4.2). This ecological approach to translation studies continues to develop, progressing from "exploration" to "construction" and "expansion" (Sect. 8.2.4). Scholars advocating this approach spare no effort to carry it forward and obtain a rich harvest in terms of the theoretical framework, backup research, international communication, base construction, etc.

Theoretical explorations and applicative studies continue to develop. The landmarks of such development and progression are as follows: (1) hundreds of pertinent publications, MA theses, and doctoral dissertations; (2) sequential explorations carried out by "cross-Taiwan Strait" scholars; (3) a program supported by China's national fund[22]; (4) columns in foreign-language or translation journals; (5) keynote reports at symposiums; (6) book reviews published domestically and abroad; (7) the establishment of an international association; (8) the convening of international symposiums; (9) the circulation of specialized journals; (10) the formation of research bases[23]; and (11) progressive international expansion.

The second decade of Eco-translatological research will be planned a decade of internationalization. On the one hand, two or three monographs have been printed in recent years; on the other hand, the operating mechanism of the International Association of Eco-translatology Research will be improved, its website (http://www.Eco-translatology.net) will be renewed, international symposiums will be held, and the *Journal of Eco-translatology* (Chinese and English editions) will be better edited.

---

[21] See Sects. 7.2–7.4 for the contents and comments of some of these theses.

[22] See note 2 in "ecological paradigm" of Sect. 3.2.3 for locally supported programs.

[23] In addition to the national base of Eco-translatology research in Macao, China, research centers have been established at Shanghai Maritime University, Southwest University, Qufu Normal University, and Zhengzhou University and in other parts of the world Xu (2011).

The field of Eco-translatology is undeniably growing and progressing. As stated by Liu and Xu (2011, p. 56), there are definitive reasons to believe that Eco-translatology should be watched and accepted by the academic translation field, opening a brand-new world in the area of translation theory and practice.

2.5 Summary

Eco-translatology germinated in 2001 laid its foundation in 2003, consolidated in 2006, and expanded in 2009, moving along an ascending course. It integrates global trends and Chinese factors, interior and exterior components, and objective and subjective elements. Its birth and development satisfy the demands of society, culture, science, and translation studies proper. Its emergence at the turn of the century is an event of chance and inevitability.

Three arguments premise its existence, objectivity, and sustainability. "The Sequence Chain" directs investigations into the relation between translation activities (translation ecosystem) and nature (natural ecosystem); such investigations channel the Darwinian theory of "adaptation/selection," especially ecology, in natural sciences into translation studies. Without the theory of "adaptation/selection," the theoretical construction of Translation as Adaptation and Selection would not be possible; thus, studies of macro-level translation ecosystems, meso-level ontological translation theories, and micro-level Eco-translation operations are still expected. Without this "three-level" study, the construction of an Eco-translatological framework remains a castle in the air. All these aspects assume a step-by-step progression from part to whole, from randomness to systematization.

Building in the first decade of the twenty-first century, Eco-translatology harvests its findings, explicates its research methodology and develops its route, and shapes a three-level research pattern. The number of its followers is mounting, and international exchange and communication are occurring.

A good start potentiates a bright future. There are still issues that remain untouched, ranging from research subjects to core notions, research foci and academic views of Eco-translatology to what distinguishes it from other approaches to translation studies, i.e., the distinctive features of Eco-translatology. These must be the key concern of the academic translation field now and thus will be presented in the ensuing chapters.

References

Baker, M. (1992). *In other words: A coursebook on translation*. London and New York: Routledge.

Bartes, R. (1964/2005). *Semiotic aesthetic theory* (X. W. Dong & K. Wang, Trans.). Shenyang: Liaoning People's Press.

Bassnett, S., & Lefevere, A. (Eds.). (1990). *Translation, history and culture*. London and New York: Pinter.

Blackmore, S. (1990). *The meme machine*. New York: Oxford University Press.

Cai, X. (2006). A brief review of an approach to translation as adaptation and selection. *Chinese Science & Technology Translators Journal, 1*, 58–59.

Cronin, M. (2003). *Translation and globalization*. London: Roultledge.

Cui, Q. (2006). *Chinese translation market: Current situation and analysis*. Retrieved August 25, 2006, from http://www.teacherwong.com/zxxx/ShowArticle.asp?ArticleID=389

Fang, M. (1999). *Translation research and practice*. Qingdao, China: Qingdao Publishing House.

Fang, M. Z. (2009). Editor's notes on eco-translatology. *Shanghai Journal of Translators, 2*, 2–17.

Fill, A. (2001). Ecolinguistics—state of the art 1998. In A. Fill & P. Mühlhaüsler (Eds.), *The ecolinguistics reader: Language, ecology and environment*. London: Continuum.

Gabbard, G. O. (2000). *Psychodynamic psychiatry in clinical practice* (3rd ed.). Washington, DC: American Psychiatric Press.

Gorlée, L. (1994). *Semiotics and the problem of translation: With special reference to the semiotics of Charles S. Peirce*. Amsterdam and Atlanta, GA: Editions Rodopi B.V.

Gu, Z. (2004). *Linguistic culturology*. Beijing, China: Tsinghua University Press.

Haught, J. F. (2007). Evolution, design and the idea of providence. *Chinese Cross Currents, 4*, 116–128.

Hermans, T. (2000). *Norms of translation*. Oxford, England: Oxford University Press.

Hu, G. S. (2001, December 6–8). *Study on translation as adaptation and selection*. Paper presented on the third Asian Translators symposium of International Federation of Translators, Hong Kong, China.

Hu, G. S. (2002). Prosperity, perplexity and exploration in translation theory. *Translation Quarterly, 25*, 80–104.

Hu, G. S. (2003). Translation as adaptation and selection. *Perspectives: Studies in Translatology, 4*, 283–291.

Hu, G. S. (2006, August 14–16). *Understanding eco-translatology*. Paper presented at the international conference on translating global cultures: Towards interdisciplinary (re)construction, Beijing, China.

Hu, G. S. (2008). Eco-translatology: A primer. *Chinese Translators Journal, 6*, 11–15.

Hu, G. S. (2009a). An "ecological turn" signified by the "sequence chain". In G. Hu (Ed.), *Translation and cross-cultural communication: Integration and innovation* (pp. 26–39). Shanghai, China: Shanghai Foreign Language Education Press.

Hu, G. S. (2009b). Eco-translatology: Trans-disciplinary integration on translation studies. *Shanghai Journal of Translators, 2*, 3–8.

Jakobson, R. (1970). *Main trends in the science of language*. New York: Harper and Row.

Ji, X. (2005, March 17). Listen to Mr. Ji Xianlin in Translation (interviewed by Li Jingrui). *Guangming Daily*.

Katan, D. (1999). *Translating cultures*. Manchester, England: St. Jerome Publishing

Keesing, R. (2001). *Cultural anthropology: A contemporary perspective* (G. Zhang & J. Yu, Trans.). Taipei, Taiwan: Ju Liu Books Ltd.

Kerridge, R. (1998). Introduction. In R. Kerridge & N. Sammells (Eds.), *Writing the environment: Ecocriticism and literature* (pp. 1–9). London: Zed Books Ltd.

Lance, H., & Jacky, M. (1991). *Redefining translation: The variational approach*. London and New York: Routledge.

Laplace, P. (1835/1978). *Cosmological system theory* (Y. Li et al. Trans.). Shanghai: Shanghai Translation Press.

Li, G., & Ren, D. (1989). *Introduction to interdisciplinary science*. Wuhan, China: Hubei People Press.

Li, Y., & Huang, Z. (2005). A unique theoretical build-up—reviewing Hu Gengshen's *Approach to Translation as Adaptation and Selection*. *Foreign Language Education, 6*, 95–96.

Lieberman, P. (1975). *On the origins of language: An introduction to the evolution of human speech*. New York: Macmillan.
Liu, A. (2010, November 9–10). *On translation eco-environment*. Paper presented at the first international eco-translatology symposium, Macau, China.
Liu, Y., & Xu, J. (2004). A new translation book with spirits of exploration—book review of *Translation as Adaptation and Selection*. *Chinese Translators Journal, 6*, 40–43.
Liu, Y., & Xu, J. (2011). On emerging eco-translatology in recent ten years. *Journal of Eco-Translatology, 2*, 52–56.
Lv, J., & Hou, X. (2006). *Translatology—a constructive perspective of translation study*. Shanghai, China: Shanghai Foreign Language Education Press.
Marx, K. (1972). *Marx and Engels collected works* (Vol. 3, p. 492). Beijing, China: People's Press.
Meng, F. (2009). The new turn of translation studies after post-modernism. In H. Gengshen (Ed.), *Translation and interculture: Integration and innovation* (pp. 48–54). Shanghai, China: Shanghai Foreign Language Education Press.
Meng, F. (2011a). Contemporary translation studies from eco-translatology perspective. *Journal of Eco-translatology, 1*, 73–79.
Meng, F. (2011b). On present translation studies from the perspective of eco-translatology. *Journal of Eco-translatology, 1*, 73–77.
Meng, F. (2012, November 23–25). *Studies on metaphysics of translation eco-environment*. Presented at the third international eco-translatology symposium, Chongqing, China.
Mühlhaüsler, P. (2003). *Language of environment, environment of language: A course in ecolinguistics*. London and New York: Paul & Co Pub Consortium.
Neufeldt, V. (1995). *Webster's New World College Dictionary*. New York: Macmillan.
Newmark, P. (1988). *A textbook of translation*. London; Toronto, ON; Sydney, NSW; Tokyo: Prentice-Hall.
Nida, E., & Taber, C. (1969). *The theory and practice of translation*. Leiden: E. J. Brill.
Nystrom, J. (2005). The definition of partnering as a Wittgenstein family-resemblance concept. *Construction Management and Economics, 23*(5), 473–481.
O'Hagan, M. (2005, May 2–6). Multidimensional translation: A game plan for audiovisual translation in the age of GILT. In *Proceedings of the Marie Curie Euro conferences MuTra 'challenges of multidimensional translation'*, Saarbrücken, Germany.
Parham, J. (2002). *The environmental tradition in English literature* (pp. 1–8). Hampshire: Ashgate Publishing.
Pawley, A. (1983). Two puzzles for linguistics theory: *Nativelike Selection and Nativelike Competence*. In J. C. Richards & R. W. Schmidt (Eds.), *Language and communication* (pp. 191–227). London: Longman.
Phillipson, R. (2006). Book review. *Language Policy, 5*, 227–232.
Pöchhacker, F., & Kaindl, K. (Eds.). (1994). *Translation studies: An interdiscipline*. Amsterdam and Philadelphia: John Benjamins.
Qian, G. (2002). *The theory of language holography*. Beijing, China: Commercial Press.
Qian, J. (2001). *The selection of existence*. Beijing, China: Chinese Social Science Press.
Sapir, E. (1921). *Language: An introduction to the study of speech*. New York: Harcourt, Brace.
Schleicher, A. (1863). *Die darwinsche Theorie und die Sprachwissenschaft: Offenes Sendschreiben an Herrn Dr. Ernst Hackel*. Weimar, Germany: Böhlau.
Snell-Hornby, M. (1988). *Translation studies: An integrated approach*. Amsterdam and Philadelphia: John Benjamins Publishing Company.
Song, Z. (2007). *Choice and adaptation: A study of translation process from the pragmatic perspective*. Changchun, China: Northeast Normal University.
Steiner, G. (1975). *After babel: Aspects of language and translation*. Oxford, England; New York: Oxford University Press.
Verschueren, J. (1999). *Understanding pragmatics*. Beijing, China: Foreign Language Teaching and Research Press & Edward Arnold (Publishers) Limited.

Wang, N. (2005). Confucian thought and western ecological holism. In D. Le & B. Li (Eds.), *Cross-cultural conversation* (Vol. 15, pp. 88–96). Shanghai, China: Shanghai Culture Press.

Wang, R., & Zhou, H. (2004). *Human and ecology*. Kunming, China: Yunnan People's Press.

Warren, R. (Ed.). (1989). *The art of translation: Voices from the field*. Boston: Northeastern University Press.

Westling, L. (2002). Introduction. In J. Parham (Ed.), *The environmental tradition in English literature* (pp. 1–8). Hampshire, England: Ashgate Publishing Limited.

Wilss, W. (1996). *Knowledge and skills in translator behavior*. Amsterdam and Philadelphia: John Benjamins Publishing Company.

Wittgenstein, L. (1955). The role of theory in aesthetics. *Journal of Aesthetics and Art Criticism, 16*, 27–35.

Wu, F. (2006). *Translation: Conceptual integration process*. Changsha, China: Central South University.

Xu, J. (2009). *Translation ecology*. Beijing, China: Three Gorges Publishing House.

Xu, J. (2011, November 11–13). *Insistence and construction—a speech at the closing ceremony of the "second international eco-translatology symposium"*, Shanghai, China.

Xu, J., & Tang, J. (2002). Translation: A noble enterprise (Preface). In J. Xu & J. Tang (Eds.), *Babel Translation Series*. Wuhan: Hubei Education Press.

Yang, H. (2004). Revelation on system theory to study of translation theory. *Journal of Changshu University, 18*(6), 38–39.

Yang, P., & Xu, J. (2011, November 11–13). *New progress of eco-translatology. speeches. Delivered at the second international eco-translatology symposium*, Shanghai, China.

Yang, X. R. (2010). Editor's notes on eco-translatology. *Foreign Languages Research, 4*, 62–67.

Yu, X. (2007, August 14). Construction of a nature-friendly ecological civilization. *Guangming Daily*.

Zeng, W. X. (2006). Reflections on the "cultural turn" of translation studies. *Foreign Languages Research, 3*, 45–47/80.

Zhan, Z. (2002). *Construction: Theoretical foundation and educational applications*. Taipei, Taiwan: Chang Chun Group.

Zhang, L., & Jin, W. (2011). Eco-translatology and its three ecological dimensions—a discussion with Prof. Hu Gengshen. *Journal of University of Shanghai for Science and Technology (Social Science), 4*, 261–266.

Zhang, M. (2005). Cultural symbiosis on binary opposition of translation conception. *Journal Beijing International Studies University, 2*, 6–9.

Zhang, S. (2002). Process category and process philosophy. *Academic Exchange, 5*, 17.

Chapter 3
Eco-Translatology: Research Foci and Theoretical Tenets

3.1 Introduction

Eco-translatology, a totally novel, or "formerly nonexistent," perspective of translation studies, is a systematic approach with newly coined terms and concepts designed to "strike out a new line" that is utterly different from the previous perspectives. What, then, makes it "novel," and what is the "new line"? On what basis is it defined theoretically, and what does it study? What are its core notions, rational characteristics, and ethical tenets? The author Hu (2011a) of this book intends to present his answers in this chapter.

The past decade, the blink of an eye in history, is crucial to the discourse and paradigmatic construction of Eco-translatology. What arguments or views of this approach have been accepted? Moreover, what elements of this approach has been undergoing modification, and what has been weeded out? All these issues will be explicated in this chapter.

3.2 The Eco-Translation Paradigm

3.2.1 *The Scientific Research Paradigm*

A "paradigm" is a complex of beliefs, values, techniques, etc., shared by members of a community. The notion and theory are contributions of Thomas Kuhn, an American philosopher of science, in *The Structure of Scientific Revolutions* (Kuhn, 1962). As the theoretical basis and operational norm on which the natural sciences rely, a "paradigm" sets the values and behavioral patterns followed by researchers of a given science. The notion of a "paradigm," the core of Kuhn's paradigm theory, is essentially a theoretical system.

(Hugs) G. S. Hu, *Eco-Translatology*, New Frontiers in Translation Studies,
https://doi.org/10.1007/978-981-15-2260-4_3

As an umbrella term, it covers such general issues as the "orientation" and "tenets" of a specific field, embodying the overall discipline, value judgment, and research methodology of a research approach or mode. It is defined as "a world view underlying the theories and methodology of a particular scientific subject" in *The New Oxford English-Chinese Dictionary* (2007, p. 1540).

In *The Structure of Scientific Revolutions*, Kuhn (1962) points out that research conducted by Nicolaus Copernicus, Isaac Newton, Charles Darwin, Henri Poincaré, Albert Einstein, etc. broke down the research paradigms of their times and thus triggered scientific revolutions. He states that scientific research does not pile up facts, theories, and methods to develop a simple assembly of information but rather is a process of changing paradigms via constant revolutions. A "paradigm" setting up guidelines for translation studies is the way to investigate a given problem, as Hermans (1999, pp. 9–10) puts it. It is thus clear that a paradigm integrates the theoretical hypotheses, research patterns, research methodologies, value standards, and metaphysical tenets shared by the members of a given academic community and represents the totality of the community members' worldviews, values, and methodologies.

A paradigm is not only a prerequisite for scientific research but also a sign of a mature science; as Kuhn (1962, p. 12) stated, "the successive transition from one paradigm to another via revolution is the usual developmental pattern of mature science." It is not until a paradigm is established among at least some of the researchers in a discipline that the discipline progresses from its pre-scientific stage to its scientific stage. For instance, the history and development of Western translation studies is rightfully a vivid picture of the evolution and transition of a paradigm.

The development of scientific research is a process from doubts to the criticism and transcendence of the traditional paradigm.[1] The choice of a paradigm on the part of researchers is conditioned by the basic rules of a given theory (i.e., accuracy, consistency, universality, simplicity, and effectiveness) on the one hand and the theoretical tradition, research methods, and cultural values shared by all disciplines on the other hand.

3.2.2 Translation Research Paradigms

Since the late 1950s, Western translation studies have entered a cross-disciplinary stage. Translation studies have been claimed as an agglomeration of paradigms from a diversity of disciplines. As stated by de Waard and Nida (1986, p. 185), "Translation ... is a kind of act that can be systematically described, and [is] closely related to many other disciplines.... Translation is built up on the basis of Psychology, Linguistics, Communication Studies, Anthropology, Semiotics, and other disciplines."

---

[1] See the discussion of paradigm in Liao (2001b, pp. 14–18).

Gentzler (2001, p. vi), as noted by Susan Bassnett in "Series Editor's Preface" to *Contemporary Translation Theories*, also "enthusiastically" advocates "the closer relations between related disciplines such as literary studies, linguistics, history, ethnography, anthropology, and sociology." Gorlée (1994, p. 133) echoes Gentzler in his *Semiotics and the Problem of Translation*: "Translation is an 'interdiscipline' (or rather 'transdiscipline') combining an approach from (general and applied) linguistics with an approach from (general and comparative) literary studies, in addition to contributions from such disciplines as information theory, logic, and mathematics on the 'scientific' side, and social anthropology, sociology, and theology, on the more humanistic side."

The notion of translation theorists' classification of translation history mirrors the evolution of the research paradigm. Peter Newmark considers the rise of the linguistic theory of translation the boundary between the prelinguistic model stage and the linguistic model stage of translation; he believes that translation studies in the first stage disregarded the connection of translation with meaning, thought, and linguistic universality and thus were confined to debates on priority choices between the source text and the target text, literal translation and free translation, elegance and faithfulness, at the cost of theoretical improvements (Liao, 2001a, 2001b, p. 166). Since the 1950s, Catford, Nida, Fedorov, Mounin, and Wilss usher in to translation studies systemic-functional grammar, transformational-generative grammar, linguistic structure, and textual function to analyze and establish rules of linguistic transfer in translation activities, giving a new complexion to translation studies.

Nida (1984, pp. 9–15) divides translation studies into a philosophical school, linguistic school, communicative school, and sociosemiotic school. These paradigms of translation studies appeared in succession and featured gradual innovation and transcendence in theoretical hypotheses, research models, and methodology: the linguistic school differentiates itself from the philosophical school by denying the isolation of texts from their context and situation; the communicative school differentiates itself from the linguistic school by rejecting the overreliance on linguistic structure and disregard of communicative function; and the sociosemiotic school endeavors to differentiate itself from the communicative school by taking into consideration the interaction between the signs involved in translation. According to Gentzler (1993), translation studies fall into five schools: "the North American translation workshop," "the 'science' of translation," "early translation studies," "polysystem theory," and "deconstruction," representing the evolution of translation studies paradigms (Liao, 2001a, 2001b). Paradigmatic innovation and evolution are associated with a renewal of academic notions and holistic changes of the world outlook and science philosophy of researchers (Lv, 2008, p. 56).

Generally, the basic paradigm pattern of Western translation studies has shown no ultimate change in recent years. Theoretical systems and research paradigms coexist and "co-evolve," depicting a pluralistic pattern of contemporary progression. Diversity is highly valued in Chinese academia, based on a combination of Chinese and Western elements, ancient and modern thought, and the pursuit of unity in diversity and integration of the parts. Integration and the examination of merit give rise to innovation and creativity (Jiang, 1993, p. 47). "Diversity" embodies the individuality and distinctive features of theories; "unity" embodies the universality and holistic connection of theories. The worldview, values, and methodology deeply rooted in this traditional cultural thought of "unity in diversity" naturally have an impact on translation theory explorations in China. Translation notions and theories arrive at "unity in diversity" or "pluralistic integration" through mutual emulation, blending, adaptation, penetration, substitution, and evolution. Based on the above discussion, we are emboldened to declare that from the viewpoint of paradigmatic transition, translation studies, after progressing through "intuitionism" to "structuralism," are now at the stage of "pluralism" and will end at "holism."

The generalizations of translation theoreticians from diversified perspectives clearly delineate the developmental course of translation studies while revealing the confrontation and substitution of paradigms. Among these are the expansion of the research horizon and alternation of the research perspective, sublation of traditional translation notions and shift of research foci, introduction of research approaches, and innovation of research methods.

Notably, the "paradigm shift" of translation studies, in contrast to from the incommensurability of the sciences and philosophy, features the complementarity, mutual fusion, and coexistence of paradigms. Significance illumination and feature discrimination are of great importance to paradigmatic explorations in translation studies.

3.2.3 The "Ecological Paradigm" of Eco-Translatology

An "ecological paradigm," or an ecological approach to translation, is gradually taking shape and accumulating concern and approval with such concepts as the basic notions, value judgments, research methods, terminology, and findings of many translation investigations with a strong ecological tint. A number of researchers, mainly from the Eco-translatological "academic community," have approved the Eco-translatological description; accepted the Eco-translatological guidelines; observed the Eco-translatological methodology of interdisciplinary study, analogical transplant, and systematic synthesis; and adopted the Eco-translatological criteria. Such investigations of diversified types and themes are not isolated cases but cooperative research under the Eco-translatological paradigm.

Eco-translatology has made several attempts to position its paradigm. For instance, it intends to identify a translation paradigm with universal theoretical properties and basic rules to be observed in translation (Hu, 2004a, 2004b, 2004c, pp. 9, 181–182). As an interdisciplinary approach to translation, it is an ecological paradigm that regards Translation as Adaptation and Selection to delve synthetically

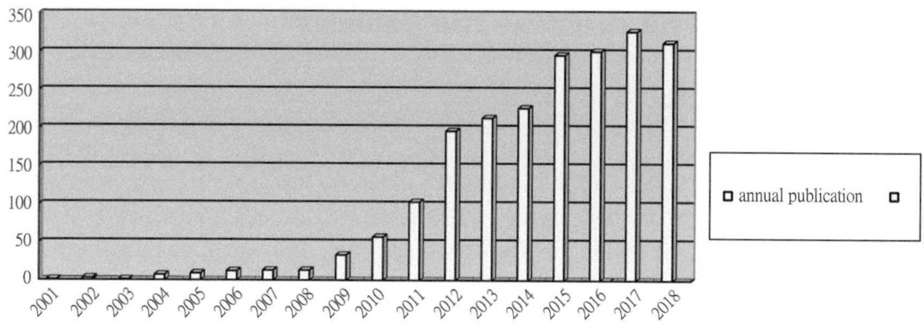

Fig. 3.1 Publications on or relevant to Eco-translatology

into translation via ecological rationale (Hu, 2003, p. 278, 2004a, 2004b, 2004c, p. 41, 2006, 2008, 2009a, 2009b).

Relevant studies of such an "ecological paradigm" orientation range from project-driven explorations[2] to thesis discussions, for example, "Translator-centered Translation Theory and Choice of Source Text: Lin Yutang's Choice of Translating *Six Chapters of a Floating Life*" (Zhai & Zhang, 2005), "Ku Hongming's Translation of *Lunyu* in the Light of Translation as Adaptation and Selection" (Bian & Yao, 2008), "An Eco-translatological Perspective to the Translation of Public Signs—A Case Study of the Slogans of Shanghai Expo" (Shu, 2010), "On Present Translation Studies from the Perspective of Eco-translatology" (Meng, 2011), "Translation Pedagogy Ecosystem: An Eco-translatological Perspective" (Song, 2011a, 2011b), "On Translational Eco-environment" (Fang, 2011), "Eco-Literature and Eco-translatology: Deconstruction and Reconstruction" (Wang, 2011) and "An Eco-translatological Approach to Literary Translator Criticism" (Liu, 2011).

Incomplete statistics show that over two thousand journal articles or academic dissertations of Eco-translatological theories or applications had appeared as of 2018. The following figure clearly shows an ascending number of publications on or relevant to Eco-translatology (Fig. 3.1).

Note that the statistics presented above are from only the first decade of Eco-translatology. In the early years of its second decade, starting in 2010, the "International Association for Eco-translatology Research" was established, the *Journal of Eco-translatology* began to circulate, and international symposiums on Eco-translatology were convened. With the increasing number of eco-translatological researchers, the second decade will surely witness the rise of an "ecological paradigm" featuring "integration and transcendence."

[2]For example, "Eco-Translatology: ecological approach to translation studies" [08YBB007], supported by China's National Social Science Fund; "Studies on Translation Schools: insight into 'ecological school of translation studies'" [RP/ESLT/2009], supported by Macao Polytechnic Institute; and a number of programs supported by provincial funds, universities or institutes.

3.3 Theoretical Bases of Eco-Translatology

3.3.1 Eco-Holism

Ecol-holism[3] contains at its core a supreme value of the holistic interests of the ecosystem over human interests, with a basic standard and ultimate scale of assessing human living modes and technological, economic, and social developments on the basis of their benefits to the integrity, harmony, stability, balance, and sustainable existence of the ecosystem.

Given that ecology is founded on holism and values holistic interrelation and interaction in methodology, it is rational for Eco-translatology, an ecological approach or perspective of translation studies, to have as its primary guideline a methodology of holism and as its foci synthetic studies on translation ecosystems. In this sense, Eco-translatology is to some extent a translation approach of metaphorical analogies, a comprehensive view, and an integrative study.

Eco-translatology considers it accurate to view translation as an integrative and harmonious system. Regarding the interaction among the system elements, this system, as a unified entirety, conceives of a power much stronger than that of the simple functional addition of its elements. This trait of integrity and interlocking fully expresses the theory of the group effect, which considers that the appearance of one ecological action results from multiple factors within the group.

The translation ecosystem, which involves society, communication, culture, language, etc., allows for spatial and temporal changes and features automatic regulation and openness, as do other ecosystems. Regarding a definition, the translation ecosystem can be defined as a functional unit of translation in which languages, translation essentials, and nontranslation essentials (e.g., society, communication, culture) interact and correlate via a constant material circulation and energy flow within given spatial and temporal ranges. In its narrow sense, it is interpreted as the coordinated development of external manipulation and internal linguistic mechanisms in studies of the translational environment, which permit it to cover investigations into all activities pertinent to translation by expanding the horizon of translation studies. Studies of translation history show that isolationism is irrational and unreal; in contrast, holism from the perspective of the translation ecosystem tells the whole truth, whether in the shaping of translation criteria; in an individual translator's choice of texts to translate and a translation strategy or a particular historical period of a nation; or in the formation of the holistic translation traits, translation policies, translation effects, etc. of a given time.

Overall, because the ecological perspective of translation studies stresses "holistic synthesis," it focuses on holistic investigations into the translation ecosystem and the internal structure of this system. This synthetic and integrative holism of ecological translation studies is conducive to the formation of a relation of mutual benefit and cooperation within the translation ecosystem that promises harmonious

[3] See Wang (2004).

ecological beauty. This will undoubtedly influence translation theory studies to move towards the "pluralistic unity" and "holistic synthesis" of translation theories via mutual learning, transplantation, adaptation, penetration, confrontation, substitution, and evolution through comparison and synthesis across times and national boundaries.

3.3.2 *Eastern Eco-Wisdom*

"Eastern" here refers mainly to China, centering on traditional Chinese eco-wisdom. One key reason that Chinese scholars pioneered the proposal of Eco-translatology is that there are abundant relics of eco-wisdom in Chinese culture. "Life is what has been focused on at the initial stage of Chinese culture" (Mu, 1997, p. 43). The "understanding of life" is deemed the mainstay of traditional Chinese cultural thought, which naturally contains an understanding of "life," "existence," or "ecology." Classic eco-wisdom in traditional Chinese culture constitutes, therefore, a crucial theoretical fulcrum and ideological principle. Profound philosophical concepts and critical thought can be extracted from this eco-wisdom featuring "the unity of man and Nature," "the doctrine of the golden mean," "the principle of human-centeredness," and the "integrative approach." These concepts are based on Chinese cultural wisdom, life wisdom, and eco-wisdom.

"The unity of man and Nature,"[4] focusing on harmony, is one of the basic notions of classical Chinese philosophy, with "harmony" referring to a combination of "natural harmony," "human harmony," and "mental harmony." It is embodied by the pursuit of the harmony and unity of the translator and translation ecosystem in Eco-translatology. The harmony between the translator and the translation ecosystem refers to the need for direct translators to adapt to the translational eco-environment and to make choices that comply with the dynamic rules of the translation ecosystem to achieve "moderation" and "balance" between the source and target texts, translator and author, translator and readers, target language and target culture, and translator's adaptations and selections. For the sake of harmony between the translator and the translational eco-environment, a translator, regularly employing different translation theories and diverse translation strategies and techniques, brings human creativity fully into play to make optimal adaptations and selections, blending his or her work into the translational eco-environment. Consequently, the pursuit of harmony and coordination between elements in the translational eco-environment to realize "the unity of heaven and humanity" is associated with "ST-centeredness," "faithfulness," "spiritual similarity," "sublimation," "beauty in sense, form and sound," "accuracy, or fluency and speed" (of interpretation) (Hu, 2009c).

---

[4] Initially proposed in the *Book of History* and modified by philosophers in Chinese history.

"The doctrine of the golden mean," formerly the model of Confucian ethics,[5] is applicable and feasible in the translation process or studies of translation theories when it is employed as a methodology of optimizing decision-making. Overly "literal" or "liberal" translations, overly "foreignized" or "domesticated" translations, and overrepresentation or underrepresentation are all undesirable, as is extremism in translation theories.

"The principle of "human-centeredness" represents the original form of Chinese political and cultural thought. Humankind and nature are one entity since humans live in nature, in the natural environment. They will change along with any minute changes in the natural environment. Harmony between humans, between humankind and nature, and between mind and body is precisely what Confucian culture values (Ji, 2008). The description and interpretation of the translator's role in the translation process are fundamental issues in translation theories. Thus, the translator issue is an eternal topic in translation studies; "the relation between the translator and the translational eco-environment" is one of the objects of Eco-translatological studies, and the translator issue is central in Eco-translatology. "The principle of person-orientation" is embodied in translation studies as the notion of "translator-orientation." "Translator-centeredness" not only proposes positioning the translator as the translation subject but also provides an answer to the question of "who translates" and, moreover, advocates "translator-centeredness" to label the translator the essential factor of a successful translation act.

Traditional Chinese cultural values contain holism, an integrative view, an organismic perspective, and a harmonious perspective,[6] believing that all human and other lives, even those of animals and plants, rank as links in the immense biological chain and thus links in the immense material and ethical cause-and-effect chain. Every step that we take will have corresponding effects on us and even our surroundings, direct or indirect, temporary or long-term, immediate or distant. As has been stated in the traditional Chinese cosmic view, "the cosmos is composed of directions and times," meaning that the cosmos is not a closed system but an entirety of openness, interaction, and interrelation and that it is not isolated, resting, invariable, or mechanically aligned but changing and progressive. The quintessence of traditional Chinese culture lies in its holistic view, its emphasis on change and changing laws. That view is enlightening in analogy with a dynamic translation ecosystem.

All these classic eco-thoughts of "nature," "life," "existence," "moderation," "the principle of person-orientation," "harmony," and "holism" in Chinese culture give rise to and nurture Eco-translatological studies. These classics esteemed by

[5]In Chinese history, there are diverse interpretations of this doctrine, which can be summarized as synonymous with "harmony in diversity," or "excess is just as bad as deficiency."

[6]Prof. Cheng (2005, p. 140) has generalized Chinese philosophy and culture from a methodological perspective in three principles, namely the "principle of wholeness," the "principle of internality," and the "principle of organicity," and adds that "an integrative view of all these three principles will make a typical methodology of Chinese thought."

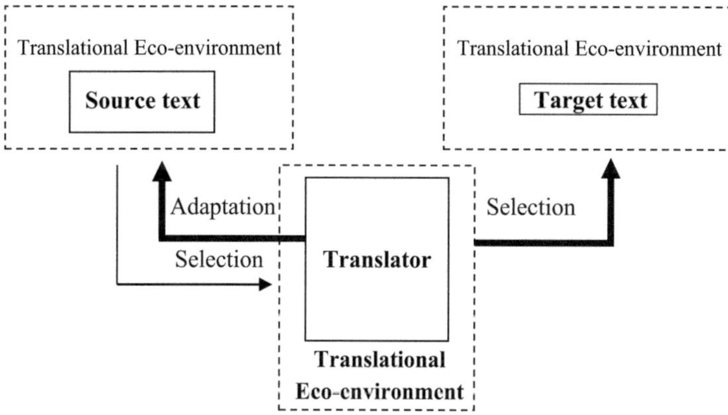

Fig. 3.2 Translation process of translator's adaptations and selections

philosophers and thinkers from all nations have proven to be a crucial fulcrum inspiring Chinese scholars in pioneering Eco-translatological studies.

3.3.3 Translation as Adaptation and Selection (TAS)

Translation as Adaptation and Selection is proposed by assimilating the metaphorically Darwinian theory[7] of "adaptation/selection" to probe into material translation issues. Based on the theory of "adaptation/selection," the author of this book defines translation as "a selection activity of the translator's adaptation to fit the translational eco-environment via textual transplants, dominated by the translator to targeting of cross-cultural transmission of messages." Because of this understanding of translation (see Sect. 5.3), the translation process is hence sketched in light of "adaptation/selection" theory, as shown in Fig. 3.2.

The dotted lines around "source text" and "translator" in the figure represent the "translational eco-environment." This schematic diagram of the translation process attempts to convey the idea that the translation process in Eco-translatological terms consists of the translator's adaptation to the translational eco-environment with the ST as a typically important component and the selection of the TT by the translational eco-environment with the translator as a typically important component, and that translation act in this translation process is translator-centered.

Selective adaptation and adaptive selection in the translation process feature (1) "adaptation"—the adaptations of a translator to the translational eco-environment and (2) "selection"—the selections of the TT by a translator who stands for the

[7]Translation as Adaptation and Selection picks up just the key notions in the adaptation elements of Darwinian theory, which also contains mutation theory and evolution theory, i.e., "natural selection" and "survival of the fittest."

translational eco-environment. Translation in this translation process is depicted as a cycling process of adaptations and selections on the part of the translator.

In the first stage of the translation process interpreted by the theory of Translation as Adaptation and Selection, a translational eco-environment with the source text as a typically important component selects the translator. The arrow to the translator under the gridlines is a filament, representing the selection of the translator by the translational eco-environment with the ST as a typically important component, to highlight the thick lines representing the adaptations of the translator to the translational eco-environment with the ST as a typically important component. In the second stage, the translator, having accepted the selection of the translational eco-environment and thus standing for it, selects the form of the final target text as the translational eco-environment (see also in Sect. 5.5.1).

It is often said that choices or selections are the basis of diction, writing, and even life. In academia, it is often said that translation consists of choices or selections. Answers have been provided here to such questions as what mechanism lies behind these choices or selections and why the translation process is interpreted as "a selection activity of the translator's adaptation to fit the translational eco-environment via textual transplants" (see details in Hu, 2004, pp. 119–127).

3.4 Research Areas of Eco-Translatology

3.4.1 Translational Ecologies (Studies on Translational Environments)

The eco-environment, the environment of ecological relations, means "a complex of all exterior factors acting on the existence and development of human and animals." Translational ecologies here denote the interrelation and interaction between translation subjects and their environment. In other words, translational ecologies are the existential and working state of translation subjects in their environment.

In light of dictionary definitions and analogical interpretations of ecology, the translational eco-environment is thus defined as a complex of all exterior factors acting on the existence and development of translation subjects. Subjects here in their broad sense cover all lives involved in translation activities, or the "translation community," including the author, translator, readers, initiators, patrons, publishers, agencies, and editors. The exterior environment contains a natural economic environment, linguistic, and cultural environment, social and political environment, etc. The translational eco-environment is a collection that integrates a diversity of natural or human factors relevant to the occurrence, existence, and development of translation activities.

In "Preliminaries," the author of this book has pointed out that translational ecology refers to a "complex" of the existential and working state of translation subjects in their environment, whereas the translational eco-environment refers to an "aggregation" of multiple exterior factors relevant to translation. In this sense, translational ecology stresses the state of "entirety" and "integration," while the translational eco-environment stresses "multiple" and "concrete" environmental elements.

Translational ecology and the translational environment are integrated as an entirety. A translator acts in a specific ecological environment that is checked by other translation subjects; a TT must comply with the norms of the target culture and is subject to the constraints of the target culture's social and political powers. The translational eco-environment, an integrative whole for any translation subject, is to be adapted to but not ignored or surmounted. Ignorance of this integrative requisite will result in a violation of translation ethical norms by any disturbance of the order of the translational eco-environment and translational environment, for example, "ghostwriting," "over-rewriting," or "plagiarism," in pursuit of individual or group interests.

The eco-environment of translation is hierarchical and is split into macroscopic, mesoscopic, and microscopic environments. Until now, what has been discussed is mainly the macroscopic environment, or the environment in general. A macroscopic view of the translational eco-environment tells us that nations have distinctive social and political systems and that language policies and language communities have distinctive translation policies; a mesoscopic view shows that the translational eco-environment of literary translation will differ from that of applied text translation even within a nation; a microscopic view reveals the interior structure of translation studies, namely, theories, applications, criticism, history, etc. More specifically, the individualistic translational eco-environment contains various minute differences.[8]

What deserves special attention here is that "translational eco-environment" differs from "context" in terms of connotation, denotation, scope, and perspective. "Context is the environment of using languages" (Wang & Chen, 2001, p. 37). Context starts from using languages while excluding language itself and language use. The "translational eco-environment," comprising the source language, source text, and target language system, is nevertheless the overall environment covering the existential state of the translator and the target text. It is an aggregation of various factors that serve as checks for the optimal adaptations and selections of the translator and also as the prerequisite and basis of his/her multidimensional adaptations and adaptive selections. In particular, in light of the "natural selection" principle, the translator, having accepted the selection of the translational eco-environment and thus standing for it, selects in the second stage the form of the final target text, whereas context is the form of this specific function. The translational eco-environment is, therefore, broader in connotation and denotation than the context of translation. It can thus be concluded that the translational eco-environment interprets and synthesizes translation acts by viewing translation as a holistic "translation ecosystem."

[8] See Fang (2011, pp. 1–5).

The translational eco-environment is a key concept of Eco-translatology. It refers to the worlds of the source text; the source/target languages; an internally interacting entirety comprising the linguistic, communicative, cultural, and social aspects of translation; and the author, the readers, the clients, etc. (i.e., the translation community) in light of earlier Eco-translatological explorations describing translation as a selection activity of the translator to fit the translational eco-environment via textual transplants and the translation process as the translator's adaptations and selections. "Worlds," "entirety," and "collection" all denote an "aggregation" of eco-environments relevant to translation. The translational eco-environment integrates, therefore, not only the macro-environment, meso-environment, and micro-environment but also the exterior environment and interior environment, not only the objective environment (for example, the source text, target text, textual function, translation strategies, and translation norms) and the subjective environment (the translator, author, readers, publishers, negotiators, editors, etc.) but also the material environment and spiritual environment (Hu 2011b).

It can thus be claimed that everything exterior to the translator is accumulated into the eco-environment of translation and that each translator is a component of the others' translational eco-environment.

The complexity of the ecosystem is displayed in its indwelling multidimensionality; i.e., the overall system is split into subsystems and subsystems into further subsystems to end with a translation ecosystem of indefinite vertical divisions, with each constituting the environment of the others horizontally. Given the relation between translational ecology, the translational eco-environment and texts, the topic of "micro-level textual studies" is elaborated on and interpreted in Chap. 6 of this book.

3.4.2 Textual Ecologies (Studies on Source–Target Texts)

Textual ecologies are about the state of the ecological environment and life of texts. In Eco-translatological terms, the source language is one textual ecosystem, and the target language is the other textual ecosystem. The textual ecosystem of the source language concerns linguistic ecology, cultural ecology, communicative ecology, etc., in the source language system, and the textual ecosystem of the target language concerns those ecologies in the target language system.

Linguistic ecology, cultural ecology, and communicative ecology can all be divided into macroscopic and microscopic levels. Macroscopic linguistic ecology refers to the harmonious coexistence of dominant languages, minority languages, and even endangered languages; microscopic linguistic ecology refers to the harmonious relation among linguistic elements within a translation. Macroscopic cultural ecology denotes the harmonious coexistence of dominant cultures, weak cultures, and endangered cultures; microscopic cultural ecology denotes the harmonious relation among cultural elements within a translation. Macroscopic communicative ecology concerns the connection and intersection of international, regional, and interindividual exchanges; microscopic communicative ecology concerns the interaction between communicative purposes and communicative acts within a translation.

Centering on "textual ecologies," Eco-translatology probes into the features of the textual ecosystems of the source/target language and the differences between them; explores the rules and mechanisms of transplanting and transforming from the source language ecology to the target language ecology; and investigates the existential state of the translations, the reasons for their "short life" or "survival" and ways for translations to survive in order to offer an ecological perspective and theoretical basis for choices of translation strategies and interpretation of (un)translatability, bringing into full play the effectiveness and values of translations.

3.4.3 "Translator-Community" Ecologies (Studies on Translator Acts)

The translation community refers to the collection of human beings relevant to translation activities and interacting with the happening, development, operation, products, functions, effects, etc. of translation activities. In other words, it refers to the author, target readers, translation critics, translation editors, publishers, agencies, patrons, or clients, etc., with the translator as their representative.

Members of the translation community differ in their modes of thought, educational background, interests and hobbies, translation views, aesthetic standards, translation experiences, etc., and there are also differences in textual types, reader expectations, receptor culture, circulating channel, norms, etc. The distinction between these subjective and objective factors and differences between the interior and exterior environments will naturally give rise to the diversification of adaptations and selections on the part of "translation community" subjects, who are obliged to adjust themselves dynamically via dialogue and negotiation to fit and nourish the holistic translational eco-environment. Meanwhile, ecosystems of translation should be mutually adaptive for effective interaction and co-evolution. The holism of the "translation community," with the translator as its representative, is fueled by the holistic, correlative, dynamic, and balanced "ecological rationality" of the translation ecosystem,[9] constituting the Eco-translatological trait and merit of emphasizing the translator and the human factors.

[9] First, because ecology studies the relation between humans and nature and Eco-translatology studies the relation between the translator and translational eco-environment, the translator (i.e., the representative of the translation community) is naturally one of the major two essential aspects of Eco-translatology and is thus the undeniable subject of translation. Second, only living beings are provided with a self-adjusting capacity in a natural ecosystem, and in the translation ecosystem, particularly in the translation process, only the translator is equipped with a self-adjusting capacity and the power of dynamic "selective adaptation and adaptive selection". Third, Eco-translatology carries forward the theory of Translation as Adaptation and Selection, in which the translation process features "translator dominance" and "translator-centeredness".

From the Eco-translatological viewpoint, only the "translation community," with the translator as its representative, or only the translator is capable of coordinating the relationship between the "translational environment," "translation," and "translation community" to realize, via "translator responsibility," the interrelation, interaction, and balanced and harmonious holism of the "environment, text, and humankind" in translational ecology. In other words, in the "translation community" ecosystem, it is the translator who has the responsibility for coordinating relations, practicing ecological rationality, maintaining eco-balance, and protecting ecological harmony. It may also be said that the translator can integrate a sense of responsibility in a broader sense into translation activities only when he/she assumes responsibility for "the otherness" ranging from texts and the "translation community" to the translational eco-environment and manages his/her relationship with "the otherness" from an ecologically holistic and rational perspective (see also Sect. 5.4.2).

3.4.4 Relationships Among the Three Ecologies

The "three ecologies" here refer to translational ecology, textual ecology, and translator ecology as a thread of the studies and argumentation of Eco-translatology. The relationship between the "three parties" is also concerned in Eco-translatology, namely, the relationship between the translational eco-environment, translation, and the translator, illuminating this approach as a "relationship" science of the three parties involved. Given that they share similarities in research subjects, principles, and methods, the "three ecologies" and "three parties" are mutually applicable, although they differ in standpoint and orientation.

On the one hand, the translator will survive and develop if he/she is adaptive to the translational eco-environment, and his/her translations[10] will survive and last. The survival and development of the translator mean the "existential" state on the part of the translation practitioner, and the survival and durability of translations mean the "life" state on the part of the translation products. The environment for the survival of translators and translations is what the "ecology" of translation concerns. The existential state of a translator is tested by the life state of translations; both rely on translational ecology as a component. In this case, the "existence" of the translator, the "life" of translations, and the "ecology" of translations (i.e., "3 themes") combine to show a picture of correlation and interaction, mutual reliance, and dynamic balance (i.e., "3 relationships between").

On the other hand, the sequence "environment, text, and translator," acting as the constraints on, basis of and center of the translation process, respectively, embodies logicality and practicality since it complies with the orientation of "the

[10] "Translations" is used here to cover translation products of all length.

Sequence Chain" (language, culture, society/communication, etc.), with the real-life translation process (it is usual for the translator to adapt to the "environment" and comprehend the source "text" before adapting to the "environment" and selecting the final form of the target "text"), with the current "turn" of translation studies (from "translation-centeredness" to the "linguistic turn" and then to the "cultural turn"), and with the research subjects and core content of Eco-translatology (i.e., explorations of the translational ecology, textual ecology, and ecology of the "translation community").

Furthermore, "3 translations as" [translation as eco-balance (translational eco-environment), translation as text transplants (translations), and translation as adaptation and selection (translator)], which will be elaborated on in Sects. 5.3.3–5.3.5, integrate with "3 ecologies" and "3 parties" to form a panoramic view of Eco-translatology. This ecological approach to translation studies can be generalized as a study of the relationship between "3 parties" centering on the theme of "3 ecologies" by adopting the technical route of "3-level" studies from translation studies at the macrolevel and translation theories at the mesolevel to translations at the micro-level to arrive at the notion of "3 translations as." All these, together with the "3 paradigmatic traits" of practicality, openness and universality probed into in the last chapter of this book, are not only interrelated, interactive, and interdependent organic components of the theoretical system of Eco-translatology but also knitted together in a manner of internal connection for the logical and corresponding relation between their research contents or subjects.

3.5 Methodology of Eco-Translatology Research

Eco-translatology, a holistic study of translation from an ecological perspective, is something of an "individualistic" or "specific" study compared with general translation studies. Seen from the methodological aspect, all standard, common, or "universal" methods applicable in general translation studies are also applicable to the "individualistic" or "specific" studies of Eco-translatology. Eco-translatology is, however, fundamentally different from other translation studies approaches and therefore contains some distinctive methodological methods to reflect its "individualistic" feature.

3.5.1 Transdisciplinary Overview

Because Eco-translatology incorporates both translation studies and ecology, a transdisciplinary view of Eco-translatological investigations is naturally adopted, and this approach is distinguished from others (see also Sect. 1.4).

This transdisciplinary or cross-disciplinary method of Eco-translatology complies with not only the academic norms and trends of modern scientific research but also the essential "cross-disciplinary" trait of translation studies proper (see Sect. 4.2.3).

Generally, this ecological perspective of translation studies synthesizes and describes translation activities in a holistic manner by borrowing and applying ecological theories and notions, findings, and methods. This book constructing and interpreting the ecological discourse system is both a practical "interdisciplinary" case of intermingling and integrating translation studies and ecology and an example of realizing the "intersection of humanities and the sciences" and employing a cross-disciplinary method to conduct translation studies.

3.5.2 Metaphorical Analogies

"Metaphorical analogy" forms one of the most crucial methods in Eco-translatological studies. The feasibility of metaphorical analogy in Eco-translatology lies in the similarities, at least to some extent, between translation ecosystems and natural ecosystems. It has been clearly demonstrated with research findings that these two systems are conspicuously similar in many aspects.

First, ecology stresses the interaction between the ecological environment and organisms, as is the case for the translation ecosystem. The picture of interrelating, interactive, and mutually overlapping relationships between the relevant factors in the translation ecosystem renders it a complicated entirety with the potential of interacting with other translation ecosystems. Second, in nature, an eco-balance exists to some extent between organisms and their existential environment via interaction, as is the case for the translation ecosystem. Owing to differences in the modes of thought, educational backgrounds, interests and hobbies, translation views, aesthetic standards, translation experiences, etc., on the part of the "translation community," with the translator as its representative, and differences in text types, receptor expectations, receptor cultures, circulating channels, normal environments, etc., discrepancies in these subjective and objective environments and interior and exterior environments are bound to give rise to different adaptations and selections on the part of "translation community" subjects who are obliged to adjust dynamically to adapt to the holistic translational eco-environment (Hu, 2009a, p. 6, 2009c). Meanwhile, we should bear in mind that translation itself is a complex ecosystem, provided likewise with the trait of balance as in a natural ecosystem. Third, in a natural ecosystem, the conscious and purposeful activities of human beings are capable of improving or confining, innovating, and reshaping ecological relations in a natural environment; in a translation ecosystem, the conscious and purposeful activities of the "translation community," with the translator as its representative, are capable of improving or confining, innovating, and reshaping ecological relations in the translational eco-environment. Fourth, the two ecosystems share some working principles. For instance, in natural ecology, the competition and rejection principle means that the more similar two species' utilization of the same resource, or the more they overlap ecologically, the more fierce the competition that exists between them is, as is the case for the academic translation field. For instance, a lower degree of comparability will exist since translation criteria differ significantly between text genres of translations, whereas a higher degree of comparability will result if the translations fall into one text genre, ending in intense competition. One example is the current translation industry, in which translators compete in an increasingly aggressive manner owing to the professionalism and commercialization of translation services. Fifth, these two ecosystems contain similar phenomena and operating patterns. For example, natural ecology involves "the beauty and infinite complexity of the adaptations between all organic beings,"[11] and "variation of organisms results from impact from environment."[12] In translational ecology, there are many phenomena of "adaptations," "selections," "existence," "elimination," etc., in the translation process. The adaptive optimal selections of the translator in the translation process can be considered conscious or subconscious translator acts. Translation consists of a succession of decision-making based on optimal selections, and the translation is the product of the translator's selection of adaptations to the translational eco-environment. Last, the two ecosystems all concern "human beings," and their acts, similarities, and universalities in this aspect are self-evident.

In summary, there are obvious similarities between translational ecology and natural ecology. This suggests similar rules in these two ecosystems and in turn indicates that some rules of natural ecology are likewise applicable to translational ecology; meanwhile, similarities between these two ecosystems promise the employment of metaphorical analogy in conducting studies on translational ecology.

3.5.3 Conceptual Borrowings

Since metaphorical analogy is feasible in Eco-translatological studies, conceptual borrowing, another essential research method, naturally follows suit. Borrowings of ecological concepts contain borrowings of both ecological concepts and theories or principles as well as terminology. The term "conceptual borrowings" is adopted here because all these borrowings at different levels are essentially borrowings of ecological concepts. Borrowings of ecological concepts in Eco translatology include borrowings of natural ecology, the ecological environment, eco-balance, ecological aesthetics, and ecological harmony, whereas those of ecological theories or principles in Eco-translatology include borrowings of system theory, variety theory, "scale" theory, and "feedback" theory, and those of ecological terminology in Eco-translatology include "ecological environment" (→ translational eco-environment), "ecological chain" (→ translation chain), "ecological community" (→ translation community), "survival (of organism)" (→ survival of translator), and

[11] For these two quotations, see Chap. 4 in Darwin's (1859).

[12] "Men are organisms by nature, hominid of primates in mammalian order" (see Huang, 2009).

"natural selection" (→ translator's selective adaptation and adaptive selection). Metaphorical analogy and conceptual borrowing, seen from a methodological point of view, are also deemed embodiments of transdisciplinary or cross-disciplinary traits of Eco-translatology.

3.5.4 Holistic Integration

The adoption of "holistic integration" as a research method is triggered by the general theoretical notion of "Eco-translatology as a holistic study." Because ecology is a science based on holism, the core concept of current ecological theories, and because ecological research methods highlight holistic ideas of interrelation and interaction, systematic studies of a holistic and integrative nature based on ecological holism rank as the main interest, primary content and principal methods of Eco-translatology, regardless of whether it is labeled an ecological approach to translation studies or an ecological perspective of translation studies.

For example, when investigating the "translation ecosystem," the "ontological translation ecosystem" should not be deemed the only research focus of Eco-translatology (in contrast to the restrictive horizon of past endeavors featuring "scholasticism"). Instead, attention should be paid to the "translational education ecosystem," "translational market ecosystem," and "translational management ecosystem," and even the "translational ecosystem," on which the first three rely, to examine and inspect in a synthetic and integrative manner.

When investigating "translation history," "the Sequence Chain" is proposed in Eco-translatology to study and synthesize translational ecology, the language ecology, cultural ecology, social ecology, and natural ecology of translation activities synchronically or diachronically. If only a single aspect, phase or dimension is concerned, a holistic view is not applied, and such research will be criticized as "partial" or "negligent."

It should also be pointed out first that holism and systematicity are the quintessence of traditional Chinese culture and that this philosophy of "holism" will be embodied as a methodology in the research endeavors of Chinese scholars.

Second, a holistic view or systematic balance view may indeed move towards the horizon of other approaches, but in translation studies, the holism of translation ecosystems is innately designed by ecosystems as "natural" and inevitable, whereas that of other approaches is determined by the competence and awareness of translation researchers. In other words, whenever translation studies are conducted from perspectives of ecosystem or ecological rationality, interrelation and interaction, balance and coordination, holism, and harmony within the system will be the natural and inevitable consequences. Otherwise, the ecological nature will be eradicated, and the ecological perspective will be a case of self-denial. The reason is that, once more, the holism and systematicity of the ecosystem are determined by the ecosystem itself as an "existential prerequisite" to establishing itself as rational.

3.6 Ecological Principles for Eco-Translatology

When approached from an ecological perspective and an ecosystem perspective, translation studies display an intense color of ecological rationality, regardless of whether a study is of a natural ecosystem or translation ecosystem and whether the system is large or small or at the superior or lower level. The rational features of these ecosystems can be labeled as follows: (1) emphasizing holism and relevance, (2) seeking a dynamic balance, (3) reflecting ecological aesthetics, (4) identifying the "translation community," (5) adhering to Eco-translation ethics, and (6) highlighting unity in diversity (Hu, 2011b). These features or principles of ecological rationality are illuminative of translation studies in terms of conceptualization, technical route, and methodology.

3.6.1 Emphasizing Holism and Interrelation

Ecology stresses holism and relevance, highlighting the reliance of individuals on the holistic environment. As noted by Merchant[13] (Merchant, 1999, pp. 99–100):

Along with current challenges to mechanistic technology, holistic presuppositions about nature are being revived in ecology's premise that everything is connected to everything else and in its emphasis on the primacy of interactive processes in nature. All parts are dependent on one another and mutually affect each other and the whole. Each portion of an ecological community, each niche, exists in a dynamic relationship with the surrounding ecosystem. The organism occupying any particular niche affects and is affected by the entire web of living and nonliving environmental components. Ecology, as a philosophy of nature, has roots in organism—the idea that the cosmos is an organic entity, growing and developing from within, in an integrated unity of structure and function.

The interrelation and interaction of different components within the system make the system an interrelated entirety, conceiving, moreover, a power much stronger than the simple functional addition of its elements. A change of any component will trigger changes of other components within the ecosystem since they are interrelated and interact. This "butterfly effect" is an embodiment of the group effect, which denotes the appearance of one ecological action resulting from multifold factors within this group; meanwhile, this trait also reveals that everything in natural ecology is a link in the interlocking and integrative network of multifold dimensions and hierarchies.

It is the case for a natural ecosystem, and the translation ecosystem follows suit. Interaction and overlapping prevail within the translation ecosystem, which makes it a complex entirety. In this case, studies of the translation ecosystem confined to

[13] Carolyn Merchant, Ph.D., is professor of environmental history, philosophy and ethics in the Department of Conservation and Resource Studies at the University of California, Berkeley.

a certain subsystem (e.g., ontological translation ecosystems) or a certain component (e.g., patronage of translation activities) leave much to be desired. From the perspective of ecological rationality, the interrelation and entirety features of different ecosystems should remain the field of interest. Specifically, discussions on constructing the theoretical framework of Eco-translatology should take into account the holistic interrelation and interaction between translational management ecosystems, translational market ecosystems, translational education ecosystems, and ontological translational ecosystems. Ecological rationale decides that interrelation.

3.6.2 Seeking Dynamic Balance

In nature, organisms and their existential environment interact to reach a certain eco-balance. On the one hand, differences in environmental conditions will give rise to variations in an organism's morphology, physiological activities, chemical elements, genetic characteristics, geographic distribution, etc.; on the other hand, to adapt to different environmental conditions, organisms must adjust constantly.

When interrelated components within the ecosystem remain stable in composition and proportion, and the input and output of energy and material remain balanced, this ecosystem displays a state of balance. In other words, when an ecosystem is relatively balanced, its population structure and quantitative proportions display no evident change, and its input and output in terms of energy flow and material circulation approximate equilibrium. Whenever changes occur within this ecosystem, feedback and self-adjustment arise to maintain a relative balance. An ecosystem in a balanced state arrives correspondingly at the relatively stable stage, where this ecosystem, comparatively speaking, conceives organisms of maximum quantity and productivity and hence attains a much greater capacity for self-adjustment. Meanwhile, the more complicated its interior structure is, the more powerful its capacity of self-adjustment or vitality will be.

Translation is a complicated ecosystem. Owing to the interrelation, similarity, and isomorphism between translational ecology and natural ecology (see Sects. 2.3.1 and 2.3.2), this trait of balance in natural ecology is also a characteristic of translational ecology. A system of dynamic balance with interdependent translation-oriented ecologies will take shape only through interaction between translation subjects and objects, or translation subjects and their exterior eco-environment. Generally, the self-adjustment capacity of the translation ecosystem is in direct proportion to the competence of the "translator community," with the translator as its representative within the system.

3.6.3 *Reflecting Ecological Aesthetics*

Nature is beautiful and rational. Its beauty is voiced by a multiplicity of colors, patterns, sounds, etc., and its rationality lies in its order and harmony. The beauty of nature is colorful: the iridescent clouds, the red sun and the blue sky over verdant forest, yellow earth, and blue sea. The beauty of nature takes many shapes: expansive horizons, straight cliffs, lofty plateaus, sea waves, lake ripples, lightning curves, the roundness of the sun, and lozenges of snow. The beauty of nature is musical: the roar of the sea, the bubbling of a spring, the pattering of rain, the whistling of wind, the rustling of leaves, human voices, animal calls, and birds warbling. With these colors, patterns, and sounds, nature arouses within humans images of beauty, reflections on beauty, joy from beauty, and the pursuit of beauty.

These images of beauty, reflections on beauty, joy from beauty, and pursuit of beauty are given full play in translational ecology, particularly in the translation process. Translation studies show that some scholars in the academic translation field are indeed pursuing "semantic beauty," "formal beauty," and "phonetic beauty," and others are insisting on "the truth," "the good," and "the beautiful." Beauty on the lexical level, the syntactic level, and the logical level has been discussed, and the beauty of accuracy or obscurity and the beauty of difference or conciseness or even verboseness have also been sought. It can be stated that the pursuit of beauty, symmetry, balance, contrast, order, rhythm, meter, etc., at a macroscopic or microscopic level has never been asserted in the translation process. Rather, those elements are categorized as ecological aesthetic elements and principles.

3.6.4 *Identifying the "Translator Community"*

As stated in Sect. 3.5.2, eco-balance in nature is obtained through interaction between organisms and between organisms and their existential environment. The same is the case for translational ecology, to analogize this interaction between organisms and between organisms and their existential environment.

The "translator community" refers to the humans involved in translation activities, including the translator, readers, author, patrons, publishers, and critics, with the translator as their representative. Owing to differences among the modes of thought, educational backgrounds, interests and hobbies, translation views, aesthetic standards, translation experiences, etc. in the "translator community," with the translator as its representative, and differences in the text types, receptor expectations, receptor cultures, circulation channels, norms, etc., discrepancies in these subjective and objective, interior and exterior environments are bound to give rise to different adaptations and selections by "translator community" subjects who are obliged to adjust themselves dynamically to adapt to the holistic translational eco-environment. Meanwhile, ecosystems pertinent to translation should also adapt to each other to achieve effective interaction and co-evolution. Ecological rationality requires a holistic view of the "translator community," with the translator as its representative, and of the translation ecosystem as an interactive, dynamic, and balanced entirety. At the same time, a characteristic or merit of Eco-translatology is that it highlights the translator and "human" elements.

3.6.5 Adhering to Eco-Translation Ethics

An analogy of ecological ethics in translation indicates that Eco-translatological ethics contains four basic principles:

1. The principle of "balance and harmony";
2. The principle of "multiple integration";
3. The principle of "symbiosis and diversity"; and
4. The principle of "translator responsibility."

Owing to their significance, these four principles will be elaborated on in Sect. 3.7.

3.6.6 Highlighting Unity in Diversity

Unity in diversity represents the law of the unity of opposites in human life and nature. The whole cosmos is harmonious unity in diversity. Diversity represents distinctions, or individuality, while unity represents interrelation, or universality. Unity in diversity denotes abundance, order, and vitality. This ecosystem characteristic denotes multiple changes; multifold factors; and ecological aesthetic principles incorporating symmetry, balance, contrast, rhythm, meter, etc (Hu 2011c).

This ecological notion of "unity in diversity" is not only directive of microscopic textual operations (see details in Sects. 6.2–6.7) but also illuminative of mesoscopic theoretical systematization (see details in Chap. 5) and even guides macroscopic translation studies construction (see details in the next chapter).

In regard to the existence of universal laws for general translation issues, scholars in the academic translation field are divided in their opinions, and some have declared that there is no such thing as a universal translation theory. However, in light of this law of "unity in diversity," the potential of a relatively universal translation theory cannot be denied, at least theoretically.

The "structure" map of translation studies presented by Holmes (1972) displays theoretical translation studies as divided into general translation theories and partial translation theories (see also Venuti, 2000, p. 172). Holmes (1988, p. 73) adds in another article that "many translation theories to date are not general but specific in scope and deal with only some of the various aspects of translation theory," and "most of the theories that have been produced to date are in reality little more than prolegomena to…a general translation theory." Steiner (1975, p. 235) also divides translation theories into "universalist" and "relativist" theories and believes that this classification is somewhat similar to the two primary mode of existence of human society, i.e., "complete adaptation or partial adaptation to the environment" (Lance & Jacky, 1991, p. 34). As stated by Lefevere (1993, pp. 229–230), "translation studies will draw greatly from a much more integrative discourse, and it is necessary to bring home to all researchers that this discourse is, if not their research focus, closely related. In recent years, endeavors towards this much integrative discourse have emerged in translation academia."[14] Real-life translation studies also make this clear: in contemporary investigations of translation theories, there appear such progressive trends as universality of theories, integration of disciplines, panoramic view of the horizon, diversity of method, compatibility of system, multiplicity of paradigm, compounding of recognition, multidirectional criticism, synthesis of viewpoint, and creativity of orientation (Hu, 2009a, 2009b). These trends can serve as examples of the realization of "unity in diversity" in a theoretical system.[15]

In sum, owing to its nature of the emphasis on entirety/interrelation, the pursuit of dynamics/balance, the identification of ecological aesthetics, adherence to Eco-translation ethics, and the highlighting of unity in diversity as well as the coevolving, progressive, and integrative nature of ecosystems,[16] ecological rationality[17] emerges as the macroscopic guideline for constructing the Eco-translatological discourse system. Therefore, Eco-translatology, under the guidance of ecological rationality, displays itself as an integrative mode of thought, an organic mode of thought, a correlative mode of thought, and a process mode of thought. It stresses the integration and unity of analysis and synthesis as well as the impacts and restrictions of the translational eco-environment on translator acts. It highlights the integration of "translation studies," "translation theories," and "translations" and pursues the integrative coordination of translational ecology at the "macro-," "meso-," and "micro-level."

---

[14] Many researchers have touched on "multidisciplinary" or "integrative" translation studies, such as *Translation Studies: An Integrated Approach* (Snell-Hornby, 1988), *Redefining Translation: The Variational Approach* (Lance & Jacky, 1991), *Translation Studies: An Interdiscipline* (Pöchhacker & Kaindl, 1994), and *Multidimensional Translation: A Game Plan for Audiovisual Translation in the Age of GILT* (O'Hagan, 2005), etc.

[15] As has been pointed out by Prof. Tan (1998, pp. 12–16), Chinese translation studies are in their fourth stage, and Western translation studies are in their fifth stage; moving from interdependence towards interaction and integration, they will finally arrive at "unity in diversity" in terms of their theoretical system.

[16] Regarding ecological traits, an ecosystem is balanced in that it follows the principles of integrative co-evolution, self-cycling, eternality of material and energy conservation; it is progressive in that it promotes competition and symbiosis, co-evolution, survival of the fittest; it is integrative in that it pursues healthy coordination in physiology, psychology, and ethics (Wang & Zhou, 2004, p. 5).

[17] "Ecological rationality", as stated by Xue Xiaoyuan in *Why Are We in Want of Ecological Rationale*, "ranks as the supreme value in our time" (see http://theory.workercn.cn/contentfile/2009/04/07/101822834372772.html. Retrieved on March 22, 2012).

3.7 Principles of Eco-Translation Ethics

Translation, as a cross-language, cross-cultural "systemized social act with the translator as its subject" (Chesterman, 2001), is inseparable from ethics. Along with the deepening of contemporary translation studies and translation investigations, translation ethics is a subject of increasing concern. Studies on translation ethics have been conducted by Berman (1984), Pym (1997), Venuti (1998), Chesterman (2001), Christiane (2001), etc. in Western countries and Sun (2007), Lv and Hou (2009), Zhu (2009, 2010), etc. in China.[18]

These studies are mostly generalizations of aspects of culture, ideology, operation, professional ethics, etc., and discussions of scholars from different countries contribute to the development of translation ethics study. With the development of translation studies, nevertheless, new aspects of translation ethics studies are expected.

On the one hand, Eco-translatology, founded on the isomorphic analogy of natural ecology to translational ecology, is a paradigm of the ecological perspective of translation. Based on traditional Chinese eco-wisdom and the Darwinian theory of "adaptation/selection," it highlights ecological holism to delve systematically into translational ecology, textual ecology, "translation community" ecology, and their interaction and interrelation as well as the integration and synthesis of the translational ecology entirety and ontological translation theories from an ecological perspective.

On the other hand, ecological ethics refers to a series of human moral codes to address the relationship between humankind and the eco-environment, ranging from the surrounding animals to nature. It usually denotes the ethical relations and adjusting principles of human activities relevant to natural ecology. All ethical aspects of humans in natural ecology constitute the realistic contents of ecological ethics, covering the rational direction of natural eco-activities, protecting eco-balance and organism diversity, protecting and utilizing natural resources rationally, and making scientific decisions on significant events relevant to natural ecology and eco-balance and on the moral qualities and responsibilities of protecting natural ecology and organism diversity.

The present section, based on translation realities, will focus on the basic principles of translation ethics in Eco-translatology that are pertinent to the research objects and main contents of this perspective by analogy with ecological ethics.

3.7.1 Principle of "Balance and Harmony"

Balance and harmony here refer to the holistic balance and harmony of a combination of factors ranging from translational ecology and textual ecology to "translation community" ecology, as well as integration and balance between languages and

[18] Berman (1984), Chesterman (2001), Pym (2001, pp. 139–154), Pym (1997), Venuti (1998), Zhili (2007), Lv and Hou (2006), and Zhu (2009, 2010).

cultures; interior and exterior factors; and macroscopic, mesoscopic, and microscopic thinking.

The balance of textual ecology contains a balance of linguistic ecology, cultural ecology, communicative ecology, etc. of the texts in translation. Regarding the balance of linguistic ecology, the translator is responsible for maintaining a semantic balance between the source and target languages, a balance of sentence meaning, a balance between the "spiritual similarity" and "expressiveness" of the source and target languages, a balance between the practical values and aesthetic values of the source and target languages, a balance of the source language style and target language style, etc.

In regard to translation practice, a balance of the source language and target language ecologies for any recognized or influential translation is usually needed.

Concepts such as "equivalence," "correspondence," "symmetry," and "nondiscrimination" from "equivalent" theories have all been recognized and accepted in translation theories to date. All these concepts can be reduced to the pursuit of "balance" between the source and target languages in a variety of aspects, regardless of whether they interpret the nature and products of the translation in terms of linguistic forms, semantic function, textual message, information volume, communicative purpose or relation between different parties of the source and target languages.

In regard to the requirements of translation studies proper, on the one hand, in Eco-translatological studies, "balance" ranks at the top of the essential traits of any ecosystem and is therefore a core notion of Eco-translatology. The impact of the translational eco-environment on translations is self-evident, and, as suggested by the declaration "No context, no text," if there is no translational eco-environment, there is no successful translation. Harmony and balance within translational ecology as a whole are therefore required, and without balance between the ecosystems of translation studies, there will be no healthy development of Eco-translatology, let alone fulfillment of the academic mission of maintaining language and cultural diversity on the part of Eco-translatology. On the other hand, in translational ecology, the balance of translational ecology, also embodied by negotiation and compromise between different parties of the translation ecosystem, will take into account coordination and balance between the author, the readers, the source text, the target text, etc. The translator is expected to conduct a dialogue of equals with the author by transcending temporal and spatial limits, overcoming possible barriers, and determining the equilibrium point between the author and the readers, with a full understanding of contemporary readers' real demands and receptivity, to realize the fusion of horizons between the author, the translator, and the readers and to achieve a mutually beneficial, healthy, and orderly ecological circulation.

Take, for example, the balance between the source-text ecology and the target-text ecology. The translator, via "selective adaptation" and "adaptive selection," is responsible for not only maintaining and shifting the linguistic ecology, cultural ecology, and communicative ecology of the source text but also for the "survival" and "longevity" of the shifted text in the new linguistic ecology, cultural ecology, and communicative ecology, i.e., the translational eco-environment of the target language.

As has been stated, linguistic ecology, cultural ecology, and communicative ecology are all of different sizes. Macroscopic linguistic ecology refers to the harmonious coexistence of dominant languages, minority languages, and even endangered languages; microscopic linguistic ecology refers to the harmonious relation between linguistic elements within a translation. Macroscopic cultural ecology denotes the harmonious coexistence of dominant cultures, weaker cultures, and endangered cultures; microscopic cultural ecology denotes the harmonious relation between cultural elements within a translation. Macroscopic communicative ecology concerns the connection and intersection of international, regional, and interindividual exchanges; microscopic communicative ecology concerns the interaction between communicative purposes and communicative acts within a translation.

It is thus clear that the translator is committed to maintaining the "balance" and "harmony" of the source text and the target text in terms of linguistic, cultural, and communicative ecologies. In Eco-translatological terms, Eco-translation is somewhat a "balancing act" or a "harmony theory" concerning textual ecology, translational ecology, and "translation community ecology." That is the purport of Eco-translatology and the guidelines for the translation ethics of the "ecological paradigm."

3.7.2 Principle of "Multiple Eco-Integration"

Multiple eco-integration refers to the criteria of translation assessment. What matters here is not fidelity to the source text or catering to the readers but a degree of holistic adaptation and selection on the part of translations to enable them to realize their "survival" or "longevity" in the new linguistic, cultural, and communicative ecologies on the basis of maintaining balances in textual ecology.

The "degree of holistic adaptation and selection" includes both the degree of multidimensional "selective adaptation" (in linguistic, cultural, communicative, or other dimensions) and the degree of "adaptive selection" in terms of other factors or components of the translational eco-environment when the translator produces a translation. In general, the higher a translation's degree of "selective adaptation" and "adaptive selection" is, the higher its "degree of holistic adaptation and selection" will be; relatively speaking, the best translation is the one with the highest "degree of holistic adaptation and selection."

The translation methods of Eco-translatology are accordingly summarized as "multidimensional transformations," i.e., transformations in linguistic, cultural, and communicative dimensions, for the following reasons:

1. From the theoretical viewpoint, the integration of the linguistic, cultural, and communicative translation approaches will give rise to a systematic study based on translation realities since language, culture, and communication continue to be the research foci of translation theoreticians. For example, in terms of functional linguistics, the language expression of texts in translation is the linguistic

dimension, contextual effects are the cultural dimension and interpersonal intentions are the communicative dimension, related to a certain degree to Halliday's ideational, interpersonal and textual functions and the register theory of field, tenor, and mode.

2. From the viewpoint of practice, language, culture, and communication continue to be the research foci in the academic translation field and the key perspectives of transformations in the translation process; a translator usually makes adaptive selections in the order of language, then culture and then communication.
3. From the viewpoint of logic, translation is the shift of language, which embodies culture as a depository of communication; hence, there exists an interior, logical connection between language, culture, and communication, constituting the essential contents of translation shifts.
4. From the viewpoint of maintaining "textual ecology," the translator, via "selective adaptation" and "adaptive selection," assumes the responsibility of maintaining and transforming to the greatest extent possible the linguistic, cultural, and communicative ecologies of the source text and guaranteeing to the greatest extent possible the "survival" and "longevity" of these transformed linguistic, cultural, and communicative ecologies in the target translational eco-environment. The maintenance of balance and coordination between the linguistic, cultural, and communicative ecologies of the source text and target text, nevertheless, corresponds to "'three dimensional' transformations" in the operational methods of translation to arrive at the "balance" and "harmony" of the linguistic, cultural, and communicative ecologies of the source text and target text.

Components of the translational eco-environment have impacts of varying degrees on the shaping of the target text. Therefore, the inability to embody, display, or integrate these components will be noncompliant with the translation realities and hence result in unfairness. Therefore, the ethical principle of "multidimensional integration" is needed to highlight the theoretically and holistically ethical and moral responsibilities of the evaluative criteria of translations and operational steps of shaping them.

3.7.3 *Principle of "Symbiosis and Diversity"*

"Symbiosis and diversity" refer to the diversity of translation theories and the symbiosis of translations. In light of ecological theories and principles, symbiosis is a basic state of organism existence, i.e., the state of interdependence and co-evolution between organisms. Diversity and symbiosis display the differences and coexistence of things, as with the diversity and symbiosis of organisms in natural ecology. Likewise, Eco-translatology, under the guidance of eco-holism and eco-rationality, advocates diversity in translation theories and the symbiosis of different translations, which are expected to be the trends and norms of translation studies. Meanwhile, diversified translation theories and different translations will develop and evolve constantly in the translational eco-environment via compliance with the natural principle of "survival of the fittest."

On the one hand, translation studies are academic investigations "seeking sameness but keeping difference." As stated in Chinese classics, "Men tread different paths that lead to the same destination" [from "Introduction of *The Book of Changes*"] and "A man of noble character seeks sameness with difference" [from "Annotations of 'the Khwei Hexagram' (in *The Book of Changes*)"]. Diversity in translation theories is compliant with both the realities of translation studies and the tradition of Chinese classical academic ethics. Even Venuti (1998, p. 81) advocates the concept of an "ethics of difference" and believes that "foreignization" is moral and differences respect cultural otherness.

It can thus be said that a diversity ethics of translation studies displays respect for translation researchers' right to construct theories.

On the other hand, the eco-environment of textual ecology, translational ecology, and "translation community" ecology is dynamic and changing; to adapt to changes in the eco-environment at different levels, or to maintain balance and coordination between textual ecology, translational ecology, and "translation community" ecology, the coexistence and symbiosis of different translations are "natural phenomena," norms of translation acts.

In this case, the principles of ecological translation ethics rightly justify the following understanding: the natural law of "survival of the fittest" works in the humanities in a somewhat different manner than in nature. That is, in the adaptation of species in nature (animals or plants) to the natural environment, their "selection" and even "obsolescence" are absolute, synonymous with "extinction," "elimination," and "disappearance" in the ecological sense, as in the cases of the dinosaurs, polar wolf, and seed fern. In contrast, the "obsolescence" of translators'/translations' adaptation to the translational eco-environment and "selection" in translation are relative, synonymous with "dejection," "failure," "abandonment," "substitution," "disappointment," etc., in the sense of human acts. That is, the so-called "fittest" or "unfit," "strong," or "weak" translators/translations are not absolute but relative. Different translations, owing to their adaptation to different translation purposes and different readers, are likely to coexist in a symbiotic relationship. "Survival of the fittest" and "coexistence and symbiosis" all comply with fundamental laws of ecology. A "symbiosis" ethics of translation can be regarded as respect for "the right to translate."

3.7.4 Principle of "Translator Responsibility"

"Translator responsibility," relative to "otherness" responsibilities, refers to the notion that the translator takes full responsibility for the translation process, translation act, and translation activities. It is the translator who is responsible for coordinating the interrelation between "translation community," "translation environment," and "translations" to achieve the translational eco-holism of interaction, balance, and harmony.

In regard to translator responsibilities, relevant discussions can be found in the academic translation field. Venuti (1995) raises the issue of translator responsibilities in his book *The Translator's Invisibility: A History of Translation*. Chesterman (2001, pp. 139–154) interprets "translational ethics of commitment" in his article "Proposal for a Hieronymic Oath." Christiane (2001) states that translation *Skopoi* are conditioned by the translator's responsibility for his/her partners in translation activities, meaning his/her responsibility for the other participants in this communicative action of multilateral interaction and cooperation. Xu (2003, p. 294) points out that a responsible translator will give full play to his/her subjective initiative to turn conflicts into harmony in seeking a balance between extremes of domestication and foreignization. Sun (2007) has even published an article on translation ethics and translator responsibilities.

It is true that in translation activities, the translator is just a routine member in the translational ecology who is expected to conduct "dialogues" on an equal basis and exchanges with different forms of otherness and is to be influenced, "manipulated," "conditioned," and "intervened with." He/she is to make choices of his/her own in translation activities and, nevertheless, to make decisions compliant with ecological rationality on any issue related to translation and to assume responsibility for all otherness. Translation is not a unidirectional language or cultural input but a dialogue of differences between two languages and cultures. The translator has the responsibility to safeguard the balance and "neutrality" of the source language ecology and the target language ecology for a unique and harmonious coexistence of the source language and culture in the target language and culture.

Likewise, the translator, although a member of the "translation community" in a relation of "dialogue" on an equal basis with the other members of this community,[19] differs from the other members primarily in that he/she is the only one to participate directly in the translation process and to practice translation acts. It is the translator as the representative of the "translation community" who assumes the specific responsibility for planning and coordinating the interrelation between "translational environment," "translations," and "translation community" and for displaying, via "translator responsibility," the translational eco-holism of interrelation and interaction, balance, and harmony. In this sense, "translator-centeredness" and "translator dominance" in the translation process embody the Eco-translatological ethical principle of "translator responsibility" in aspects of the translation process and translator acts.

The reason is that "eco-holism" can be practiced only by or via the translator, and eco-rationality can be highlighted only by or via the translator. From the viewpoint of translation ethics, "translator-centeredness" and "translator dominance" at the

[19] "Dialogue on an equal basis" denotes an attitude or mode, while "translator responsibility" denotes implementation and outcomes. Regarding translation, though many participants are needed in dialogues who can be "intervened with" or "manipulated," the translator is the one in charge who assumes the ultimate "responsibility."

level of the translation process or translation act will be embodied mostly by "translator responsibility" at the macroscopic or ethical level.

In the "translation community" ecosystem, the translator has the responsibility to coordinate relations between different parties, to practice eco-rationality, to maintain eco-balance, and to protect ecological harmony. He/she has the responsibility to adapt to an ecosystem, to nurture the target language ecology, to keep a watchful eye on the reception and spread of translations, and to strive for the "survival" and "longevity" of translations. In other words, the translator is to display an eco-holistic view of translational ecology with balanced and harmonious ecosystems via reliance on the ethical principle of "translator responsibility." It can also be said that the translator is capable of integrating a more powerful sense of responsibility into translation activities only by assuming responsibility for all otherness, including translations, the "translation community" and translational eco-environment, and by censoring his/her relationship with all "otherness" from the viewpoints of eco-holism and eco-rationality.

What Eco-translatology pursues is the holism and relevancy of translational ecology, and what it concerns is the balance and harmony of translational ecology. After all, who is to implement and practice a translation activity and maintain the holistic and related state of a translation activity? It is the translator as the representative of the "translation community" who implements and practices a translation activity, and it is only through the translator's coordination with other agents that the holistic and related state can be maintained. That is the translator's responsibility! In this sense, ontological translation and rational discussions of translation issues, all designs and plans, all arguments and opinions, and all wishes and expectations should be configured as the translator's consciousness of being meaningful, the translator's competence in being effective, the translator's duty to be practiced, and the translator's responsibility to be realized.[20] Otherwise, effectiveness will be reduced, and any project may become a castle in the air. In a certain sense, Eco-translation ethics is a new ethics of "translator responsibility." Based on this, Eco-translatology defines "translator responsibility" as an ethical principle of crucial significance.

All the above discussions show that the ethical principles of Eco-translation are closely and related to and coordinated with the identity, nature, research objects, and contents of Eco-translatology. The author of this book believes that there is still room for Eco-translation ethics to develop and expand. Studies of Eco-translation ethics are multidimensional and open-ended, and the principle of "diversity and symbiosis" is likewise applicable in these studies.

[20]Actually, what is emphasized is the translator's responsibility for a variety of aspects, no matter whether it is Andrew Chesterman's "ethics of representation," "ethics of service," "ethics of communication," "norm-based ethics," and "ethics of commitment" or Sun Zhili's "reproduction of the source text," "accomplishment of requests from the initiator," "compliance with the norms of target society and culture," "satisfaction of target readers' demands," and "abidance of professional ethics" (Sun, 2007, pp. 14–18).

3.8 Other Research Foci and Theoretical Tenets

As a translatological paradigm of interdisciplinary traits, Eco-translatology employs eco-rationality to synthesize translation as a whole. The research foci of Eco-translatology will be presented, including the Sequence Chain, eco-rationality, translation as textual transplants, and translation as eco-balance. Doing things with translations, the translational eco-environment, the translation ecosystem and the post-event are not presented in detail in the present section since they will be elaborated on in the ensuing chapters.

1. "The Sequence Chain"

   "The Sequence Chain," a linked chain indicating the logic and horizon expansion from "translation" to "nature," has been discussed prior to the basic theories of Eco-translatology to interpret the connection and interaction between translation activities (translational ecology) and nature (natural ecology) (see details in Sect. 2.3.1). "The Sequence Chain" from Eco-translatology explicates the interrelation, interaction, and progression of translation activities (translational ecology) to nature (natural ecology). This chain of "translation ←→ language ←→ culture ←→ humans/society ←→ nature," a logic and relation chain of recognition expansion and logic connection, lends systematization and authenticity to not only macroscopic translation studies but also microscopic textual transformation and assessment.
2. "Eco-rationality"

   "Eco-rationality" has been generalized by the author of this book (Hu, 2009a, 2009b, 2010a, 2010b, 2010c) as "emphasizing holism and interrelation, seeking dynamic balance, reflecting ecological aesthetics, identifying 'translation community', adhering to Eco-translation ethics and highlighting unity in diversity" (see details in Sect. 3.6). It should be pointed out that Eco-translatology, under the guidance of ecological rationality, manifests as a kind of holistic thinking, organic thinking, correlative thinking, or process thinking. What is emphasized includes not only the integration of analysis and synthesis and the impact and constraints of the translational eco-environment on translator acts but also the coordination of "translation studies," "translation theories," and "translations" and the pursuit of the integration of these three.
3. "Translation as Textual Transplants"

   What is Eco-translation? At the macroscopic level, Eco-translation, synthesizing translation from an ecological point of view, bestows an ecological significance of balance and harmony on translation activities and translation studies. At the microscopic level, particularly from the textual level of the Eco-translatological perspective, however, translation is deemed textual transplants (Translation as Textual Transplants, TTT). Specifically, Eco-translation transplants texts from linguistic ecosystem into another linguistic ecosystem.

   Connotations of different linguistic ecosystems have been touched on in the "eco-balance" section: the textual ecosystem of the source language concerns linguistic ecology, cultural ecology, communicative ecology, etc., in the source language system, and the textual ecosystem of the target language concerns those in the target language system (see details in Sect. 3.4.2). Therefore, textual transplants in Eco-translation commence from the ecological structure within the source text[21] to select a text to translate and reproduce the source text by following the intrinsic ecological structure within the source text in the translation process. In other words, the translator of textual transplants in Eco-translation is committed to maintaining, coordinating, and balancing the source/target language ecologies and reconstructing the source language ecology in the target language ecology.

In the source/target language ecologies, "maintenance" ranks as the primary demand, and when maintenance is inaccessible, "coordination" will fill the gap; "coordination" purports to arrive at a balance, and when "balance" is inaccessible, "reconstruction" is targeted. "Reconstruction" here means to create in the target language ecosystem an eco-environment adaptive to the source language ecology to "revive" the world of the source language system covering the source text via the translator's efforts, or even certain special techniques, and to ensure the survival and longevity of the translation in the target language eco-environment.

4. "Translation as eco-balance"

The true essence of Eco-translatology is an emphasis on the balance of translational ecology. It is, as it were, a "Translation as Eco-balance" (TEB) view; translation studies and even the strategies and techniques of translation can all be reduced to "the art of balance."

Balance here means a holistic balance of comprehensive factors to cover the balance of textual ecology, translational ecology, and "translator community" ecology. Specifically, the balance of textual ecology refers to the linguistic ecology, cultural ecology, communicative ecology, etc. of texts. For instance, regarding the balance of linguistic ecology and the balance of textual ecology, the translator is expected to exert him-/herself to maintain the semantic balance between the source and target languages, the balance between the "spiritual similarity" and "expressiveness" of the source and target languages, the balance between the practical values and aesthetic values of the source and target languages, the balance between the styles of the source and target languages, etc. Regarding translational ecology, holistic harmony and balance are expected and desired (as stated in the previous section). In this sense, from the viewpoint of Eco-translatology, Eco-translation is synonymous with "the art of balance" within and without textual ecology, translational ecology, and "translator community" ecology.

Relevant discussions on textual ecology can be found in Sect. 3.4.1, those on the principles of "balance and harmony" can be found in Sect. 3.6.1, and those on "translatability" and "translation of styles" in Sects. 7.5.1–7.5.4.

[21] That is, the automaticity of texts themselves is emphasized in exploring the interior rules of texts, particularly rules of language "transplantation" and "shifts," in addition to the exterior environment and conditions.

5. "Doing things with translations"

   "Doing things with translations" in Eco-translatology is in essence embodied in two aspects: (1) a translator has his/her purpose or aim of conducting translation activities (subjective motive), and (2) translations can do things (objective effects). Regarding the former, a translator may translate for clothing and food, for prestige and interests, for emotional expression, for ethical appeal, for religious belief, or for an intellectual pursuit. In this case, the translator's subjective motive for "doing things with translations" is to (a) "make a living," (b) "realize ambitions," (c) "satisfy interests," (d) "empathize" with others, and (e) "compete" with others. The effects, merits, and achievements of translation are too marvelous to be denied or neglected. The functions of translation from the viewpoint of "doing things with translations" are (a) promoting exchanges and communication, (b) innovating languages, (c) stimulating cultural progress, (d) facilitating social transformation, (e) accelerating eco-civilization, and (f) driving the development of translation studies (see details in Sect. 5.6).

6. "Translational eco-environment"

   The "translational eco-environment" is proposed for the first time in "An Initial Exploration into an Approach to Translation as Adaptation and Selection" (Hu, 2001). The reasons for employing this notion instead of "context" are (1) its correspondence with the "adaptation/selection" law in the Darwinian theory of evolution and (2) its differences from the notion of "context" in connotation, scope, reference, etc. Context refers to the environment of language, with the usage of the language as a reference while excluding language itself and language use. The "translational eco-environment," nevertheless, is an umbrella term to collect the existing states of translator and translations as a whole, integrating the source language, the source text, and the target language, and thus has a wider scope than "context."

   In recent years, the scope of the "translational eco-environment" has been expanded further to integrate the texts involved, cultural context, and the "translation community," as well as a complex of spiritual and material factors. It falls into the macro-environment, meso-environment, and micro-environment, exterior and interior environment, objective and subjective environment, or material and spiritual environment. In translation, all factors exterior to the translator are components of the translational eco-environment; meanwhile, a translator is one component in the translational eco-environment of his/her peers (Hu, 2006; 2008). As a key notion of Eco-translatology, the translational eco-environment has been attracting interest from academics, particularly for its scope expansion, conceptual adjustment, etc.

   Overall, as denoted by "No context, no text," the translational eco-environment is of great significance to the production of translations. For any specific translation event, regardless of the translation of style or content, choice of diction or syntactic structure, the translational eco-environment should be involved and considered in different manners (directly or indirectly, materially or spiritually) and to different degrees (closely or loosely). Therefore, relevant and further explorations are still desired.

7. "Translation ecosystem"

   In light of ecological rationality in Eco-translatology, the translational ecosystem as a whole involves the translational management ecosystem, translational market ecosystem, translational education ecosystem, translational ecosystem itself, and overall translational environments (see details in Sect. 4.2). In general, relevant studies are underway, for example, thematic studies on translation market ecosystems (Cui, 2006; Wu, 2008) and translation education ecosystems (Li, 2010; Song, 2010, 2011a, 2011b; Tao, 2011, 2012). Arguments on the construction of the Eco-translatological discourse system in the present book are essentially studies of translation theories and, in this sense, studies on the ontological translational ecosystem.

8. "Post-event penalty"

   The "post-event penalty" refers to advice or guidelines for translation theories or translator acts instead of any concrete or specific translation strategy, technique, or method. It is employed figuratively to highlight the translator's adaptive selections, i.e., his/her active endeavor to "manipulate/dominate/rewrite" at every stage of the translation process, instead of the pre-event "requirements," "warnings," or "limitations" imposed on the translator. In addition, it points to the assessment and processing of the translator's selections or the translations after the selections of every stage or the whole translation. Specifically, the law of "survival of the fittest" governing the eco-environment at the post-event stage of translation will select and judge the selections of the translator. The "post-event penalty" is therefore consistent with "survival of the fittest." The absence of pre-event or during-event "requirements" or "penalties" leaves room for "translator-centeredness" and "translator dominance." It is a different case, however, for the post-event stage—here, it is "natural selection" that selects and judges. The "post-event penalty," or punishment, comprises obsolescence or elimination in different forms or to different degrees, i.e., criticism, poor sales and even the withdrawal of unfit or inferior translators/translations.

   Cases of the "post-event penalty" are not uncommon in translation activities. The mechanism of the "post-event penalty" for translator adaptations and selections remains, however, a field that is less trodden for the time being and thus is a potential area for further exploration. Case analyses of translation "errors" from the viewpoint of the "post-event penalty" are of some significance as a research theme, especially those on the grave losses (in politics and diplomacy, economy and trade, science and technology, culture and education, interpersonal relations, etc.) resulting from defects in the translator's adaptations and selections.

In addition to the above, there are other research foci or theoretical tenets of Eco-translatology, such as the "holistic degree of adaptation and selection," the "translator community," and "cross-disciplinary integration and interaction." Since they will be elaborated on in the ensuing chapters or sections, we will not go into detail in the present chapter.

In general, among the research foci and theoretical tenets mentioned above, there are macroscopic ones, such as eco-paradigm, ecological rationality, translational eco-environment, translational ecosystem, translation as textual transplants, and translation as eco-balance, and microscopic ones, such as the Sequence Chain, adaptation/selection, "three-dimensional" transformation, "doing things with translations," translator-centeredness, and "post-event penalty." Literature reviews tell us that some theoretical tenets, such as adaptation/selection, "three-dimensional" transformation, "doing things with translations," translator-centeredness, eco-paradigm, and translational eco-environment, enjoy a wider acceptance in the theoretical investigations and applicative attempts of translation studies, whereas ecological rationality, the Sequence Chain, and the "post-event penalty" remain somewhat neglected, leaving much room for further explorations.

The author of this book stops with the major contents and general remarks of the items listed in the present section; regarding their theoretical props and processes of argumentation, readers are advised to turn to the relevant chapters for further reading.

3.9 Summary

The present chapter focuses on the "ecological paradigm" of Eco-translatology, displays its theoretical basis, interprets its objects and methods, summarizes its ecological rationality, proposes its ethical principles, and clarifies its research foci and academic perspectives. The literature review tells us that the research foci and theoretical tenets illustrated in this chapter are core notions of Eco-translatology and the theoretical framework of a number of postgraduate theses, and these theoretical arguments and research perspectives are adopted or commented on by a group of translation researchers.

The essentials of the Eco-translatological discourse system are illustrated as follows (Table 3.1).

Based on all the above, the following further inquiries may occur:

1. What are the latest developments and further investigations in this field on the basis of these Eco-translatological findings?
2. How can Eco-translatology be carried further by its "three-level" study, and what constitutes these three levels?
3. What roles do these three levels play, and how are they coordinated?

All these inquiries are expected to be satisfied in the ensuing chapter, The Macrolevel: An Entire Translation Ecosystem.

Table 3.1 Essentials of the discourse system of Eco-translatology

Essentials	General description
Name	Eco-translatology
Translator	An emerging paradigm of systematic studies on textual ecologies, translational eco-environments, and "translation community" ecologies as well as their interactions and interrelationships. Regarding the scene of translation as a holistic ecosystem, it describes and interprets translation activities in terms of ecological principles of eco-holism, Eastern eco-wisdom, and *Translation as Adaptation and Selection*
Philosophical mainstays	Eco-holism, Eastern eco-wisdom, and *Translation as Adaptation and Selection*
Object of study	Textual ecologies, translational eco-environments, and "translator community" ecologies
Research methodology	Metaphorical analogies, eco-conceptual borrowings, eco-holism
Research subjects	Textual ecologies, translational eco-environments, "translator community" ecologies, and their relationships and interactions; macro-translational ecosystems, meso-ontological theories, and micro-textual Eco-translation operations; overall interpretation and description of both the ontology of translation theories and the entire translational ecosystem
Core concepts	Translation as Eco-balance (TEB); Translation as Textual Transplants (TTT); *Translation as Adaptation and Selection* (TAS)
Theoretical tenets and key terms	"The Sequence Chain," eco-paradigm, eco-reason, translational eco-environment, translational ecosystem, Eco-translation ethics, translational eco-balance, textual ecology, textual transplants, multi-symbiosis, multiple eco-integration, translator responsibility, selective adaptation and adaptive selection, "three-dimensional" transformations, "doing things with translations," post-event penalty, holistic degree of adaptation and selection, etc.
Stages of theoretical development	1. A period of initial exploration from 2001 to 2004, marked by the paper "An Initial Exploration into an Approach to Translation as Adaptation and Selection" presented at the 3rd FIT Asian Translators Forum in 2001 and by the monograph titled *An Approach to Translation as Adaptation and Selection* published in 2004 2. A period of deep integration from 2005 to 2009, indicated by the paper "Understanding Eco-translatology" presented at the International Conference on Translating Global Culture: Towards Interdisciplinary (Re)Construction in 2006 and the paper "Eco-translatology: A Primer" published in the *Journal of Chinese Translators* in 2008 3. A period of all-around expansion since 2009, witnessed by the founding of the International Association for Eco-translatology Research in 2010, the successive holding of international symposiums on Eco-translatology since 2010, and the launching of the *Journal of Eco-translatology* (JET) in 2011

References

Beman, A. (1984). *L'etranger: Culture et Traduction Dans L'Allemagne Romantique*. Pairs: Gallimard.

Bian, L. H., & Yao, Z. F. (2008). Ku Hongming's translation of the analects in the light of translation as adaptation and selection. *Journal of Lishui University, 3*, 18–20.

Cheng, Z. Y. (2005). *Standing up in C-E translation-Chinese philosophy and new positioning of Chinese culture*. Beijing: China Renmin University Press.

Chesterman, A. (2001). Proposal for a Hieronymic Oath. *The Translator*, (1), 139–154.

Christiane, N. (2001). Loyalty re-visited: Bible translation as a case in point. In A. Pym (ed.), *The return of ethics* (pp. 185–202). Manchester: St. Jerome.

Cui, Q. L. (2006). *Chinese translation market: Current situation and analysis*. Retrieved August 25, 2006, from http://www.teacherwong.com/zxxx/ShowArticle.asp?ArticleID=389

Darwin, C. R. (1859). *On the origin of species by means of natural selection, or the preservation of favored races in the struggle for life* (1st ed.). London: John Murray.

de Waard, J., & Nida, E. A. (1986). *From one language to another: Functional equivalence in bible translating*. Thomas Nelson: Nashville.

Fang, M. Z. (2011). Translation ecology. *Shanghai Journal of Translators, 1*, 1–5.

Gentzler, E. (1993). *Contemporary translation theories*. New York: Routledge Inc.

Gentzler, E. (2001). *Contemporary translation theories* (Rev. 2nd ed.). Clevedon: Multilingual Matters Ltd.

Gorlée, L. (1994). *Semiotics and the problem of translation: With special reference to the semiotics of Charles S. Peirce*. Amsterdam: Editions Rodopi B.V.

Hermans, T. (1999). *Translation in systems: Descriptive and system-oriented approaches explained*. Manchester: St. Jerome Publishing.

Holmes, J. S. (1972). *The name and nature of translation studies. Unpublished manuscript*. Amsterdam: Translation Studies Section, Department of General Studies.

Holmes, J. S. (1988). *Translated! Papers on literary translation and translation studies*. Amsterdam: Rodopi.

Hu, G. S. (2001, December 6/8). S*tudy on translation as adaptation and selection*. Paper presented on the Third Asian Translators Symposium of International Federation of Translators, Hong Kong.

Hu, G. S. (2003). Translation as adaptation and selection. *Perspective: Studies in translatology, 4*, 283–291.

Hu, G. S. (2004a). *Translation as adaptation and selection*. Wuhan: Hubei Education Press.

Hu, G. S. (2004b). From "Translator's Subjectivity" to "Translator-centeredness". *Shanghai Journal of Translators, 3*, 9–15.

Hu, G. S. (2004c). An empirical study on evaluating the "Approach to Translation as Adaptation and Selection" through questionnaire surveys. *Foreign Languages and Their Teaching*, (4), 40–44.

Hu, G. S. (2006). Understanding translation as adaptation and selection through case studies. *Foreign Language Education, 4*, 50–54.

Hu, G. S. (2008). Interpreting translation process from the perspective of adaptation and selection. *Journal of Sichuan International Studies University, 4*, 91–95.

Hu, G. S. (2009a). An "Ecological Turn" signified by the "Sequence Chain". In *Hu Gengshen's translation and cross-cultural communication: Integration and innovation* (pp. 26–39). Shanghai: Shanghai Foreign Language Education Press.

Hu, G. S. (2009b). On Fu Lei's translation philosophies: An eco-translatological perspective. *Journal of Foreign Languages, 32*(2), 47–53.

Hu, G. S. (2009c). Eco-translatology: Trans-disciplinary integration on translation studies. *Shanghai Journal of Translators, 2*, 3–8.

Hu, G. S. (2010a). Eco-translation vs. natural ecology: Relevance, similarity, isomorphism. *Shanghai Journal of Translators*, (3), 1–7.

Hu, G. S. (2010b). Translation eco-system in context of eco-translatology. In *Ren Dongxing's theoretical construction of translation as a discipline* (pp. 296–315). Shanghai: Shanghai Foreign Language Education Press.

Hu, G. S. (2010c). *Studies on translation theory* (pp. 10–30). Shanghai: Shanghai Foreign Language Education Press.

Hu, G. S. (2011a, November 11/13). *Eco-translatology: Where is "Different" and "Special"*. Paper Presented on the Second International Eco-Translatology Symposium, Shanghai.
Hu, G. S. (2011b). Eco-features of eco-translatology and the implications to translation studies. *China Foreign Language, 6*, 96–99/109.
Hu, G. S. (2011c). Eco-translatology: Research foci and theoretical tenets. *Chinese Translators Journal, 2*, 1–5.
Huang W. (2009). *Recognizing characteristics of life science disciplines*. Retrieved January 23, 2010, from http://2009.cersp.com/article/browse/35826716.jspx
Ji, X. L. (2008). *Memory muse and emotion expression*. Xi'an: Shaanxi Normal University Press.
Jiang, K. Y. (1993). *The new theory of aesthetics*. Beijing: People's Literature Publishing House.
Kuhn, T. S. (1962). *The scientific revolutions*. Chicago: The University of Chicago Press.
Lance, H., & Jacky, M. (1991). *Redefining translation: The variational approach*. London: Routledge.
Lefevere, A. (1993). Discourses on translation: Recent, less recent and to come. *Target*, (2), 229–241.
Li, G. S. (2010, November 9/10). *Comparative study on martial theory and translation theory from eco-translatological perspective*. Paper presented by at the first international eco-translatology symposium, Macau.
Liao, Q. Y. (2001a). *Contemporary translation studies in U.K.* Wuhan: Hubei Education Press.
Liao, Q. Y. (2001b). Research paradigms and translation studies in China. *Chinese Translators Journal, 5*, 14–18.
Liu, A. H. (2011). On eco-environment of eco-translatology. *Journal of Eco-Translatology, 2*, 57–64.
Lv, J. (2008). Paradigmatic criticism and problematic consciousness-a critical discussion on the two principle approaches to translatology. *Journal of Foreign Languages, 5*, 55–63.
Lv, J., & Hou, X. Q. (2006). *Translatology — A constructivist perspective of translation study*. Shanghai: Shanghai Foreign Language Education Press.
Lv, J., & Hou, X. Q. (2009). Cultural criticism in translation studies. *Foreign Language and Literature, 1*, 122–126.
Meng, F. J. (2011). Contemporary translation studies from eco-translatology perspective. *Journal of Eco-Translatology, 1*, 73–79.
Merchant, C. (1999). *The dead of nature—Woman, ecology and the scientific revolution* (Wu Guosheng et al., Trans.). Changchun: Jilin Peoples Publishing House.
Mu, Z. S. (1997). *Nineteen lectures of Chinese philosophy*. Shanghai: Shanghai Guji Press.
Nida, E. (1984). Approaches to translation in the Western World. *Foreign Language Teaching and Research, 2*, 9–15.
O'Hagan, M. (2005, May 2–6). *Multidimensional translation: A game plan for audiovisual translation in the age of GILT*. Paper presented at the Marie Curie Euroconferences MuTra "Challenges of Multidimensional Translation" – Saarbrücken.
Pöchhacker, F., & Kaindl, K. (Eds.). (1994). *Translation studies: An interdiscipline*. Amsterdam: John Benjamins.
Pym, A. (1997). *Pour Une Ethique du Traducteur*. Arrras: Artois Presses Universite et Presses de l'Universite d'Ottawa.
Pym, A. (Ed.). (2001). *The return of ethics* (pp. 139–154). Manchester: St. Jerome.
Shu, H. J. (2010). The translation of public signs from eco-translatology perspectives——Taking the signs of Shanghai World expo as an example. *Shanghai Journal of Translators, 53*(2), 41–44.
Snell-Hornby, M. (1988). *Translation studies: An integrated approach*. Amsterdam: John Benjamins Publishing Company.
Song, Z. P. (2010). *Translation pedagogy ecosystem: An eco-translatological perspective*. Paper presented at the First International Symposium on Eco-translatology, Macao.
Song, Z. P. (2011a). Eco-system of translation teaching in view of eco-translatology. *Journal of Eco-Translatology, 1*, 59–65.

Song, Z. P. (2011b). Issues on ecological holism and translation teaching. *Journal of Eco-Translatology, 2*, 72–78.
Steiner, G. (1975). *After babel: Aspects of language & translation*. Oxford: Oxford University Press.
Sun, Z. L. (2007). The translator's responsibilities. *Chinese Translators Journal*, (4), 14–18/94.
Tan, Z. X. (1998). Pay attention to comparative studies of Chinese and Western translation theory in translation studies. *Chinese Translators Journal*, (2), 12–16.
Tao, Y. L. (2011, November 11/13). *On translation textbooks studies: Perspectives from eco-translatology.* Paper presented at the second international eco-translatology symposium, Shanghai.
Tao, Y. L. (2012). Translation textbook design in translation and interpretation: Perspectives from eco-translatology. *Foreign Language World*, (3), 81–88.
Venuti, L. (1995). *The translator's invisibility: A history of translation*. London: Routledge.
Venuti, L. (2000). *The translation studies reader*. London: Routledge.
Venuti, L. (1998). *The scandals of translation: Towards an ethics of difference*. London: Routledge.
Wang, D., & Chen, C. (2001). *Modern rhetoric*. Shanghai: Shanghai Foreign Language Education Press.
Wang, N. (2004). Ecological holism. *Reading Journal*, (2).
Wang, N. (2011). Eco-literature and eco-translatology: Deconstruction and reconstruction. *Chinese Translators Journal, 2*, 10–16.
Wang, R. S., & Zhou, H. (2004). *Human and ecology*. Kunming: Yunnan Peoples Press.
Wu, Y. H. (2008). The golden time for chinese translation industries. Retrieved August 6, 2008, from http://news.mainone.com/hangyefenxi/htm
Xu, J. (2003). *On translation*. Wuhan: Hubei Education Press.
Zhai, H. M., & Zhang, D. R. (2005). Translator-centered theory and choice of Lin Yutang's Translation of Fu Sheng Liu Ji. *Journal of Anhui Normal University (Humanities & Social Sciences)*, (1), 115–119.
Zhu, C. W. (2010). The translator's duties: A reading from the ethics of translation. *Foreign Language and Literature, 6*, 77–82.
Zhu, Z. Y. (2009). Translation studies: Norms, description and ethics. *Chinese Translators Journal*, (3), 5–12/95.

Chapter 4
The Macrolevel: An Entire Translational Ecosystem

4.1 Introduction

The theoretical structure of Translatology as a discipline of translation studies has already been mentioned in the section titled *Translatology* in the Preliminaries. Studies on Translatology, universally relevant domestically and abroad, in the form of either an illustration or a description of its theoretical structure, differ from one another in the divisions of pure or applied theory, of exterior system or interior systems, of general or peculiar sections, or of macro- or microlevels, but no distinct and unified doctrine acts as a general principle among all the theoretical structures of translation studies. In other words, all the higher level theoretical structures of translation studies require a ruling doctrine, namely, the ultimate "theoretical constitution." As a result, on the one hand, translation theorists in different countries develop their translation theories based on their personal differences in recognition, which makes it difficult to form a common foundation for translation studies, and on the other hand, in the theoretical systems that have been constructed, unified coordinating criteria and relevant ties are theoretically needed between the various sections or layers of the disciplines of translation studies, which makes translation studies greatly diverse and disconnected.

However, the ecological outlook can make it possible to compensate for the shortcomings of other translation theories in the higher level of theoretical structures. According to the ecological viewpoint, unity, relevance, dynamics, and harmony are all highlighted in the ecological system. As a unity, most of the factors in ecological systems are in a relationship of co-motivation, co-benefits, co-existence, and co-development, which gives rise to an ecological unity of balanced stability and harmonious motivation. All this is determined by the ecological reasoning of the ecological organism, but it cannot be transferred or altered by the individual will, different types of human recognition, or other abilities. Thus, from the ecological perspective, a unified macro-level doctrine and a relatively universal criterion

(Hugs) G. S. Hu, *Eco-Translatology*, New Frontiers in Translation Studies,
https://doi.org/10.1007/978-981-15-2260-4_4

can be developed spontaneously in the construction of a theoretical system of translation studies by means of adhering to the ecological rationale.

In accordance with the macro-ecological rationale of Eco-translatology (see Sect. 3.6), a unified Eco-translational system of Eco-translatology will be established in this chapter, including a translation management ecosystem, translation market ecosystem, and translation education ecosystem as well as the translation ecosystem itself and the ecosystem of the translation environment as a reliable system in the outer zone. On this basis, the multidimensional and multilayered internal constituents of the Eco-translational system will be explored, the interlaced Eco-translational system will be analyzed, and studies of the subsystem of Eco-translatology will be critically evaluated. Furthermore, in accordance with the macro-level eco-principle or eco-perspective of Eco-translatology, studies in various disciplines, such as linguistics and cultural studies, that are related to translation activities, as well as the cross-disciplinary integrity between communication science, anthropology, comparative literature and ecology, namely, the return of ontological studies of Translatology, have been highlighted. Finally, the construction and maintenance of the balanced stability of the Eco-translational system will also be mentioned.

4.2 Complex Thinking, Complexity of Ecosystems, and Cross-Disciplinarity of Translation Studies

4.2.1 Complex Thinking

A complex system can be studied by means of "bridging the gap between the natural sciences and the social sciences,"[1] while the exploration method of the complex thinking system is to investigate the movement and alteration of things with the main features of nonlinear thinking, entire thinking, relevant thinking, and process thinking.[2]

The method of complex thinking[3] possesses ten fundamental points: (1) a method is valid only when it includes complexity; (2) this method will provide operative principles that can help the thought itself; (3) this method is not a design (including the whole set of projections) but an action strategy that can continuously be self-adjusted in the feedback of reality; (4) the method should be highlighted to promote "a new quality"; (5) one should understand that time is irreversible, which has been regarded as an indispensable condition for explaining the historical process as a contemporary event; (6) the systematic integrity of things and inter-relevant

[1] See Mainzer (1999), p. 64).

[2] See Peng (2003, p. 9).

[3] See *Complex Thinking* by Huang. http://www.huangjiwe.com/blog/?p=558. Retrieved on May 26, Huang, 2012.

cognition is emphasized; (7) the existential significance of self-organization is acknowledged, with a process including reciprocity of cause and effect, feedback, randomness, hysteresis, interference, disorder, coordination, lapping, bifurcation and catastrophe, and reorientation; (8) the inter-effort between the cognitive object and its environment is emphasized; (9) the inter-relevance between the observant subject and the observed object is highlighted; and (10) the limitation of formal logic in complex systems is confessed, and the contradiction or logical setback encountered in observation is regarded as an actual realm that is unknown or more complicated. The complex deduction principle includes the synchronic qualities of complementarity, competition, and contradiction of conceptions.

> The method of complex thinking requires that we should never encapsulate our conceptions in the time of thinking, but rather, we should break through the constrained boundary and, amongst the thinking diversity, we should never forget that they are a unity integrated together. It is an expansive force toward the overall realization of things, but meanwhile, attention should be paid to its contradictory features.[4]

Translation studies, which comprehensively cover the knowledge of various subjects of the social sciences or humanities, such as comparative literature, linguistics, cultural studies, communications, anthropology, media studies, philosophy, methodology, discipline-development theory, and psychology, have various relevant elements such as the original author, the source-language code system, the content conveyed in the source language, the conveyance channel, the translator, the lingual code system used by the translator, the content conveyed with the target language, and the readers of the original and the translated version. Translation, so to speak, proves to be the most complex issue in the human world. Thus, it is reasonable for us to explore it systematically from the perspective of complex thinking.

4.2.2 *Complexity of Ecosystems*

Complex science provides us with a new worldview for the translation studies discipline. The term complex system is defined[5] in complex science as a system with intelligent, self-adaptable, medium-numbered subjects who can act on the basis of partial information. Based on the above definition, the complex system can be comprehended as follows:

1. It is not a simple system or a random system.
2. It is a complex system, not a complicated system.

---

[4] See Morin (2005, p. 14).

[5] The content of *Complex System* is examined and published by the National Science and Technology Terminology Committee, which can be seen in http://baike.baidu.com/view/228782.htm. Retrieved on May 26, 2012.

3. In the complex system, there are many subsystems that are interdependent and coordinate with one another and that can co-evolve. In the complex system, the subsystems can be multilevel and multi-scale divisions.

The complex system features a multiplicity of elements, with forceful coupling effects among them. The complex system is composed of various small systems. The ecosystem, for instance, is composed of various species groups and creatures.

The Eco-translational system as a unity, which is discussed below, is a gigantic system composed of various subsystems, such as the translation management ecosystem, translation market ecosystem, and translation education ecosystem as well as the ecosystem of translation itself and the ecosystem of the translation environment as a reliable system (see Sects. 4.3.1–4.3.6). It is a compound system, one major feature of which is that it is composed of different levels and scales (see Sect. 4.4.3).

Due to the innate nature of the multiplicity of dimensions and levels of the ecosystem and to the interrelations between various levels, the large system will be composed of subsystems, which, in turn, are composed of various sub-subsystems. However, whether it is an Eco-translatological system or a different sub-ecosystem, the different interior structures and interior connections of different sub-ecosystems, or the interrelationships between various sub-ecosystems, the design and description herein are aimed at a general directive principle, that is, to adhere to the basic ecological laws and to conform to the features of the ecological rationale.

4.2.3 Cross-Disciplinarity of Translation Studies

Along with globalization, translation has developed as a medium or tool to bridge different cultures in our present-day information society. As translation practice covers a vast scope, and as different subjects in our society are increasingly obviously "overlapping" and influencing one another, the cross-disciplinarity of translation studies is more frequently coming to light, not only in translation activity itself but also in translation studies as well as translation criticism. The latest achievements of the social sciences and natural sciences have been constantly absorbed in translation studies, which has enriched and expanded the realm and scope of the field of study. Thus, the cross-disciplinarity of translation studies has become increasingly obvious.

The word *cross-disciplinarity* first appeared in New York, in the USA, in the 1920s, with the initial meaning of *cooperative research.* In 1985, the Cross-Science Conference that was held in China made the term *cross-science* or *cross-discipline* widespread in the world of science. In the early days, there was no distinction between *cross-science* and *cross-discipline*. Since the 1990s, *cross-disciplinary* began to be used by some scholars to replace *cross-science.* From the present-day viewpoint, interdisciplinary study is the initial stage of cross-science research, as such research is confined to the existing academic disciplines, which are designed mostly by certain scholars. Thus, to achieve further developments in the research, efforts must be made to break through the confinement resulting from the division of various disciplines to drive cross-disciplinary research that is broader in scope and higher in level.

Cross-disciplinarity aims mainly at the realization of integrated research on various issues by passing over the previous approaches that were used to study each academic area individually. Currently, various flourishing academic disciplines that are mostly cross-disciplinary are emerging worldwide.

Because of the complexity of translation and translation studies, which is mainly embodied in their cross-disciplinarity, the subjectivity of the translator and translation researchers, and the experience-orientation of translation studies, it is difficult to develop new research directions, which has led to unprecedented challenges for scholars of translation studies. It is impossible, so to speak, to accomplish further developments in translation studies without adopting cross-disciplinary approaches.

The traditional translation studies are confined to the thinking pattern of duality, which centers on the four fundamental elements of translation studies, namely, the author and the translator and the original and the translated versions. The development of deconstruction made it possible for translation studies to break away from the confinement of bipolar oppositions and to enrich and improve the disciplinary system of translation studies. In reviewing the history of translation studies, we can see that to develop Translatology, and to establish and improve the discipline of Translatology, cross-disciplinary and comprehensive dimensions of research must be developed.

Since the end of the 1950s, translation studies in the West have finally attained a new stage of cross-disciplinary research. It can be safely said that present-day translatological research has become a discipline with multiple academic norms.

As said by De Waard and Nida (1986), p. 185, "Translation ... is an activity which can be systematically described and is closely associated with various disciplines. ...Translation is conducted on the basis of various disciplines such as psychology, linguistics, science of communication, anthropology, and semiotics, etc."

Snell-Hornby (1988, p. 29) also points out, "...The holistic principle itself has become increasingly dominant in the study of language over the last few years, and in recant translation theory it is of primary importance." Similarly, Susan Bassnett remarks in the *Series Editor's Preface* to *Contemporary Translation Theories* by Edwin Gentzler, "Translation Studies seems to be developing for the future, enthusiastically advocating the closer relations between related disciplines such as literary studies, linguistics, history, ethnography, anthropology and sociology" (Gentzler, 1993, p. viv). Recently, in the academic lecture titled *On Approaches to Translation Studies and Multiplicity of Conceptions*, Gorlee (Gorlée, 1994, p. 133), in a down-to-earth attitude, restates the same viewpoint by pointing out that "...Translatology is a cross-discipline (or inter-discipline) in that it combines the research norm of general linguistics and applied linguistics and that of general literature and comparative literature into one, and besides this, it also refers to the norms of social science disciplines such as socio-anthropology, sociology and mythology, etc." Translation studies "involves the multiplicity of different terrains" (Duarte, Rosa, &

Seruya, 2006, p. 2). "The multiplicity revealed through the act of translation is doubled by the multiplicity of meanings attributable to the act itself, making interdisciplinary approaches necessary" (Nouss, 2005, p. 228, quote from Duarte et al., 2006, p. 3).

Through the research on the Sequence Chain (i.e., translation ↔ language ↔ culture ↔ human beings/society ↔ nature, which can be referred to in Sect. 2.3.1), the interrelationship of translational activity and ecological nature, the similar structure of translation activities and ecosystems, and the features and laws of the interaction and inter-complementarity of translation and human society have been discovered. This not only provides a theoretical basis for interdisciplinary translation studies but also opens up new alternative spaces in translation studies.

4.3 The Entire Translational Ecosystem

The theoretical framework of the translational ecosystem in the vision of Eco-translatology is constructed in accordance with the five main fundamental bases as follows:

1. It is based on the cognitive dimension of *the Sequence Chain* between translation (translational ecology) and nature (natural ecology) (Sect. 2.3.1);
2. It is based on the reference to the fundamental elements or components of which the entire translational ecosystem is composed (see the interpretation and connotation of the items *ecosystem* and *translational ecosystem* in the fourth section, titled *Understanding Eco-translatology through Terminology*, in the Preliminaries of the book);
3. It is formed on practical translation activities, namely, the relevant and interactive behaviors that connect various links of *the chain of translation*);
4. It rests on references to the related conceptions or statements in the former theoretical systems of translation studies (see the section titled *Translatology* in the Preliminaries of the book); and
5. It is based on the feature of the macro-ecological rationale of Eco-translatology (Sects. 3.6.1–3.6.6).

Undoubtedly, to understand the entire translational ecosystem, a further understanding of Eco-translatology will be helpful. As has been revealed in modern science,[6] "We always understand a thing through the changing system which contains each part of it" (Serres, 1974, p. 9), while "the thing will be thoroughly understood only if the basic units of which the thing is composed is understood" (Lu, 2005, p. 11). Therefore, the translational ecosystem in the vision of Eco-translatology,

[6] For instance, the world can be understand by means of the cognition of atom, electron, proton, neutron and field, life can be interpreted by means of the decoding of genes, etc.

including its components, structures, functions, procedures, and interrelationships, will be discussed below.

4.3.1 Translation Management Ecosystem

The translation management ecosystem, obviously, is the macrosystem of management that is used to manage, coordinate, maintain, and evaluate Eco-translation. According to the ecosystem, the *translator community,* Eco-translational environment, and cross-cultural activities of human beings are joined together for coordination. Based on a thorough comprehension of the composition, structure, functions, and procedures of the translational ecosystem, adaptive management strategies will be developed to maintain and develop the entity, validity, and continuity of the translational ecosystem.

The translation management ecosystem requires that the entire translational ecology be considered; that the production of translational ecosystems be managed by means of the principles of Translatology, ecology, sociology, and management science; and that the profits and values of the entity and validity of the translational ecosystem be maintained and developed. In addition, the best integral management of the social value, expected value, and Eco-translational potential should be carried out so that the entirety, ideal conditions, profits, products, values, and services of translational ecosystems will be long maintained.

The translation management ecosystem is related to certain administrative organs as well as certain laws, regulations, assessments, and tests. According to the social requirements, for instance, the *Temporary Regulations for Translation Qualification (Level) Test* was developed by the former Personnel Ministry of China for China's translational ecology. Since 2003, *The National Translation Qualification (Level) Test* has been carried out throughout China. In September of the same year, *The Regulations for Translation Service* were published by the General Administration of Quality Supervision, Inspection and Quarantine of the PRC to regulate the market of translation services and to expand the professional functions, which made it possible for consumers to choose the objective criteria of translation services.

The translation management ecosystem is also related to the organization and administration of various professional associations. The administration is somewhat different from the functions mentioned above in that a joint translator community's main functions are maintaining the translator community's profits, confining the professional activities of individual translators, promoting exchanges between translators, expanding the activity space of the translator community, and enhancing the social status of the translator community to highlight the collectivity and sociality of the translation. Although such an administration cannot enforce restrictions on the behaviors of individual translators, it can exert an explicit influence upon the self-discipline of the translation profession. The present situation and development of this aspect is similar to what Mr. Yang (2009, p. 6) pointed out: "We are now in a condition of dispersing backwardness in that there is in need not only of the scholars' theoretical research but also of the establishment of morale and laws by the related governmental organizations."

In addition, accompanying the development of globalization throughout previous decades, an increasing number of corporations, in a sense, have grown from the common cross-national corporations into truly global organizations, which has triggered the instant and consistent requirement for translation services and has resulted in an explosive increase in translation services and the corresponding demand for translation administration. In the complicated framework of translation administration, project management, which features the integrity of the artistic, scientific, and professional dimensions, is the critical factor of the translation process because the entire translation process is related to exchanges of and interactions in various aspects of the work. The administrator of the translation project, as the director of the whole translator community, drives the whole translator community to create a harmonious symphony of translation by means of various translation tools and a translation management system.[7] Thus, translation has developed rapidly into a cultural industry with vast potential. The market system has stimulated the translation industry to develop rapidly and, meanwhile, revealed new and higher level requirements for the administration of the translation profession.

4.3.2 Translation Market Ecosystem

As the top goal pursued in translation, cross-cultural exchange in translation serves as the primary criterion for evaluating the quality and profitability of translation. Regarding the nature and function of translation, the translator as the subject of translation activities is the producer of the translation product. The final product of translation activity should be consumed and enjoyed by the readers of the translated version. In other words, market services should be delivered through cross-cultural communication and exchange between different languages.

The emerging translation market has become a newly developed modern service industry in the cultural economy, second only to the education profession. According to the investigation of the world translation market by an American organization, the translation market throve in 2005, with a general value of 22.7 billion US dollars. With the Chinese translation market as another example, according to the statistics, there are more than 3000 corporations registered in the translation profession and more than 10,000 translation corporations registered as translation acquisition corporations or translation and publishing houses, which actually act as translation corporations. In Beijing, specifically, 100 or more corporations have mushroomed as registered translation corporations. In addition, the productive value of the translation market has been increasing rapidly, from 11 billion US dollars in 2003 to 20

[7] Refer to Translation and Translating Management from *Globalization and Localization* by Huang Xiang and Wang Weihua, pp. 12–13. Vol. III, 2012.

billion US dollars in 2005 and then to 30 billion US dollars in 2007. As pointed out by Ms. Sheryl Hinkkanen, secretary general of the International Translators Association, in her plenary lecture titled *On the Global Translation Industry, Past and Present* at the China International Translation Industry Forum in 2006, cosponsored by the Chinese Translators' Association and Tongji University, decent progress has been made by the international translation circle in establishing the criteria of the translation industry. At the same time, the structure of the translation industry has also experienced rapid and fundamental changes. However, the 1990s still saw the translation industry developing mainly in the local markets of various countries, with linear relationships between most or all links in their service and supply chain. Since then, however, the translation industry has developed quickly into a global market, including various computer tools and international competition as well as long service and supply chains among many countries. One action taken to deal with the changing operative environment is standardization, with an aim of not only directing the translation industry in the coordination of its activities but also assisting customers by guaranteeing service quality. National-level criteria were first published, with the intention to expand their coverage (Shao, 2006). As has been reported, more than one hundred translators were needed during the Beijing Olympic Games 2008, and more than seven thousand news reports and lectures required translation. More than one hundred forty languages were used at the press conferences and award-presentation ceremonies, and more than one hundred fifty interpreters were also needed. As pointed out in an analytic report on the requirement for talented professionals published by the Beijing foreign-funded firm Talented Professional Service Co. Ltd., after the opening ceremony of the Olympic Games, the productive value of translation amounted to more than seventy million yuan, covering ten languages. These data cover only a fraction of the translation services needed because several hundred sports delegations and thousands of foreign tourists would also have needed translation services (Wu, 2008).

Importance should also be assigned to the development of the version-publishing market. Since the establishment of China Translation and Publication Corporation in 1973, for instance, various documents with tens of billions of words or characters have been translated. This corporation has developed from a document translation service solely for the UN to an overall translation service for various organizations domestically and abroad, from an exclusively domestic translation service to a long-/short-term international translation service providing professional translators or interpreters, from a one-foreign-language-oriented translation service to a multiple-foreign-language-oriented translation service, and from providing written translations to providing alternative or simultaneous interpretations. Professionalism and internationality have been emphasized in the translation services of the corporation, with online translation and remote translation as the primary means of delivering translation services.

The specific feature of the translation industry lies in the fact that it is not only an industry but also the core link in related global multilingual industrial chains with abundant relevant industrial links. Unlike the traditional industrial chain, the global multilingual chain is a multidimensional network rather than a linear connection.

The translation industry is the core of the global multilingual network, which drives advanced relevant industries such as the cultural industry, publication industry, press industry, information processing industry, and training industry.

Furthermore, as reported in *English Education Weekly of the 21st Century* (Wen, 2009), translation circles domestically and abroad have witnessed a new, significant tendency to establish large-scale multilingual information-transferring bases to organize translation industrial centers. These multilingual information bases, which form a distinguished high-level translation service industry independent of the translation procedure, are related not only to simple translation services but also to services of various broader scopes, such as multilingual consulting services and multilingual software development. On such information platforms, international databases of terms and lingual materials, together with coordinated translations and standardized administration of translation, enable the operative systems to work simultaneously, with significant savings of translation time and costs. With the publication and development of a multilingual information platform, the productive value amounted to two hundred thirty billion yuan (RMB) in the overall translation market in China. As pointed out by Mr. Huang Youyi, Vice President of the Chinese Translators' Association, in the 18th World Translation Conference held in Shanghai, the translation industry in China has encountered new opportunities against the background of reform, opening up, and globalization, whereas various problems have arisen in the aspects of marketing, laws, and services. He held that a permit system to enter the translation market should be advocated to maintain a high quality of translations and to enhance the entire level of the translation market. Mr. Tang Jiaxuan, honorary president of the Chinese Translators' Association, also pointed out at the Symposium on All China Translation Work on the Thirtieth Anniversary Celebration for the Establishment of the Chinese Translators' Association held on December 6th, 2012, that epoch-making changes have taken place to the contents, conditions, and means of translation work during this time of globalization and information explosion, and the traditional mode of individual, manual translation activity has been replaced by a new, modern industrial pattern with a numerical, Internet-oriented tendency featuring group participation.[8] That is, significant changes have taken place to the translation career since the campaign of reform and opening up, especially since the turn of the twenty-first century, in that the objects, genres, requirements, means, and markets of translation are utterly different from those of previous years, that translation work has experienced an alteration from an individual activity to the industrialization of lingual services, and that translation has moved to a new stage of information as a systematic profession by intermingling with the tide of globalization.[9]

The translation market ecosystem, with its functions of cross-cultural and cross-social communication and exchange, is significant to cross-cultural existence. The significance of cross-cultural existence will be affected if the cross-cultural products

[8] See *Perspective and Interview* in The Chinese Translation Internet, Vol. VII, 2012.

[9] See *Perspective and Interview* in The Chinese Translation Internet, Vol. I, 2012.

of translation stray from the expected high-level goals fail to attain the goal, or even cease to function. In the present situation, more efforts should be made to study how to improve the translation market ecosystem, including the publication of translated versions,[10] and to promote the sound, orderly, and skillful development of the translation market.

4.3.3 *Translation Education Ecosystem*

The translation education ecosystem is a system of multiple ecological environments that, with translation education at its center, functions to regulate and administer the establishment, management, and development of translation education.

The translation education ecosystem is divided into three layers. First, there is the individual or collaborative education ecosystem composed of all-inclusive environments, including the exterior translation environment, social environment, and regulative environment,[11] with translation education at its core. The second level constitutes and is centered on a particular school or educational level, reflecting the interior interrelationships of the education system. The third level constitutes a system for the study of the exterior environment, including national and international factors, with the individual cultivation of students or trainees as its primary dimension.

The complexity of the ecosystem is embodied in the multidimensional and multilayered overlapping relationships of the ecosystem; that is, the entire general system is composed of subsystems, each of which is composed of sub-subsystems; therefore, the translation ecosystem comes into being with boundlessly divided translation subsystems interlacing in the translation environment. For instance, the translation education ecosystem is composed of attached subsystems such as the international translation education ecosystem, national translation education ecosystem, regional translation education ecosystem, and school translation education ecosystem. Of course, subsystems can also be composed of their sub-subsystems. For instance, the school translation education ecosystem contains a discipline construction system, text construction system, teaching method system, and assessment and testing system. The components of the systems are connected, confined, and affected by one another. The more integrity the ecosystem has, the less structure it has. Under certain conditions, an ecosystem can join a more extensive system as an element. Integrity is the fundamental core of the ecosystem.

The structure of the translation education ecosystem has two facets: the macro-facet and the micro-facet. The macro-level ecological research on translation

---

[10] It includes various issues such as choice of topics, sponsors, supporters, version readers, translator's stance, social profitability, social reaction, and cultural effect.

[11] The regulatory environment can include culture, science and technology, language, nationality, ethics and morals, philosophy, democratic and legal systems, social customs, religion, arts, and sports and games.

education refers to various environmental systems (i.e., the eco-environment of translation domestically and abroad) centered on translation education, analyzing its function, the cross-cultural interaction between its function and translation education, and the human beings involved in order to determine the right direction for educational development, for the best system of translation education, and for various strategies that should be undertaken. The micro-level ecological research on translation education refers to the influence of schools, classrooms, and facilities or even the location of jobs upon teaching. It refers not only to the analysis of microsystems, including curriculum, general principles for teaching, teachers, teaching strategies, teaching methods, and teaching evaluations, but also to the influence of the affinity relationships of families, teacher–student relationships at schools, relationships between students, and students' private lives and attitudes towards education.

The teaching structure of the translation education ecosystem ranges from the undergraduate stage to the MTI stage, doctoral stage, and postdoctoral stage, which reflects not only the different layers of education (in 2006, for instance, the Education Ministry approved establishing a translation major among undergraduate students in the universities or colleges of China, and undergraduate courses on translation are now taught at 15 universities or colleges of China. Systematic teaching of translation has been established in the universities or colleges of China, covering various learning stages, such as the undergraduate stage, master's degree, doctoral degree, and the postdoctoral stage) but also the translation education process, from simplicity to complexity and from junior stage to senior stage.

A country, a province, or a municipality is usually regarded as a large translation education ecosystem with many subsystems attached to it. Meanwhile, the translation education ecosystem includes the original translation education system as well as the translation environment system. The translation education system contains three "functional groups," namely, the professors, students, and education administrators. In the translation education ecosystem, translation education and its attached organs or layers are regarded as central, with the three "functional groups" as ties and with the central aim not only of training professional translators or interpreters but also of creating translation products. Regarding other layers of the eco-environment, multiple elements have been influenced, and an interactive system has emerged.

4.3.4 The Ecosystem of Translation Itself

The ecosystem of translation itself, with the ontological research of Translatology as its core and related research from the perspectives of linguistics, cultural studies, communications theory, and anthropology as its support, is intermingled with the transdisciplinary complex of research (Hu, 2009, pp. 12–13). Emphatically, the group body serves as the center of the whole translator community, embodying the ultimate orientation of the pattern of translation activity, the absolute thinking about translation, and the cultural value of translation. In Eco-translatology, translation is dominated by the translators and supported by cross-cultural texts dealing with the transfer of information based on the doctrine of Eco-translatology. In other words, translation is an activity in which the translator chooses to adapt the translation eco-environment and to transplant the text from the source-text ecology to the target-text ecology.

The research on the ecosystem of translation itself refers, certainly, to fundamental problems of translation theory, such as the nature of translation, the process of translation, and the value of translation.[12] Specifically, research on the ecosystem of translation itself refers not only to philosophical thinking patterns, including those of philosophy, esthetics, logic, and the science of thought, but also to lingual symbols, including those of semantics, grammar, pragmatics, stylistics, and rhetoric, and to social and cultural aspects, including sociology, anthropology, social customs, religion, natural language, and artificial language.

In the ecosystem paradigm, the translator community, with the translator as its representative, realizes and generalizes translation affairs, translation skills, and translation opinions in a reasonable way; abstracts and crystallizes a series of translation knowledge, such as the conceptions, general principles, methods, criteria, and criticism of translation; thinks scientifically about the logic and manner of translation activity from the past (i.e., the translation history and the history of translation studies) to the present and from the present to the future; logically analyzes the laws and aims of the contradictory movements between the translation subject and object; and systematically researches and realizes the cultivation and promotion of the necessary skills and qualities of translation. The translation ecosystem can, to the maximum degree possible, include independent translation thoughts as well as the peculiarity of the thinking patterns of translation studies, which makes the translator's recessive thought exterior and the exterior translation theory interior. The translation ecosystem demonstrates to society the existence of the science of translation, which supports the developing framework of the independent knowledge system, bearing the significance of theoretical promotion and practical direction in the existence and development of the disciplinary system of translation studies.

Research on the ecosystem of translation itself features theoretical, explosive, and argumentative dimensions. It is embodied not only in colleges and scientific research organs but also in related academic researchers producing studies for various academic symposiums and academic journals. Its vitality exists in the exploration of the structures and laws of translation activity and translation phenomena, together with reasonable explanations, the constant discovery of new phenomena, and the continuous development of new ideas. Thus, the variety of the rapidly increasing academic schools with different styles is the very mark of the ecosystem, which is full of vitality and animation.

[12]The related content has been mentioned by various scholars domestically and abroad (see Holmes, 1972; Bassnett, 1980; Tan, 1988; Liu, 1999, p. 18; Xu, 2003a, 2003b, pp. 21–23; Yang, 2007, p. 4; etc.).

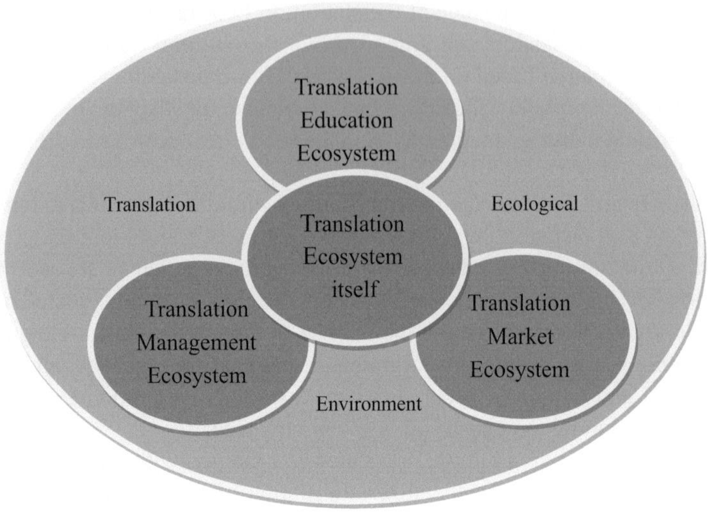

Fig. 4.1 Illustration of the overall translation ecosystem

The ecosystem of translation itself, as the complex body of the thinking of translation science and rational intelligence, carries the responsibility, conscience, and duty for the cross-cultural communication of the "translator community," which serves as the director and supporter of cross-cultural development in various societies. Henceforth, compared with the other ecosystems mentioned above, the ecosystem of translation itself proves to be the most active and most complicated ecosystem, which, for the most part, embodies the wisdom and value of translation activity and translation studies as a discipline.[13]

4.3.5 Overall Translational Environments

Regardless of whether a system is the translation management ecosystem or the translation market ecosystem, or even the translation education ecosystem or the ecosystem of translation itself, apart from their interrelationship and interaction and the relation of one translation ecosystem to another, it will be associated with and influenced by, directly or indirectly, actively or inactively, to a greater or smaller degree, the outside environments related to translation activities, such as the natural-economic environment, the lingual-cultural environment, and the social-political

[13]The components of the ecosystem of translation itself and the interrelationships of the components, together with the related studies, can be seen in one of my theses titled *Eco-Translatology, the Cross-Disciplinary Integrity of Translation Studies*, published in *Shanghai Journal of Translators,* vol. II of 2009, pp. 1–7.

environment.[14,15] Because of the vast coverage of the outside translation ecological environment and its many variants, it can usually be described with such words as "unity," "integrity," "the general collection of all," and "the world," which makes it unnecessary to accurately classify and distinctly define it in specific translation operation. Therefore, the outside translation ecological environment, as a micro-framework of translation studies, is regarded as an entire supporting system in which various single sub-ecosystems of translation can be included and intermingled (see Fig. 4.1).

The translation ecological environment is one of the main objects of Eco-translatological research.[16] It can be viewed on the basis of the above statement that the sub-ecosystems, which to some extent have an organic unity independent of and identical to each other, constitute a complete translation ecosystem. A harmonious and active translation ecosystem will, in its kernel structure, embody rationality, legality, fairness, and suitability. Its rationality refers to its conformity with the general laws and practices of translation; its legality refers to its conformity with the related laws or regulations of the translation business in the local setting; its fairness refers to its conformity with the hobbies and professional capacities of the translators; its suitability refers to its conformity with the related intentions of the commissioner, supporter, and other profit-related figures of the translation version. The chart of the overall translation ecosystem can be illustrated as follows:

The above "4+1" illustration of the translational ecosystem (i.e., the four sub-ecosystems plus the general supporting ecosystem) shows that the ecosystem of translation itself is placed at the center as the core of the maintenance and development of the general translational ecosystem; that the translation education ecosystem, which is around the core, serves as the basis for the translational ecosystem to be maintained and developed; that the translation market ecosystem is the platform for the translational ecosystem to be maintained and developed; and that the

[14] For instance, the translation ecological environment can be defined as "the general collection of all the outside conditions to influence the existence and development of translation subject. The translation subject here bears the general meaning, referring to all the human beings taking part in translating activity, including the original author, the translator, the reader, the sponsor, supporter publisher, dealer and editor of the translation work, etc. Whereas the outside environment includes natural-economic environment, the lingual-cultural environment, and the social-political environment, etc., which are related to the translating activities" (Fang, 2011).

[15] For instance, it refers to the micro-environmental elements; supporting environmental elements; and elements of the author, the translator and the reader. The elements can be further divided. That is, the micro-environmental elements include the political environment, the cultural environment, the economic environment, and the natural environment; the supporting environmental elements include translation resources, professional environments, translation studies, translation skills, and the elements of certain rights; the author elements include the author's intention, the author's language, and textual elements; the translator's elements include the translator's intention, motivation, feelings, professional capacity, lingual capacity, and creative capacity; the reader's elements include the reader's intention, language, and feelings (Qian, 2011).

[16] Regarding the connotation and interrelationships between translational ecology and the translational ecological environment, references to Sects. 3.4.1 and 5.4.3; no further comments will be given here.

translation management ecosystem guarantees the maintenance and development of the translational ecosystem. In addition to their inter-associative and interactive functions as the translation of one to another, each unique ecosystem is typically supported by the outside ecosystem. The sub-ecosystems mentioned above are maintained in a relationship of coordination and balance.[17] If the layers of the translational nature and the cultural value are analyzed, apart from the outside supporting translational ecosystem, other systems can be listed in order, that is, the translation management ecosystem—translation market ecosystem—translation education ecosystem—and ecosystem of translation itself.

Along with the continuous development of Eco-translatology, importance is attached to the translation management ecosystem, the translation education ecosystem will develop more rapidly, and the ecosystem of translation itself will be more stable. In addition, the ecosystem of translation itself, the translation education ecosystem, the translation market ecosystem, and the translation management ecosystem will be researched independently, not only to improve each ecosystem but also to trigger and influence specific translation processes and translation activities as parts of the translational ecosystem; to unite with each outside translational ecosystem as its support; and, finally, to promote the sustainable development of Translatology as a whole.

4.3.6 Other Studies on Translational Ecosystems

At present, the translational ecosystem is divided into the aspects of composition, structure, function, and framework. We believe that the translational ecosystem can also be divided into other aspects, for example, the exterior system and interior system; the microsystem and macrosystem; the major environment and minor environment; or even social, economic, and political factors. For instance, from the viewpoint of a systematic network based on a biosphere, some scholars have analyzed the interrelationships between various ecological environments and translation and also between the functioning principles and laws of various elements of the ecological environment of translation. In accordance with the layer-analytic method, the translation ecosystem can be roughly divided into the micro-level translation ecosystem and the macro-level translation ecosystem, and it can be further divided into the individual translation ecosystem, group translation ecosystem, translation ecosystem, etc. (Xu, 2009a, 2009b, p. 5). Another scholar holds that the translational ecosystem is composed of five dimensions, namely, the micro-level environmental elements (including the political environment, social culture, economic environment, technological environment, and natural environment), the supporting environmental elements (including the translation resources, professional environment, translation studies, and power elements), the author elements (including the author's doctrines, the author's language, and the textual elements), the translator's

[17] Further discussion can be found in Sect. 4.5.

elements (including the doctrine of value; the translation motives; and the translator's feelings, professional capacity, lingual capacity, and creative capacity), and the reader's elements (including the reader's doctrine, the reader's language, and the reader's feelings) (Qian, 2011). The translational ecosystem can also be divided into three layers: the internal translational system, the social translational system, and the natural translational system. The internal translational system contains the system of the translator and the system of translation. The former consists of cultural ideology, political standpoint, mind-set, and esthetic outlook, and the latter consists of subsystems such as the languages, genres, and textual structures of the original text and the translated text. The social translational system contains nonsubstance concept systems, such as the language, politics, economy, ethics, culture, and poetics in the translator's social environment, and the human-relationship ecosystem, which consists of the translator, the supporter, the publisher, the commenter, and the reader. The reader's natural system contains the minor natural system, including the climate conditions, geographical location, species of animals and plants and their distribution, in which the translator lives, and the major natural system consists of the cosmic elements surrounding the translator (Bian, 2011). Although various scholars have proposed different classifications and descriptions of the translational ecosystem, we can see that they have something in common; that is, different studies of ecological translation have provided various observational perspectives and multidimensional considerations for research on the translational ecosystem in the realm of Eco-translatology.

4.4 Cross-Disciplinary Integration of the Ecosystem of Translation Itself

This section deals mainly with the discussion of the cross-disciplinary integration of the ecosystem of translation itself, as the ecosystem of translation itself is the core of the translational ecosystem as well as the intensified embodiment of the thinking pattern of Translatology and the rational wisdom, and as the translational ecosystem will ultimately lead to ontological studies of translation. According to the Sequence Chain, Eco-translatology requires cross-disciplinary integration, which, in terms of the cross-disciplinary integration of the ecosystem of translation itself, can serve to some extent as a significant starting point and expansion point.

4.4.1 Interdisciplinary Interplay and Cross-Disciplinary Studies

With the Sequence Chain as a clue, related studies will be conducted correspondingly, and by utilizing distinct studies and interrelated coordination, it is possible to generalize and integrate the ecosystem of translation itself from the perspectives of related disciplines.

On the one hand, as we all know, ecology is the science of the interrelationship between living creatures and their environment. Therefore, translation studies starting from the perspective of ecology will include research on the association, similarity, and coconstitution of the translational ecosystem and natural ecosystem; the interrelationships between the translator and translational ecosystem; the interrelationships between the author/reader/financial supporter/publisher/commenter and the translational ecosystem; the internal structures of the translational ecosystem; the interrelationships between the nature of translation, the translation process, the translation principles, the translation methods, and the translation criteria; the status and functions of translation in the natural ecosystem; the interrelationships between translation and other disciplines; and related studies on the translational ecosystem from the perspective of ecology.

Ecology is also a science based on holism (Krohne, 2001, p. 11), with a research method emphasizing the integrity of interactivity.[18] Thus, the comprehensive proof and analytic case proof will be combined in research on Eco-translatology. That is, detailed studies on certain topics will be coordinated and classified, and the relevance and interaction of fundamental research will be emphasized. In the studies of various specific topics mentioned above, attention should be paid to the theories and achievements of various disciplines such as Translatology, linguistics, semiotics, cognitive science, sociology, cultural anthropology, cross-cultural communication, and ecology. By means of the comprehensive analysis and illustration, the conclusion can be scientifically and persuasively generalized.

On the basis of the features of commonness, interaction and progressiveness in the Sequence Chain, the general direction of Eco-translatology in researching the ecosystem of translation itself is that in the framework and perceptive of ecology, the research and descriptions have been conducted from the perspectives of language, culture, and human communication, which are closely related to translation activities. Detailed cross-disciplinary research on the ecosystem of translation itself includes the following:

Research from the Perspective of Applied Linguistics In the ecosystem of translation itself, and "in view of translation as a lingual transfer" (according to "the Sequence Chain"), research from the perspective of applied linguistics should include ecology-oriented functional/cognitive/pragmatic analytic studies of the text (source text and target text), studies on relationships between the languages and the translational ecology, studies on the translation of the ecological vocabulary, and studies from the perspective of linguistics that are related to the translational ecosystem.

[18] One scholar points out that the core of Confucian humanism lies in the emphasis on holism, with a perspective of holistic vision (Tu, 2001, pp. 247–248). The author also translates"天人合一", which is the cream of Chinese culture, into "the unity of Man and Nature," which illustrates the emphasis upon "ecological unity" and "natural unity." This, however, is the core viewpoint of ecological holism, which is the mainstream of the philosophy at present.

Research from the Perspective of Cultural Studies In the ecosystem of translation itself, "in view of translation as a lingual transfer while the language is a part of culture" (according to "the Sequence Chain"), research from the perspective of cultural studies should include ecology-oriented studies of cross-cultural differences, conflicts, combination and constraint; cultural contextual studies of the translational ecosystem; studies on cultural diversity and the translational eco-environment; and studies from the perspective of cultural science that are related to the translational ecosystem.

Research from the Perspective of Anthropology In the ecosystem of translation itself, "in view of translation as a lingual transfer while the language is a part of culture, which is the collection of human activities, while human beings are a part of nature" (according to "the Sequence Chain"), research from the perspective of anthropology should include evolutionary studies on translation and human cognition, studies on human memory and translation (particularly oral interpretation), studies on the translator's requirements/feelings/desires/abilities, studies on the translator's living conditions/the development of the translator's capacities, studies on translation and human communication, studies on the mission of translation and human civilization, studies on translation and globalization, and studies from the perspective of anthropology that are related to the translational ecosystem.

Research from the Perspective of Ecology In the ecosystem of translation itself, "in view of translation as a lingual transfer while the language is a part of culture, which is the collection of human activities, while human beings are a part of nature [the natural ecology]" (according to "the Sequence Chain" and based on the fact that ecology is the science of the interrelationship between living creatures and their environment), research from the perspective of ecology should include studies on the interrelationships between the translation and the translational eco-environment; studies on the interrelationships between the author/reader/financial supporter/publisher/commentator and the translation eco-environment; studies on the internal structures of the translational ecosystem and their interrelationships; studies on the interrelationships between the nature of translation, the translation process, translation principles, translation methods, and translation criteria; studies on the status and functions of translation in the natural ecosystem; studies on the interrelationships between translation and other disciplines; and related studies on the translational ecosystem from the perspective of ecology.

Research from the Perspective of Translatology In the ecosystem of translation itself, "in view of translation as a lingual transfer while the language is a part of culture, which is the collection of human activities, while human beings are a part of nature [the natural ecology]", we can shift our research perspective from the natural ecosystem back to translation (according to the inverted "Sequence Chain"); thus, research from the perspective of Translatology should include integral studies of the translational ecosystem; coordinating studies between translation, language, culture, human beings, and nature; best-effect studies of the translational ecosystem;

comparative studies of the ecological dimension of translation; translation studies from the perspectives of other disciplinary dimensions; and related studies from the perspective of Translatology that are related to the translational ecosystem.

As studies from the perspectives of anthropology, ecology, and Translatology have been comparatively belittled or ignored in previous translation studies, studies from such perspectives are the key research realms of Eco-translatology.

Obviously, with the deepened research with the Sequence Chain as the clue, with the disciplinary perspective as the emphasis, and with the comprehensive integrity of studies as the goal, the contents of Eco-translatology and the significance of the research can be illustrated distinctly. The basic ideas of Translation as Adaptation and Selection, therefore, can be further illustrated in that what the translator adapts is the translational eco-environment; that the author, translator, reader, and other elements of the translational eco-environment, too, will need "harmony"; and that the translated version should be close to the original text. To accomplish all these aims, the translator is required to strive more strongly for a better process of translation, which inarguably should be defined as the translational ecosystem with "integral harmony."

4.4.2 Synergic Fusion of Multidimensional Studies

Ecology is an "ontological discipline," which is of contending and leading significance toward other disciplines; the orientation of ecological research is that of a comprehensive discipline, which makes it possible for Eco-translatology to integrate the translation-related disciplines comprehensively in a cross-disciplinary manner. In other words, the ecological perspective can be relied upon as the basis of integral comprehensive translation studies or as the starting point and ending point with a turn towards improved significance.

Based on the directive features of commonness, inter-reaction, and graduality in the Sequence Chain (Hu, 2003, p. 298), from the ruling perception of ecology, Eco-translatology is open to research and descriptions from the perspectives of language, culture, and human communication, which are closely related to translation activities, and will ultimately revert to ontological research from the perspective of Translatology.

Specific translation studies from the perspectives mentioned above include research from the perspectives of linguistics, cultural studies, anthropology, ecology, and Translatology. The cross-disciplinary studies are relevant and interactive in that research from the perspective of linguistics is oriented towards studies on the interior text (i.e., with the emphasis on the version); that research from the perspective of cultural studies is oriented towards studies on the exterior text (i.e., with the emphasis on the translation context); that research from the perspective of anthropology is oriented towards studies on the translator community (i.e., with the emphasis on the translator); and that research from the perspective of ecology, as the academic perspective with ultimate significance for the existence of human beings, is oriented towards domination and coordination in translation studies. Finally, research will, in turn, revert to the absolute integrity from the perspective of Translatology.

The main structure of the ecosystem of translation itself and the cross-disciplinary fundamental research mentioned above can be illustrated as follows:

In the above illustration, research from the perspective of Translatology is at the center, with the linguistical, cultural, and anthropological viewpoints around it, along with their relevance and interrelationships. Research from the perspectives of the various disciplines can ultimately be intermingled within the scope of ecology, through which a complete comparative ecosystem of translation itself can be constructed, and, on that basis, cross-disciplinary integral research can be conducted.

Additionally, the comprehensive integrity[19] of the ecosystem of translation itself from the perspective of ecology is the integrity of a multidimensional unity. It embodies not only the integrity of thinking modes (i.e., the integrity of reason and perception and of abstractness and concreteness) but also the collection of research perspectives (i.e., the traverse perspective and reverse perspective, the prospective perspective and retrospective perspective) and the convergence of cross-disciplinary research (i.e., natural sciences and social sciences, science and art). Meanwhile, the comprehensive integrity from the perspective of the ecosystem of translation itself also conforms to the ecological notion and the eco-esthetic principles of "multiple unities" and "unity of diversity." Eco-translatology itself is "a kind of worldview and methodology." Therefore, the cross-disciplinary integrity of research on the ecosystem of translation itself is also methodologically significant in terms of Holism.

4.4.3 *Other Studies on the Ecosystem of Translation Itself*

In this subsection, other sub-ecosystems of translation, including the ecosystem of translation itself, together with the related studies, will be illustrated as follows.

1. Studies on the ecosystem of translation and other sub-ecosystems

[19]According to the definition of "integrity" in *Modern Chinese Dictionary* (the fifth edition) (the Commercial Press, 2005, p. 1737) and *Modern Chinese Dictionary of New characters and Phrases* (edited by Lin Zhiwei and He Aiying, Commercial Press International Co. Ltd., 2005, p. 837), "integrity" means "reorganization through reform and coordination." The purpose of integrity is to reorganize different things by means of readjustment so that it, as a unity, can mutually improve functions and effects. Meanwhile, the "settlement of integrity can be conducted not by the delivery of new information, but by the management of something that is already known" (a saying by Wittgenstein, quoted from Liu, 2005, p. xxxii).

It is held in Eco-translatology that the ecosystem of translation itself is centered on the ontological studies of Translatology, supported by related studies from the perspectives of linguistics, cultural studies, communication science, and anthropology; interacting with and interdependent on studies from the perspectives of related disciplines; and, finally, intermingled with the transdisciplinary unity of the eco-environment of translation.

In the ecosystem of translation itself, studies from the perspective of each discipline consist of the following sub-ecosystems, illustrated as follows:

Ecosystem of translation itself—
- —Ecosystem from the perspective of Translatology
- —Ecosystem from the perspective of linguistics
- —Ecosystem from the perspective of cultural studies
- —Ecosystem from the perspective of anthropology
- —....

Because the ecosystem of translation itself is at the center of the ecosystem of translation and is the centralized embodiment of translation studies thought and rational wisdom, and as there are many important interdisciplinary issues in the system, we have discussed such issues above.

2. Studies on the translation education system and its sub-ecosystems

It is held in Eco-translatology that the translation education system is a system of multiple ecological environments with the functions of controlling and adjusting the emergence, existence, and development of translation education.

The subsystems of the translation education system can be illustrated as follows:

Translation education ecosystem—
- —International translation education ecosystem
- —National translation education ecosystem
- —Regional translation education ecosystem
- —College translation education ecosystem
- —...

The translation education ecosystem is important because it serves as the basis of the development and maintenance of the entire ecosystem of translation. The ecosystem of translation itself serves as an important source of manpower and intelligence.

3. Studies on the translation market ecosystem and its sub-ecosystems

It is held in Eco-translatology that the translation market ecosystem, which is an important part of the translation industry, is related to services for translation products. In other words, it directly provides services for cross-cultural communication and exchange between different languages. The translation market ecosystem is an important platform for maintaining and developing the entire translational ecosystem. The subsystems of the translation market system can be illustrated as follows:

Translation market ecosystem—
- —International translation market ecosystem
- —National translation market ecosystem
- —Regional translation market ecosystem
- —College translation market ecosystem
- —…

Studies on the translation market ecosystem are somewhat weak in the previous translation literature. However, it is of immediate, foregrounded significance in relation to the cultural existence of translation and the development of translation studies as a discipline. Thus, the development of and research on this ecosystem should be increasing concerns of various scholars in the circle of translation studies, particularly those in the theoretical fields.

4. Studies on the translation management ecosystem and its sub-ecosystems

It is held in Eco-translatology that the translation management ecosystem, in accordance with its name, refers to the micromanagement system for the management, coordination, maintenance, and evaluation of Eco-translation. The subsystems of the translation management system can be illustrated as follows:

Translation management ecosystem—
- —International translation management ecosystem
- —National translation management ecosystem
- —Regional translation management ecosystem
- —Municipal translation management ecosystem
- —…

The translation management ecosystem is important in the maintenance and development of the translational ecosystem as a whole. However, like the translation market ecosystem, the translation management ecosystem has been little studied in previous research. Thus, it should be an increasing concern of scholars in the circle of translation studies and the theoretical field.

The goal of the construction of the translation ecosystem lies in the promotion of a sustained translation ecosystem with a sound environment, economic management, coordinated disposition, and systematic harmony.

Translation, as a complicated intelligent activity, possesses a "complex system" (Docherty, 1993, p. 30).

As mentioned above, because of the multidimensional and multilayer internal nature of the ecosystem and in accordance with the ranking relationship, the general system consists of subsystems, and the latter consist of sub-subsystems. Theoretically speaking, therefore, the translational ecosystem that comes into shape is infinitely divisible vertically and is inter-environmentally horizontal. In other words, in the subsystems of the translational ecosystem mentioned above, the translation ecosystem itself, translation education ecosystem, translation market ecosystem, and translation management ecosystem will again be embedded in their respective sub-subsystems. For example, the research system (Fig. 4.2) from the perspective of Translatology, which serves as the center of the translation ecosystem itself, can be illustrated as follows.

- Ecosystem of translation itself—
  - —Ecosystem from perspective of Translatology
    - —Translation process research
      - —Reader adaptation
      - —Reader selection
      - —...
    - —Translation principle research
    - —Translation method research
    - —Translation criteria research
    - —...
  - —Ecosystem from perspective of linguistics
  - —Ecosystem from perspective of cultural studies
  - —Ecosystem from perspective of anthropology
  - —...

If the longitudinal subsystems in the translation ecosystem itself are ordered from top to bottom or from large to small (which is confined to the section from the

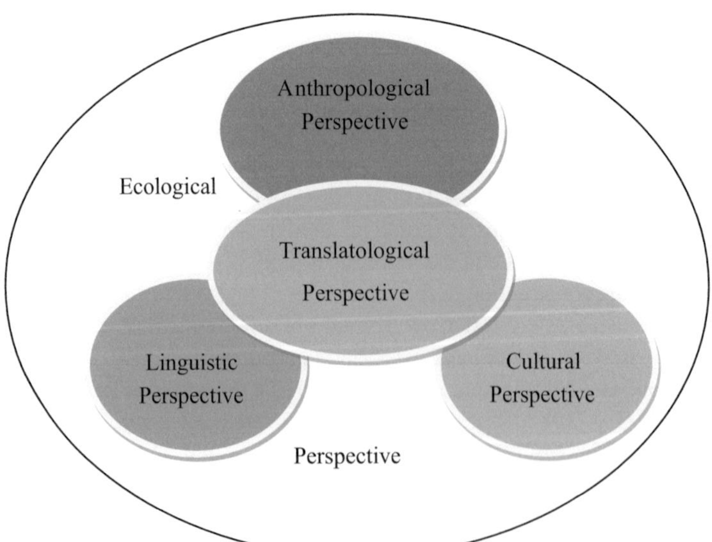

Fig. 4.2 Illustration of the structure of the ecosystem of translation itself

perspective of Translatology and the subsystem of the translation process), the structural layers can clearly be seen as follows (Fig. 4.3):

If the different-leveled systems of the entire translational ecosystem are considered as a whole, a complicated system comes into shape, consisting of the different systems and their layer structures, which is illustrated as follows (Fig. 4.4):

The above texts and figures illustrate that specific structures exist in the translational ecosystem; that they are structured in specified ranks, levels or layers; and that they are interrelated. According to such relationships, subsystems exist in the general system and sub-subsystems in each subsystem. Moreover, there are influences between and among the various levels and layers of all the systems.

Regarding translation activity, the lower the translation layer is, the closer it is to the translation operation. In other words, "there are more confinements at the bottom layers, and, in turn, there are less on the top" (Xu, 2009a, 2009b, p. 2), and vice versa.

4.5 Coordination and Balance of Translational Ecosystems

The concept of ecological balance, which was first proposed in 1935 by A. G. Stanley, a British plant ecologist, refers to a stable condition obtained by cultivating and coordinating relationships within the ecosystem among the producers, consumers, dissolving elements, and non-living-creature environments. It includes stability not only in structure and function but also in the input and output of energies.

The translational ecosystem is also a dynamic balance system. In the disciplinary system of translation, as in the natural ecosystem, various elements, and especially those with environmental aspects, will possess a constant flow and circulation of knowledge and information. The disciplinary system of Translatology is continuously developing and evolving in that translation workers and researchers have increased in number, translation education and its industrial links have become

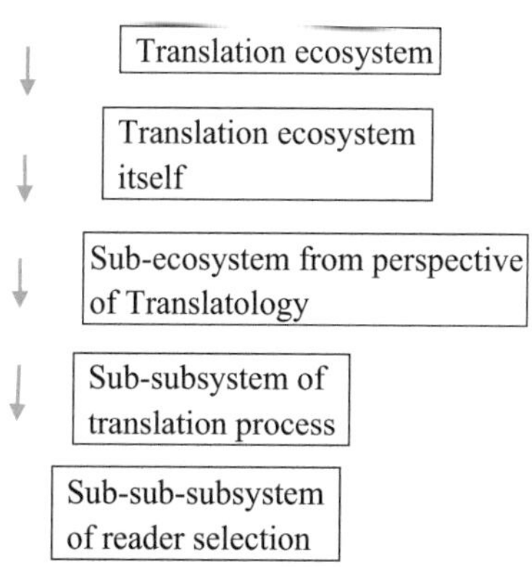

Fig. 4.3 Layer structure (partial) of the translation ecosystem itself

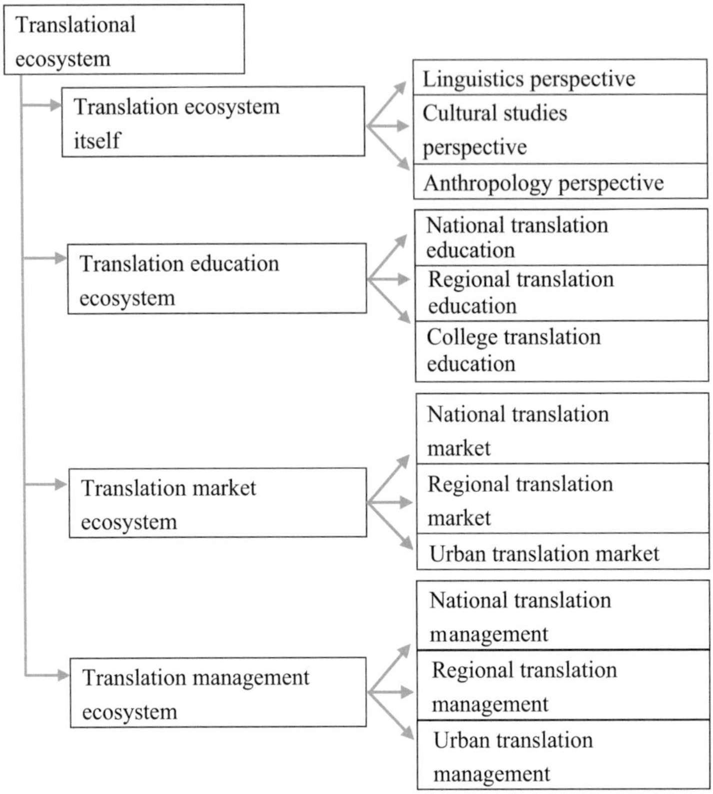

Fig. 4.4 The translation subsystems and their layer structures in the translational ecosystem

increasingly complex, and the types of translator community have also changed. Thus, the translational ecosystem is also in dynamic, constant change.

The coordination and balance of the translational ecosystem go between different ecosystems and different stages of translation activities and embody different aspects of the environmental elements of ecological translation.

4.5.1 Coordination and Balance of Ecosystems from Different Dimensions

At the microlevel, to maintain the lingual and textual eco-balance, the translator should strive for the maintenance of balance in meanings between the words and sentences of the source and the target languages, for the conveyance of the same spirit and sense in the source and target languages, and for the maintenance of the practical value, esthetic value, and style of the source and target languages.

At the mesolevel, the overemphasis upon the author and the text in traditional studies led to the *invisibility of the translator* and *the original-centeredness*; however, this approach was deconstructed and overthrown in the circle of translation studies in the 1970s because of its failure to maintain the balance of the translational ecosystem. Similarly, the theory of the *text system* and the *translational system itself* in the mode of philology has also been gradually dismissed and replaced because of its ignorance of the relation of the social system and natural system to the translational environment and its failure to meet the requirement of translation studies and social development, which has verified the basic principles of "the survival of the fittest" and "the dismissal of the weak and the prosperity of the strong."

Balance, coordination, and harmony also exist in the subjects of translation activities, such as the translator and the author, the translator and the translation initiator, the translator and the editor, the publisher and the reader, the publisher and the translation examiner, and the publisher and the bookseller as well as the version copyright holder. Balance and harmony also exist among the various elements of the translational ecosystem mentioned above and the various factors of the translational environment.

4.5.2 Coordination and Balance of Individual Ecologies and Population Ecologies

The maintenance of the coordination and harmony of the translational ecosystem includes the coordination and harmony of individual ecologies of translation and population ecologies of translation. As far as the translation individual is concerned, there exist differences in translators' thinking modes, educational backgrounds, hobbies and interests, translation ideology, esthetic criteria, translation experience, etc., and variations in translational text types, reader requirements, cultural acceptance, current conveyance channels, regulatory environments, etc. The differences and variations in both the subjective and objective environments and the interior and exterior environments will give rise to different adaptations and selections of the translators, who must adjust themselves dynamically to adapt to the entire translational ecological environment. The dynamic balance can be obtained by means of dynamic coordination, that is, the coordination and balance of individual ecologies.

The coordination and balance of the structure of the translator community should first be considered, such as the coordination and balance between written translators and interpreters and between literary translators and nonliterary translators. For a long period in China, importance has been attached to the education of literary translation talent, to commentaries on literary versions, and to research on literary translation, while education, encouragement, and research in nonliterary translation have been seriously ignored, leading to an insufficient supply of nonliterary translation talent; this issue, too, should be focused upon and adjusted. Moreover, we should maintain the adaptive interrelationship between the translator community and the eco-environment. For instance, in economically developed areas in China, such as Beijing, Shanghai, and Guangzhou, where the translational eco-environment is favorable, there is a large amount of translator community, while in West China, where the economy, education, and culture are relatively backward and the translational eco-environment has a correspondingly low capacity, there is a comparatively small amount of translator community. This balance is maintained spontaneously by the translator community. However, we can change the balance properly by exerting the subjective dynamic function to scientifically, reasonably, and moderately break the former balance and to cultivate and maintain a new balance that can meet the requirements for the development of China. In other words, we should try hard to cultivate translation personnel, to expand the translator community, and to strengthen the translation industry to stimulate the development of the local economy, culture, and education. Of course, the alteration of the amount of the translator community should be coordinated and should conform to the ecological rationale. Otherwise, it will cause an imbalance in the translational ecology and will even affect social development.

4.5.3 Coordination and Balance of the Translational Ecosystem and Other Social Systems

The translational eco-environment, as a contradicting unity, needs to maintain a balance because it will lose itself if it becomes unbalanced. For instance, in the early stage of the People's Republic of China, almost all national policies, including lingual policies and translation policies, were based on those of the USSR. In the next 10 years of the so-called Great Cultural Revolution, everything related to the so-called feudalism, capitalism, and revisionism was utterly forbidden. In the general environment of the national enclosure, the translational environment became seriously poor. Consequently, China, separated from the outside world for 30 years, was out of sync with the progressive world.

During the so-called Great Cultural Revolution, which occurred in 1966, some ridiculous actions occurred based on the political policies of "class struggle as the principle" and "politics as the top priority." The intent to translate or introduce foreign literature was considered reactionary, which struck a fatal blow to the translation circle in China. Many distinguished elderly translators were persecuted, and a large number of intellectuals were labeled "foreign culture admirer" or "reactionary authority" and forced to accept "ideological reform" and live in a "cowshed," which made translation activity impossible. For instance, the translation of Shakespeare's sonnets by Tu An was forced to stop. Some translation scriptures were confiscated and destroyed. The translated versions of *Shakespearean Sonnets* and *Faust*, together with Montaigne's *Essays,* which were translated by Liang Zongdai, were burned to ashes. The translation organizations, translator associations, and related organs were disbanded by force, and translation journals such as *Translation*

Communication and *Versions* were banned. The termination of the translation practice led to the stagnation of translation studies. Such miserable phenomena could be seen in works such as *Comments on Literary Translation by the Contemporary Scholars* by Wang Shoulan and *Historical Studies on Translation Theory in China* by Chen Fukang. More than 180 articles on translation, written from third century BC to 1982, with 700,000 characters in total, are collected in *Collected Comments on Translation,* in which, however, there is no article written during the so-called Great Cultural Revolution (Biao & Xiuming, 2000).

The translation talent pool is far behind the amount needed in China. Moreover, notably, there is a great need for Chinese-foreign language top translation talent This is a kind of severe ecological imbalance. In 2008, among the copyrighted books introduced into China, approximately 12,000 needed to be translated into Chinese. Suppose 1.5 translators were needed to translate one book, then 18,000 translators would be needed for those foreign language-Chinese translations. Let us consider the case of Chinese-foreign language translation. More than 300 books were listed in promotion or investment programs; together with the foreign-language books published by the organs attached to the China International Publishing Group and by various organizations in various regions, the number amounted to at least 3000, and more than 5000 translators were needed to translate them. Thus, without even considering interpreters, 20,000 or 30,000 translators were needed. Currently, there are no professional translators in China, and most translators are part-time. How can we achieve the amount of translators needed for the translation work? (Li, 2010).

In the same period, before 2005, there was an obvious structural imbalance in translation education in China in that translation education could be found only in the postgraduate stage. To coordinate and balance the translational ecological environment, translation-oriented colleges and schools began to be built on a large scale throughout China so that the translation talent could be cultivated to meet the translation demand. In 2006, the Ministry of Education granted that the translation major should be established in colleges or universities, and courses of translation can now be offered to undergraduate students at 15 universities. Currently, a teaching system for the translation major has been established at the undergraduate, postgraduate, and doctoral levels throughout China. By the end of September 2010, the translation major for undergraduates had been established in 31 universities in China, and the MTI degree program had been granted to 158 universities. These steps have made the translation major the most popular major throughout China (Huang, 2010). The orderly structure of translation education, from English education in primary schools and high schools to the undergraduate translation major in universities, master's or doctoral degree programs in translation for postgraduates, and postdoctoral studies in translation, reflects the different education layers and the education process, from simplicity to complexity and from the junior to the senior stage. However, attention must be paid to scientific and reasonable regulation to avoid a future ecological imbalance after the translation talent pool has expanded to the point that it surpasses the requirements of the translation market.

At present, China has seen one of the best periods of the translational ecological environment in that the economic environment has expanded, the social and political environment is stable, multiple elements of the lingual and cultural environment are harmonious, and the translational ecosystem is prosperous with professional divisions and with the production of many excellent translation works. However, another aspect is not yet being coordinated, balanced, or harmonized. From the perspective of Eco-translatology, the translational eco-environment often witnesses various inappropriate actions, such as seeking fame or profit, translating roughly, plagiarizing others' versions, and falsifying a translation. "When we look at the present translation realm, especially the literary translation realm, we may find various phenomena, such as plagiarizing others' versions in re-translation of classical works, translating some best-sellers in a rush, and translating in a rough way, etc." (Xu, 2003a, 2003b, p. 12). In the research on translation theory, "there is not a great change of fashion in which the Chinese scholars follow foreign scholars in studying norms; the interdisciplinary research is not deepened yet" (Yang & Wang, 2008, p. IV). "Nowadays, some theses are published for the sake of profit and interest. Influenced by the economic striving and the material temptation, academic studies have been distorted to a great degree" (Luo, 2009, p. 10). The research on translation textbooks is also out of balance. In the previous 30 years, for instance, research on translation textbooks (2 academic works and 30 theses published in core journals) has not been commensurate with the prosperity of translation textbook publication (851 translation textbooks in total), and research on translation textbooks is not in proportion to translation studies on the whole (8390 theses published in core journals, with only 3.5% of them research on translation textbooks) (Xu & Mu, 2009, pp. 77–87). To maintain the balance of the translational ecosystem, the proportion of translation studies must be adjusted to strengthen interdisciplinary integrity in translation textbook research. In this situation, the balance and coordination of translational ecology are of the utmost importance.

4.5.4 Coordination and Balance of Ecosystems Along the "Sequence Chain"

From the "Sequence Chain" of "translation ↔ language ↔ culture ↔ human beings/society ↔ nature" (Hu, 2003, pp. 298–299; Hu, 2004, pp. 60–61), we can see not only the interrelationship between translation activity and ecological nature but also the basic features of the co-movement and complementarity of the natural ecosystem with that of human society (see Sect. 2.3.1).

The coordination and balance of the various ecological elements mentioned above are the coordination and balance of various ecosystems related to "the Sequence Chain," both inside and outside. First, there are coordination and balance inside the translational ecosystem and the coordination and balance of the translational ecosystem with the outside lingual ecosystem and other ecosystems. Second, there are coordination and balance within the lingual ecosystem and the coordination and balance of the lingual ecosystem with the outside cultural ecosystem and translational ecosystem. Third, there are coordination and balance within the cultural ecosystem and the coordination and balance of the cultural ecosystem with the outside social ecosystem and translational ecosystem. Fourth, there are coordination and balance within the social ecosystem and the coordination and balance of the social ecosystem with the outside natural ecosystem and translational ecosystem. Fifth, there are coordination and balance within the natural ecosystem and the coordination and balance of the natural ecosystem with the translational ecosystem and other ecosystems. The coordination and balance within the related ecosystems constantly circulate and expand, oriented towards the best conditions of coordination and balance.

More coordination and balance, both in content and in scope, are required in the above ecosystems related to "the Sequence Chain." One of the core notions of Eco-translatology, i.e., translation as an ecological balance, undoubtedly covers the content of and research on the above ecosystems (see Sect. 5.3.3). That is why "Translation as Ecological Balance" is not only a core notion of Eco-translatology but also an essential principle of Eco-translation ethics (see Sect. 3.7.1).

4.6 Conclusion

From the perspective of the microecological dimension, the entire translational ecosystem of Eco-translatology has been constructed, and various translation-related ecosystems, such as the translation management ecosystem, translation market ecosystem, translation education ecosystem, and translation ecosystem itself, as well as the structure and function on which the outside translational eco-environment relies, have been discussed. In addition, the multiple-setting translational ecosystem has been demonstrated and analyzed in two directions, both vertically and horizontally. Furthermore, in the generalized perspective of the microecological notion of Eco-translatology, we have applied interdisciplinary comprehensive perception and integration to the related research on the translational ecosystem itself, which is the core of the translational ecosystem. Finally, we have discussed how to coordinate and maintain the balance of the translational ecosystem in multiple layers and dimensions.

Some primary elements of the existing systematic framework of Translatology have been embodied to a certain degree in the entire micro-framework of Eco-translatology. For instance, certain ecosystems in Eco-translatology, such as the translation management ecosystem,[20] translation market ecosystem,[21] and translation education ecosystem,[22] can be regarded as "exterior research," a "general system," or

[20] It includes the translation industry, translation programs, translatological conferences, translation evaluations, and translation politics.

[21] It includes the corpus, machine translation, multimedia translations, and internet translations.

[22] It includes translation education, translation training, translation criticism, and translation tests.

an "institutional system" in the former translatological framework, whereas the translation ecosystem itself[23] is an "internal system," "special system," or "normal system."

It should be mentioned that the traditional translation (theoretic) studies have rarely coped with the realms that are indispensably related to translation studies, such as "translation management," the "translation market," and the "translation industry"; therefore, some scholars think that such realms related to translation studies are not relevant to research on translation theory, or no importance has been attached to them. (The reason is simple: If translation studies fail to address the translation business engaged in by many translators, it is reasonable not to blame them for their indifference to translation theory!) This situation has been somewhat improved in the realm of Eco-translatological research, as, in the micro-translatological framework of Eco-translatology, the translation management eco-system and translation market ecosystem, together with their "outskirts," or the public translational environment ecosystem, have been regarded as indispensable, organic components of the entire translational ecosystem, and relevant studies have been initiated.

Since the translational ecosystem itself serves as the core and basis of the entire translational ecosystem, which embodies the ultimate orientation of the behavioral patterns and cultural values of translation, the reader will probably be more interested in the following topics: How does the ontological translation theory, which belongs to "interior research" or "special systems" in the translatological system, apply to research on Eco-translatology? How can we realize and describe the ontological translation theory from the perspective of Eco-translatology? Moreover, what kinds of problems will be solved in the ontological translation theory of Eco-translatology? These questions will be discussed and answered in detail in the next chapter.

References

Bassnett, S. (1980). *Translation studies*. London: Routledge; Revised edition, 1991.

Bian, L. H. (2011). *Studies on eco-system in translation*. Paper presented on the second international eco-translatology symposium, 2011-11-11/13. Shanghai.

Biao, J., & Xiuming, Z. (2000). Chinese translation industry during "cultural revolution". *Journal of Tianjin Foreign Studies University*, (1).

de Waard, J., & Nida, E. A. (1986). *From one language to another: Functional equivalence in bible translating*. Thomas Nelson: Nashville.

Docherty, T. (1993). Theory and difficulty. In R. Bradford (Ed.), *The state of theory* (pp. 20–34). London: Routledge.

Duarte, J., Rosa, A., & Seruya, T. (2006). *Translation studies at the interface of disciplines*. Amsterdam: John Benjamins Publishing Company.

[23] It includes translation theory, ontology, epistemology, methodology, axiology, ethics, and comparative translation studies.

Fang, M. (2011). Translation ecology. *Shanghai Journal of Translators*, (1), 1–5.

Gentzler, E. (1993). *Contemporary translation theories*. New York: Routledge Inc.

Gorlée, L. (1994). *Semiotics and the problem of translation: With special reference to the semiotics of Charles S. Peirce*. Amsterdam: Editions Rodopi B.V.

Holmes, J. S. (1972). *The name and nature of translation studies. Unpublished manuscript*. Amsterdam: Translation Studies Section, Department of General Studies.

Hu, G. S. (2003). Translation as adaptation and selection. *Perspectives: Studies in Translatology, 11*(4), 283–291.

Hu, G. S. (2004). *Translation as adaptation and selection*. Wuhan: Hubei Education Press.

Hu, G. S. (2009). An "Ecological Turn" signified by the "Sequence Chain". In *Hu Gengshen's translation and cross-cultural communication: Integration and innovation* (pp. 26–39). Shanghai: Shanghai Foreign Language Education Press.

Huang, J. (2012). *Complex thinking*. Retrieved May 26, 2012, from http://www.huangjiwei.com/blog/?p=558

Huang, Y. Y. (2010). Adventure and perseverance--The opening speech on the first international eco-translatology symposium. *Journal of Eco-Translatology*, (1), 13–15.

Krohne, T. D. (2001). *General ecology*. Brooks: Thomson Learning.

Li, J. D. (2010). Chinese Books' "Going Global" call for excellent translations. *China Press and Publishing Media Network*, 2010-07-08.

Liu, M. (1999). *Contemporary translation theory*. Beijing: China Translation & Publishing Corporation.

Liu, M. (2005). *Translation thinking: In China and in the West*. Beijing: China Translation & Publishing Corporation.

Lu, F. (2005). Ecological turn for science and technology in 21st Century. *Science Times*.

Luo X. A review on translation theory. Luo X : Structuralism, deconstructionism, and constructivism; current translation theories. Shanghai: Shanghai Foreign Language Education Press, 2009: 1-10.

Mainzer, K. (1999). *Thinking in complexity---Material, spirit and complex dynamics of human* (Zeng Guoping, Trans.). Beijing: Central Compilation & Translation Press.

Morin, E. (2005). *Complex thoughts: Conscientious science* (Chen Yizhuang, Trans.). Beijing: Central Compilation & Translation Press.

Nouss. (2005). This quotation comes from – Duarte, J., Rosa, A., & Seruya, T. (2006). *Translation studies at the interface of disciplines*. Amsterdam: John Benjamins Publishing Company.

Peng, X. (2003). *Complex thinking and social development*. Beijing: China Renmin University Press.

Qian, C. (2011). A study of constituent structure of translation ecological system. *Journal of China University of Mining & Technology (Social Science Edition), 4*, 134–139.

Serres, M. (1974). *La Traduction, Hermes*. Paris: Les Editions de Minuit.

Shao, D. (2006). *The first China International translation industry forum kicks off in Shanghai*. Retrieved May 28, 2006, from http://www.china.com.cn/chinese/zhuanti/translation/htm

Snell-Hornby, M. (1988). *Translation studies: An integrated approach*. Amsterdam: John Benjamins Publishing Company.

Tan, Z. X. (1988). On translation studies. *Foreign Language Journal*, (3), 22–28.

Tu, W. M. (2001). The ecological turn in new Confucian humanism. *Daedalus, 130*(4), 247–248. Fall.

Wen, T. (2009). Multilingual platform brings new developments in translation industry. *21st Century English Language Education Weekly Journal*, A1.

Wu, Y. H. (2008). The golden time for chinese translation industries. Retrieved August 6, 2008, from http://news.mainone.com/hangyefenxi/htm

Xu, J. (2003a). Creative treason and the establishment of translational subjectivity. *Chinese Translators Journal, 1*, 6–11.

Xu, J. (2003b). *On translation*. Wuhan: Hubei Education Press.

Xu, J. (2009a). *Translation ecology*. Beijing: China Three Gorges Publishing House.

Xu, J., & Mu, L. (2009). Translation studies in China in the past three decades (1978-2007). *Foreign Language Journal, 32*(1), 77–87.

Xu, S. (2009b). Overall perspective of complex language studies. *Foreign Language and Their Teaching, 3*, 1–3.

Yang, Z. (2007). How to promote the construction and development of translation studies. In H. Gengshen (Ed.), *Translation and cross-cultural communication: Turn and development* (pp. 3–6). Shanghai: Shanghai Foreign Language Education Press.

Yang, Z. (2009). Understanding several translation problems. In G. Hu (Ed.), *Translation and cross-cultural communication: Integration and innovation* (pp. 3–13). Shanghai: Shanghai Foreign Language Education Press.

Yang, Z., & Wang, J. (2008). *Universality, particularity, and perspective (Theories and methodologies for English-Chinese contractive studies)*. Shanghai: Shanghai Foreign Language Education Press.

Chapter 5
The Mesolevel: Ontological Theories of Translatology

5.1 Introduction

It has been acknowledged that studies on translation theory and studies on translatology are on different layers, two of which are in a partial–total relationship. As a part of translatological studies, translation theory serves as the rational understanding and high-level abstraction of a series of translation knowledge, such as the nature, process, principles, methods, and criteria of translation. In research on the construction of the Eco-translatological discourse system, systematic ontological translation theory should be labeled "mesolevel" research under macro-level research for the translatological system.

In meso-level research on ontological translation theory, recognition of the nature of translation, description of the translation process, realization of translation principles, illustration of translation subjects, categorization of translation criteria, and retargeting text through "doing things with translations"[1] will be highlighted in Eco-translatology.

Specifically, in Eco-translatology, translation is briefly described as the translator's selection in text transplants to adapt the translational eco-environment and translation activity as the translator's adaptation and selection; in addition, Eco-translatology includes the interpretation of the translation process with the alternating circulation of the translator's adaptation and selection, the demonstration of the translation idea of the "translator as the center" and the ethical translation principle of the "translator's responsibility," the retargeted text of "the survivor as the fittest" and "the stronger as the longer survivor," and the primitive motive of the "translator as stimulator."

The key elements of ontological translation theory mentioned above correspond closely to the relevant questions in ontological translation theory, such as "What is translation?" "Why to translate?" "What is translation for?" "How to translate?"

[1] See Sect. 6.4.

(Hugs) G. S. Hu, *Eco-Translatology*, New Frontiers in Translation Studies,
https://doi.org/10.1007/978-981-15-2260-4_5

"Who translates?" and "For whom to translate?" In this chapter, therefore, the interpretation of systematic translation theory and the issues of ontological translation theory from the perspective of Eco-translatology, such as "What is translation?" "Why to translate?" and "How to translate?" will be highlighted.

5.2 Cognition of Translation Theory

Although there is no universally acknowledged definition or fixed statement on what translation theory is, many studies and statements by scholars of translation theory can be used as reference frames for the construction of translation theory, which, in turn, can serve as the basis for us to observe, to think about, and to research the issues of translation studies.

5.2.1 Definition and Purpose of Translation Theory

Translation theory is the rational and general knowledge of the activities, skills and views of translation, and the high-level abstraction and distillation of a series of translation knowledge about the definitions, principles, methods, and criteria of translation. Translation theory is demonstrated mostly as the systematic statement and rational interpretation of translation activities. Many scholars of translation studies have mentioned the above topics in their studies. For instance, Stanley E. Fish holds that a theory is "a system of abstract rules used to direct and dominate practical activities at the angle of the over-practice" (Fish, 1989, p. 378). In addition, "a theory is defined as a systematically related set of statements, including some law-like generalizations, which are empirically testable. A theory should give direction to the research effort needed to solve a specific problem or shed light on a situation. Theory forces research on specific areas, as well as providing alternative views and solutions to problems" (Hunt, 1991). Ernst A. Gutt holds that translation theory is that which "can make it possible to describe a complex [translation] phenomenon in simpler language, namely, to induce and to generalize the phenomenon of translation" (Gutt, 2000, p. 11). Xie (2001, p. 135), a Chinese scholar, holds that translation theory should simultaneously possess the effects of "direction" and "realization," and he also appeals to scholars in the field of translation studies to break away from the "false realization" of translation studies and translation theory.

In short, translation theory aims to identify the most fundamental and the simplest interrelations in translation.

5.2.2 The Key Components and Functions of Translation Theory

The key components and functions of translation theory have not yet been defined. Similarly, statements by scholars domestically and abroad will provide us with only a basis for reaching an agreement.

As is held by Louis Kelly, an integral translation theory should be composed of three sections: certain function and purpose, description and analysis of operation, and evaluation of the purpose-operation relationship (Kelly, 1979, p. 1).

As stated by James Holmes, translation theory should be "an integral theory including many elements to interpret and to predict all the phenomena arising in translating work and target-texts" (Holmes, 1988, p. 73). "An integral translation theory should include at least the four sub-theories: translation process, translation products, translation function and translation teaching" (Holmes, 1988, p. 95).

As defined by Peter Newmark, translation theory is "a series of knowledge including the basic principles, directions, advices and notes of translation" (Newmark, 1988, p. 9). He also points out that "translation is no grey academic abstraction, but an attitude to translation, a set of ideas and explanations to translation, and a series of rules and hypotheses used to give rise to the target-texts" (Newmark, 1997, p. 99).

Eugene Nida points out that scientific translation theory should include "a series of general propositions as principles used to a lot of translating phenomena" and "a set of translating procedures by which to produce target-texts" (Nida, 1997, p. 102).

Albrecht Neubert holds that a true translation theory should systematically integrate every translation mode or translation route and develop several partial theories or sub-theories before integrating them into "the universal translation theory" (Neubert, 1998; Beeby, Ensinger, & Presas, 1998, p. 25).

In China, Tan Zaixi points out that an integral translation theory should have five components: (1) the illustration of the nature of translation; (2) the description of the translation process; (3) the definition of the principles and criteria of translation; (4) the description of translation methods; and (5) the demonstration of various contradictions during translation (Tan, 1988, p. 27).

Liu Miqing holds that the research on translation theory "refers to the features and complexities possessed in 'translation perception' as well as its general working law and process" (Liu, 1998, pp. 7–8).

Gu Zhengkun points out that a translation theory should conform to the "systematic," "interrelated," "unique," and "significant" meta-principles that are used to evaluate translation theory (Gu, 2002, pp. 10–12).

In summary, seen from the perspective of the key components of translation theory, a relative integral ontological translation theory should possess such theoretical elements as the philosophical basis of translation theory; statements on the animation of and approaches to the nature of translation, the translation process, the principles and methods of translation, the criteria of translation and translation criticism; and explanations of various contradictions and phenomena in translation. In other words, the ontological translation theory should be an integrated collection that states the nature, process, strategy, and criteria of translation, explains the translation phenomena, and systematically and comprehensively explores and describes the fundamental elements of the philosophical basis of translation theory.

Seen from the perspective of the function of translation theory, a systematic translation theory should be helpful for realizing the theoretical dimensions of translation, stimulating translation thought, directing translation practice, evaluating and analyzing the target text, explaining the translation phenomena, and developing translation itself. Thus, concepts such as *integrity*, *inclusiveness*, *comprehensiveness, systematicity, interpretiveness,* and *operativeness* are highlighted in translation theory.

5.2.3 The Characteristics and Judgment of Translation Theory

There are many statements on judging the criteria of translation theory. Thomas Kuhn, for instance, holds that a theory should possess five features, namely, exactness, conformity, adaptability, conciseness, and source-textity (Kuhn, 1962). W. Wilss holds that a theory that is used to describe the translation process should include at least three aspects: Pöchhacker's Entwicklungspsychologie, Handlungstheorie, and Okopsychologie (Developmental psychology, Action theory, and Biological psychology) (Wilss, 1992, p. 33). Tan Zaixi proposes three "-ty"s: objectivity, systematicity, and applicability (Tan, 1997, pp. 331–352). Edward Wilson suggests five natures of a theory: repeatability, economicality, measurability, inspirability, and integrity (Wilson, 2002). Zhu Chunshen holds that one should see "if a theory possesses the systematic elements needed in construction of the modern theories universally acknowledged in the international academic research (such as the relatively exact system of terms, relatively rational discourse norms, relatively objective theoretical frames, relatively comprehensive analytical means, and relatively stable force of interpretation, etc.), so that a theory should be a system which can be verified, falsified, operated, interacted, and sustainably developed" (Zhu, 2001, p. 17).

It can be seen from the perspective of translation theory that a universally applied translation theory should possess the qualities of universality, systematicity, and operativity. Meanwhile, the great theory should be simple and beautiful. Einstein, who was the greatest scientist of the twentieth century, for example, produced the Theory of Relativity as $E = mc^2$, which is concise and plain with a special beauty. In addition, W. Holmes mentions "all science's desire for the elegance of simplicity" (Holmes, 1988, p. 101).

From the above examples, it can be seen that there are actually very few authentic, systematic translation theories, and this is true of China as well as of foreign countries. It can also be seen that in China or in other countries, a translation theory cannot be regarded as systematic and universal unless it can be used for the whole systematic description and illustration of the various theoretical elements mentioned above. Seemingly, if a statement merely describes and illustrates a certain theoretical element or other related element of the translation theory, it is but a theoretical viewpoint rather than a systematic, universal translation theory. Therefore, one cannot "treat viewpoint of one's experience as a so-called theory, with which to replace the exploration for truly-genuine translation theory" (Xie, 2007, p. 57).

For the same reason, we should be quite cautious in proclaiming and verifying a translation theory in accordance with the fundamental features, basic functions, and ontological elements of translation theory. The excessive nominations of terms of translation theory tend not only to make the differences between translation research and studies on translation theory ambiguous but also to confuse the differences between translation viewpoint and translation theory. This ambiguity will not only affect the reputation of true translation theory but also provide excuses for misleading statements, such as "theory is nothing" or "theory is useless," which make certain scholars dismiss research on translation theory.

5.2.4 The Ten Viewpoints in the Research of Translation Theory

On the basis of the understanding mentioned above, I want to illustrate the following viewpoints and ideas:

1. Theory is extremely important, and research on translation theory should be highlighted.

It is self-evident that theory is of great importance. There was a saying in ancient China, namely, "The governor works with his wisdom, and the governed with his strength." Although we cannot belittle the workers who have strength, a soulless body cannot go beyond certain boundaries.

Regarding this aspect, a persuasive example is that of Columbus, who, more than five hundred years ago, enabled Europeans to possess the silver standard and who was directed by geographic knowledge in finding a so-called new continent (even though he misnamed the land he found), for the first time triggering the process of globalization. In contrast, Zheng He, who was far more powerful than Columbus, cannot be called the finder of a new continent, even though he covered a greater distance than Columbus, as he lacked the geographic knowledge that the earth is a sphere. Therefore, compared with that of Columbus, Zheng He's voyage was of no big significance.

The same is true, in a sense, of the significance of the research on translation theory and the relationship between theory and practice. My viewpoint on this topic is the same as the following: "Without the direction and support of theory, all translation studies would but stop at the layer of comments and experience generalization of translation issues rather than actual breakthrough. Hence, we must try hard to promote the construction of China's translatological theory by great exertion in translation theory if we wish for a new breakthrough in China's translation studies" (Xie, 2012). I am convinced that "only a nation that respects thoughts can give rise to great thought. And a country in possession of great thought will possess the force by which we can move on and on."[2]

2. Theories should be systematic, and the research on theory should be entitled.

Translation theory is a system of strong senses. Therefore, translation theory should be a systematic unified discourse rather than a single doctrine of words. Judged by this standard, a concept cannot be regarded as a translation theory but is rather considered a theoretical viewpoint if it, with mere words or phrases, fails to be a systematic statement of the nature, process, strategy, and criteria of translation; to be a systematic explanation of the activity and phenomena of translation; or to be a systematic exploration and description of the serious contents such as the philosophical basis and methodology of translation theory.

The theoretical viewpoint serves as a translator's rational summary of translation experience, as the rational induction of their understanding of translation activity or translation skills, or as a rational generalization of the translation profession (including translation teaching, interpreter training, and translation industry administration) and the practical operation of translation. As a brief summary of translation experience and a refinement of translation skills, the theoretical viewpoint of translation tends to be characterized by the statement of doctrines.

It is therefore judged that various statements on translation by scholars in China or abroad, such as "Five losses of the source-text and three nontranslations" by Shi Dao'an, "Faithfulness as well as adaptability" by Xuan Zang, "faithfulness (*xin*), expressiveness (*da*) and elegance (*ya*)" by Yan Fu, "spiritual similarity" (*shen si*) by Fu Lei, "natural transformation" (*hua jing*) by Qian Zhongshu, and "beauty in sound, sense and form" by Xu Yuanchong as well as "the maintenance of the style and force of the language" by Cicero, "the translator as the dominator of the target-text" by Horace, "complete transcript, the same style and ease in composition" by Tytler, and "translation as difference" by Derrida, should all belong to the well-acknowledged important remarks on translation, namely, the theoretical viewpoints of translation, all of which are not only "the sparks of translation thoughts" but also concise, exact generalizations of translation criteria and translation skills, and fundamental elements of translation theory. They have greatly enriched the treasure house of translation studies; however, they cannot be regarded as systematic translation theories.

Regarding this aspect, Mr. Liu Jingzhi, former President of the Hong Kong Translation Association, says that "some remarks related to translation are short and concise but never long and comprehensive. The statements of 'faithfulness (*xin*), expressiveness (*da*) and elegance (*ya*)', 'spiritual similarity (*shen si*) and formal similarity (*xing si*)' and 'natural transformation' (*hua jing*), etc., are not translation theory, but a viewpoint, a criterion, or a dimension of translation" (Liu, 2005, p. 30).

[2]A sentence taken from the interpretation lines from *China's Rise*, a full-length CCTV film.

An earlier scholar also holds that "the viewpoints of the famous scholars are some opinions with partiality and nonintegrity rather than exclusive theory focused on certain topics. That is, they are not the theory in the modern sense" (Zhu, 2000, p. 2). Therefore, the comments and essays on their own translation practice "are surely filled with acute wisdom and shining words but, on the whole, are all unsystematic subjective expressions.…Therefore, they cannot be regarded as systematic translation theory" (Meng, 2002, p. 16).

As is pointed out more explicitly in *A Brief Introduction of Translatology*, published several years ago, traditional Chinese translation theory is mostly embodied in highly condensed statements, such as "adherence to the source text," "seeking faithfulness," "faithfulness (*xin*), expressiveness (*da*) and elegance (*ya*)," and "natural transformation" (*hua jing*), rather than forming a theoretical system with exact conceptions, clear definitions, and logical intensity or illustrating a clear succession and ways of thought, which makes it difficult to identify obvious schools of translation theory (Xu & Mu, 2009, p. 550).

I hold that, as usual, the translators who have engaged with translation affairs possess certain translation experience and perceptions and, too, propose various viewpoints of translation. However, not all translation viewpoints are systematic translation theory, nor can a scholar who has expressed some translation thoughts necessarily be labeled a translation theorist.[3]

3. Theory should interact with practice.

I agree with Nida's opinion that "theories must grow out of practice" (Nida, 1997), and I also agree with Neubert, who says, "Theory without practice is empty, but equally, practice without theory is blind" (Neubert, 1998; Beeby et al., 1998, p. 26). I do not agree that theory can exist without practice or that theory can be too pure to have anything to do with practical activity. The theory–practice relationship depends on the absoluteness of theory. The more absolute a theory is, the further it is from practice, and the more indirect it is in its directional function, and vice versa. However, this is related to the far or near distance, direct or indirect relations, and degree of tightness or looseness rather than whether there are relationships and links between theory and practice. In the relationship between theory and practice, I agree with Mr. Luo (2009, p. 1), who holds that "any translation thought will prove to be worthless if it breaks away from the native translation practice; a translation theory will not go further at all if it goes away from its native source, no matter how new it is."

[3]The difference between a translator and a translation theorist lies in the fact that the translator can come into being at any time, while the translation theorist will come in due time. Without due time, a translation theorist will not work even if he or she comes into being. As translation work is needed at any time, it is necessary that someone become the translator. The task of the translation theorist lies in reflection on the translation activity of human beings through proposing theoretical structures; revealing translation rules; and generalizing the past, directing the present, and perceiving the future. Translation theorists tend to emerge in times of great social and historical change, when translation activity prospers and thoughts on translation theory thrive.

4. A theory should not need many local features, though some are inevitable.

I hold that whether a translation theory has a truly Chinese feature depends on whether it is associated with effective theoretical topics, arouses attention and discussion on theoretical topics, helps to construct translation theory systems with greater force in theoretical interpretation, embodies the spirit of China's philosophy and culture in the construction and description of its theoretical system, possesses the power of Chinese discourse, etc. Overall, when a translation theory developed by a Chinese scholar can exist securely in the world translation field, the "made-in-China" feature will certainly be highlighted naturally, which is sufficient for the Chinese characteristics. In other words, if we answer "yes" to the above questions, then the Chinese characteristics are clear. Thus, I do not want to highlight the so-called academic, theoretical, and literal Chinese "characteristics"; rather, I wish to intermingle these characteristics with the "mainstream" because, on the one hand, as is said in ancient China, "Academics are universal all over the world," and theory should not be defined according to nationalities. On the other hand, only when intermingled with the "main stream" can a theory's true academic value be demonstrated. If the local feature is overemphasized, then the theory will fail to intermingle with the "mainstream" of international translation studies.

5. There should be no partiality toward local or international translation theory; however, local translation theory should be cherished.

I hold that it is inevitable that certain scholars will apply a double standard to local translation theory and foreign translation theory. Some scholars, for instance, "fail to understand or to be interested in the new local theories, while they are apt to deny all of them before knowing what the theory is like"; "they keep their eyes on the foreign scholars, and this is why they are blind to the local academic achievements"; and "even certain crude, imperfect theories developed abroad are of active significance; how can they require that Chinese scholars should be perfect at the first step of their research? We know that too demanding criteria will serve as one of the elements that prevent our studies from being effective, and what has been abandoned therein is but our animation for academic studies" (Yang, 2004, pp. 42–43). I confess that I do not object to foreign translation theory, but I am much more concerned about the development of our "local resources" of translation studies during our learning from and referring to foreign translation theories.

6. True international exchange is actually to be likely confined to the theoretical layer.

I hold that emphasis should be placed on macro-theoretical research, on "pure theory" research, and on theoretical research that is comparatively high in absoluteness. The reason is that, in terms of international exchange, cooperation, and competition, concrete research in different countries or regions has different features and advantages. Generally, micro studies at the technological level with comparatively low absoluteness have low or no comparability. Thus, if exchange and dialogue are conducted at this level, a failure to communicate with one another may result from a lack of common topics. It is difficult to conduct a true exchange at that level. On the comparatively absolute theoretical level, however, truly equal dialogue and exchange will occur due to the comparatively high degree of commonness and ease of communication. On the whole, some Chinese scholars are comparatively weak in macro-theoretical translation research, for which efforts should be greatly strengthened.

In the long run, Chinese scholars have been deprived of their discursive power, and one reason is that we lack an independent source text and animating discourse system. In other words, we lack the "commanding elevation" of translation theory in the international field of translation studies. With its "translation army" of one million translators, China is undoubtedly a "superpower of translation." We have been absorbed in practical, concrete translation projects, and translation works that are important. However, in international realm of translation studies, what matters most is not the number of translators or of translated works but the possession of widely acknowledged systematic translation theory as well as the possession of the "commanding elevation" of translation studies. Small (in terms of territories) as Israel is, it has achieved great fame because of the "poly-system theory," which has been widely studied and rapidly disseminated in recent years; owing only to this theory, Israel has stood out above other, larger countries.

To date, China has made great efforts to spread traditional Chinese culture to the world. In the field of translation, attention has been paid to the publication of translations, to translation practice, and to introducing concrete, practical achievements in translation activities to the outside world, which are certainly important. However, a key link has long been ignored, that is, the exploration, development, and introduction of comparatively mature translation theory, especially systematic macro translation theory or pure translation theory. Hence, compensation for and emphasis on the collective introduction and publication of translation theory will undoubtedly increase the effect of the revitalization of China and the dissemination of Chinese culture.

7. There are three "challenges" for the dissemination of Chinese translation theory.

It is not easy for Chinese translation theory to be disseminated. Regardless of the quality of local Chinese translation theory, three "challenges" in relation to this issue arise mainly from foreign theorists, Chinese theorists and colleagues in domestic translation studies. In the extreme version of this situation, some foreign scholars would belittle Chinese achievements, and some Chinese scholars belittle the translation theory developed by other Chinese scholars and even by colleagues in the same translation workshop. By some foreigners, I mean mainly Westerners, who have habitually ignored scholarly work from China or by Chinese scholars. For instance, recent years have obviously seen the steady development of Chinese research on translation theory; however, in an academic work titled *The Turn of Translation Studies: New Paradigms or Shifting Viewpoints* (Amsterdam/Philadelphia: John Benjamins Publishing Company, Snell-Hornby, 2006), Mary Snell-Hornby, a famous German translation theorist, does not mention translation studies by Chinese scholars, although she provides a comprehensive summary on studies of translation theory all over the world. This situation is rather regrettable. In addition, 4 years later, in an academic work titled *Critical Readings in Translation Studies* (London and New York: Routledge, Baker, 2010) by Mona Baker, no work or paper on translation studies by Chinese scholars is included, although the author provides a comprehensive and far-reaching introduction to expand her global vision of translation studies. In a sense, the habitual "mistake" or "ignorance" of some Western scholars shows that it is of great importance for scholars from China and the West to hold equal dialogues. The dismissiveness of Chinese scholars toward their domestic colleagues or disparagement among workmates is due not only to the recognition of "the foreign" over "the domestic" but also to the double standard toward others and the self and the long-lasting traditional rivalry between scholars or colleagues.

In spite of this, however, I am still assured that the dissemination of Chinese translation theory depends on Chinese translation theory itself and that other external factors will not ultimately be decisive, though they can have more or less effect at certain times. I hold that the reason that Chinese translation theory has failed to be disseminated is that we are not sufficiently good at it. Only when we do well in translation studies will all come right in the end.

8. Blindness and key points in translation studies.

Several years ago, Xu Jun pointed out the blindness and segregation of theory from practice. He said, "The translation realized by each school of translation theorists is partial to a large degree, and what they have revealed is but an aspect of translation activity rather than the full view of translation activities" (Xu, 2003, p. 56). I agree with this viewpoint, and there is also a problem of how to identify the key points of this issue. I hold that only by means of grasping the translator can we provide the key points in the research on translation theory, which is the very core of translation studies. As Nida, the American translation theorist, says, "The real problems of translation are not technical, they are human" (Nida & Taber, 1969, p. 188). Hermans says, "What translation has told us is more about the translator than the target-text" (Hermans, 2000, pp. 12–13). Therefore, the translator serves as the decisive factor no matter whether the translation is a success or a failure. For this reason, the core idea of Eco-translatology is the emphasis on the translator in that efforts should be made to pay attention to the translator's development in order to discuss how to solve and settle various problems with texts by means of the development of the translator's abilities. Only when we grasp the very key of the construction and interpretation of translation theory can we securely succeed in our Eco-translatological research.

9. There is a difference between the cultural perspective and the ecological perspective in translation studies.

I hold that translation studies from the perspective of culture deal mainly with research on cultural phenomena and their influence upon translation. It is not only concrete translation research on the layer of culture but also concrete cultural research on the layer of translation. By comparison, ecology is an ontological science, and ecologically oriented studies are rooted in a comprehensive science. The ecological perspective of translation studies is that in light of eco-reason, or translation studies directed by eco-philosophy (such as the integral balance principle, coexistence principle, or law of unified diversity), translation research is conducted with a sense of philosophy and methodology. Translation research from the cultural perspective is cultural studies in translation; however, translation studies from the ecological perspective are not ecological studies in translation but the renewed interpretation of translation activities and phenomena in the sense of ecology and from the perspective of ecology, which serves as the new description and interpretation of the translational ecology as a whole and of ontological translation theory, which will enable translation studies to possess eco-reason. In translation studies, the cultural perspective of translational ecology and the ecological perspective of ontological translation theory should not be on the same layer. It can be said in this sense that the translation studies from a cultural perspective mostly illustrate multiple phenomenological features by which the philosophical elevation and methodological perception of translation studies obviously have certain limitations.

10. Universality is the common goal for theoretical researchers.

It is believed that each theory is universal to a certain extent; however, there should be a theory that is wholly universal. This is what we are striving for in our Eco-translatological research.

5.3 "What Is Translation?" from the Perspective of Eco-Translatology

5.3.1 Understanding the Nature of Translation from Different Perspectives

Translatology is complex in that it is connected with almost all subjects. It is the very cross-disciplinary, comprehensive, and complex features of translatology that allow multiple understandings and expressions of translation. Translation can be perceived from the perspective of translation activity (or function). For instance, translation is "one of the most important and most valuable things among all the things in the world" (Goethe, 1827). "Translation is by far the most complex activity noted in the evolution history of the universe" (I. A. Richards). It is universally acknowledged in the translation world that translation is cross-cultural communication and exchange of information as well as cross-lingual and cross-cultural communication. Translation is the activity participated in by the translator and is an interlingual transformation as well as an important means of promoting human social exchange and development.

Translation can be perceived from the lingual perspective. For instance, the following definitions are familiar to the Chinese translation world: Translation is "the replacement of the information in one language with the complete information of another language rather than the isolated lingual code units" (Jakoboson, 1959/2000, p. 233)." Translation is "the replacement of textual material in one language (SL) by equivalent textual material in another language (TL)" (Catford, 1965, p. 20). Translation is "the closest natural equivalent to the source-language message" (Nida & Taber, 1969, p. 12).

In the *Chinese Translation Dictionary*, "translation" is defined as follows: Translation is the important component of lingual activity and refers to the process or result of changing the content of a language or the variant of a language into another language or variant of another language, or the exact and complete representation of the text composed of one lingual material with another language.[4]

Before the 1970s or 1980s, the definitions given by scholars in the translation world from the linguistic perspective were mostly accepted by the dictionary compilers in Chinese society. In the *Grand Dictionary of Chinese Characters* (*Ci Hai*) (1980), for instance, translation is defined as "the expression of the meaning of one language with another language." In the *Modern Chinese Dictionary* (1978, 1983, 1996), translation is defined as "the expression of the meaning of one language with another language (or in the expression of a dialect of the common national language, the expression of one dialect with another dialect or the expression of classical Chinese with modern Chinese); or as the expression of the symbols or codes representative of language or literature with language or characters." In the Encyclopedia of China (The Volume of Chinese Language) (1988), translation is defined as "the activity of expressing the spoken or written meaning with another language."

Translation can be perceived from the cultural perspective. For instance, translation serves as the bridge by which a foreign culture can be conveyed across the gap between the languages and cultures of the two sides for open-minded communication. The nature of translation activity lies in the realization of cross-cultural communication. The object of translation transfer is cultural information. For instance, Casagrande holds that the translator is actually translating not the language but the culture (Casagrande, 1954, p. 338). This can be explained in three aspects: First, the two languages involved in translation are the product of certain social cultures; second, the animation and influence of translation activity are stimulated by the underlying culture; and third, the difficulties of translation sometimes lie not in lingual differences but in cultural distance. Cultural researchers have tried to replace the research dimension of scholars of the linguistic school by mentioning that ideology, translator subjectivity, patrons of translation activity, and the historical and cultural environments that constrain translation activity should be exposed in translation studies. The cultural turn of translation studies is highlighted by Bassnett and

[4] See *Chinese Translation Dictionary* compiled by LIN Huangtian, Hubei Education Publishing House, 1997, p. 167.

Lefevere, who actually refer to exploring and explaining translation phenomena by means of modern cultural research and from the same perspective. In China, translation is also defined as "a cultural activity of a translator presenting the meaning contained in one language in another language" (Wang, 1997).

Translation can be perceived from the perspective of cross-cultural communication. For instance, an exchange between different languages is actually that between different cultures, and the nature of translation activity lies in the realization of cross-cultural communication. Translation criticism, with translation activity as the object of perception, should take the inspection of the cross-cultural effect of translation activity as its fundamental task. As the cross-cultural communication activity is undertaken at different levels, the inspection of the cross-cultural communication effect should also be spread across different levels. Translation, too, is a cross-lingual and cross-cultural communication activity. As is said by Hans Vermeer, "Translation, in all, is a cross-cultural transfer, and the translator should be an expert at two or more cultures. As language is the indispensable part in the culture, the translator should responsibly be good at two or more languages. And then, translation is essentially a behavior. In other words, it is a cross-cultural behavior" (Liao, 2001, p. 364).

Translation can be perceived from the perspective of information. For instance, translation is a kind of cross-cultural exchange and transfer of information with the essence of the information as its conveyance (Lv, 1997). Translation is a "true-to-information" service provided for two parties who do not understand each other's language and culture. Translation is a process in which a comparatively strange form of expression is transferred into a comparatively familiar form of expression, with its content including language, characters or words, figures and symbols.

Translation can also be viewed from other perspectives. For instance, the symbolic school of translation studies regards translation as a "symbol-interpretation" activity: "Translation is symbol interpretation, and vice versa."[5] In this school, translation is "a transfer of meaning from one set of language signs to another set of language signs" (Lawendowski, 1978, p. 267). Hermeneutic researchers consider translation to be understanding. The purpose of translation is the target text's equivalence with the source text in the unity of content and form, and it is essentially the copying the source text. Mona Baker defined the essence of translation as "re-narration."[6] In addition, we find the following perspectives: "The translator is a traitor," "translation as conquest," "translation as rewriting," "translation as canonization," "translation as editing," "translation as recovery," "translation as dialogue," "translation as negotiation," "translation as mediation," and so on.

The recognition of the essence of translation and the definition and interpretation of translation show that translation is a highly complex intellectual activity that is rather abundant in content. It will be helpful for us to understand the essence of

[5] See *Modern American Translation Theory* by Guo Jianzhong, Hubei Education Publishing House, 2000, p. 102–134.

[6] www.ben-kelunwen.cn/article/2010/0102/article_5036.html. 2010-1-12.

translation no matter what perspective and commentary we adhere to. However, the following exploration of the essence of translation and its special meaning from the perspective of Eco-translatology, especially inspecting and interpreting translation activity on the basis of textual ecology, translator-community ecology, and translation-environment ecology, which are the main ideas in Eco-translatology, has historically rarely been seen.

5.3.2 *Understanding Translation from the Perspective of Eco-Translatology*

Even though various definitions of translation have been mentioned above, no studies have chosen textual ecology, translator-community ecology, and translation-environment ecology, as research objects; in other words, in the present definitions of translation, more emphasis has been placed on *language* or *text*, or the function or effect of translation, than on systematic research from the ecological perspective or on textual life or on human behavior in the *survival* and *development* of the translator-community with the translator as the representative.

Life science has become a leading science in contemporary times. According to the statistics of the ten annual most important science and technology developments from 2006 to 2008, chosen by *Science Journal* in the USA, 60% of research is related to life science. This means that it is life science that is sustainably and forcibly driving the progress of modern science and affecting the progress of present-day human civilization to a great degree.

Life science is the general term for natural sciences with life material as the research object. Its rapid development has guided many fields of the natural sciences and had far-reaching influence upon the humanities. Ecology is a branch subject of life science, and many interdisciplinary subjects exist, such as Eco-economics, Eco-ethics, Eco-anthropology, Eco-aesthetics, and Eco-philosophy. In addition, "Eco-translatology, created and directed by a Chinese scholar HU Gengshen, has been booming in recent years" (Fang, 2010).

When we talk about how the translator perceives and interprets systematically from the perspective of ecology, life, and survival, we know that some Eco-translatological concepts come from the understanding of the natural ecosystem on the basis of An Approach to Translation as Adaptation and Selection. The basic principle of *the survival of the fittest* has been borrowed, transplanted, and analogized in Eco-translatology and been integrated into the translational ecosystem, which connects the principles and *translator-community*[7] with the inborn sense of survival. In other words, the basic principle of *survival of the fittest*[8] is used as a

[7] See the item *translator-community* in the subsection titled "Sketching Eco-translatology from the Perspective of Terms" in the Preface of this work.

[8] As noted by Yang Zijian in an article titled "My Reflections on the Foundation of Translatology"

lever to consider the survival system in translation practice. This is the internal and external motivation for successful translation as well as the primitive goal for the translator-community to strive for better survival and development, as "life has been highlighted as the starting point of Chinese culture" (Mou, 1997, p. 43). Moreover, *ecology*, *survival,* and *life* are the very lifeline of ecological studies.

When translation is studied from the perspective of the survival and development of life in certain eco-environments, all the information, personnel, energy, time, and space that are required for the survival and development of the translators can be regarded as the resources of the translator and of translational ecology. In the sense of the development of *life*, *survival*, and *ecology*, the nature of translation activity is the process of the translator's utilization, evaluation, distribution, preservation, fabrication, dominance, and reproduction of the resources of translational ecology. In the process, the translator's adaptation and selection occur alternately. The relationship in the recycling process is as follows: the purpose of adaptation is to seek effective survival by means of optimizing selection, while the rule of selection is *elimination of the weak in favor of the strong* and *selection of the fittest for survival.* As said by Qian (2001, p. 278), a sociology scholar in China, "The selection for survival gives birth to the most animating diversity of human beings."

Natural selection is a two-way process, and creatures will select the environment suitable for their survival according to the law of *survival of the fittest*, while the environment, too, will select various factors as the necessary components of the environment for the attainment of balance. Because whether it is the eco-environment or the grand environment of the universe, its source text is a system of dynamic balance. The reason that the system exists as it does is that there is an inner working mechanism to maintain the dynamic balance in accordance with the natural law.

In fact, translation "is but the selective imitation, no matter what it is, a science or an art" (Blackmore, 1990, p. 50). Baldwin (1909), a US psychologist, also pointed out that "natural selection is never an ecological law; as a law, it will work in all the other sciences related to life and soul." Undoubtedly, "natural selection will be workable in imitation of meme, for the complex natural world is developing by means of natural selection. As is said by Skinner, 'Some activities have been selected in an active sense, while others been eliminated'" (Blackmore, 1990, p. 76).

In sum, the spirit of Eco-translatology is centered on *ecology*, *life*, and *survival* (see the discussion of the "three eco-themes" in Sect. 3.4 Research Areas of Eco-translatology); on maintaining the balance of textual ecology and concern for the survival and existence of the source-text life in the environment of the target-text language; on the survival and development of the translator in a certain translational ecology; and on the maintenance of the stability, coordination, balance, and harmony of translational ecology, textual ecology, and the translator-community.

[*Journal of Chinese Translators*, 1989 (4): 17–19], "Translatology is a science to research the rules of thought and methods of translation." Feng Wenkun also holds that "the being of the translator should be the primary basis" in his article titled "On the Ontological Turn of Translation Studies and Its Connotation" (see *Foreign Language Teaching* Feng, 2008(4): 82–85).

5.3.3 Translation as Ecological Balance

The core idea of Eco-translatology is the emphasis on the balance of translation-environment ecology. Therefore, Eco-translatology is related to the vision of translation, namely, *translation as ecological balance,* and translation strategies and skills are actually the *balancing methods* of translation.

The term "balance" here means the integral balance of the comprehensive elements, including not only the balance of translational ecology and textual ecology, as well as that of the ecology of the translator-community, but also the cross-lingual and cross-cultural integral balance and that of the internal and external factors as well as that of macro-level, meso-level, and micro-level thought.

The balance of textual ecology concretely includes the textual balances of lingual ecology, cultural ecology, and communicative ecology. In terms of the lingual eco-balance in the textual eco-balance, the translator should strive to maintain the balances of meanings, spirits, practical values, aesthetic values, styles, etc. between the source language and the target language.

From the perspective of the evaluation of translation practice, the universally acknowledged and influential translation works are usually adept at the management and disposition of the bilingual eco-balance. This is true of *Tian Yan Lun* by Yan Fu (Jiao, 2010; Yuan, 2011; Wang, 2012).

From the perspective of the equivalence theory, in various translation theories thus far, terms such as *equivalence, correspondence,*[9] *symmetry,* and *nondiscrimination* have been researched, and some of these terms have been accepted by consensus. The different terms are used to describe the nature and results of translation studies in the aspects of the lingual form, sense function, textual information, knowledge amount, and communicative purpose of the source language and the target language as well as the relationships of the various parts. At the core, the theory deals with the pursuit of balance between various aspects of the source-text language and target-text language. Therefore, it can be described and explained with the idea of *eco-balance*. As is held by Basil Hatim in his work titled *Teaching and Researching Translation*, "The new research focuses have been balanced: the pure theoretical research has actively been undertaken on translating product design, translating process and translating function in the projects of translation studies" (Hatim, 2001, p. 10). Obviously, the issue of *balance* can also be found in such "pure theoretical research."

From the perspective of the requirements for translation research itself, balance, on the one hand, serves as the core idea of Eco-translatology as far as Eco-translatological research is concerned, as *balance* is the most fundamental feature of any ecosystem, let alone the effect of the translational eco-environment on the

[9] James Holmes says, "What the translator actually achieves is not textual *equivalence* in any strict sense of the term, but a network of *correspondence*, or matchings, with a varying closeness of fit. These correspondences are of various kinds, formal, semantic, and/or functional, mimetic, or analogical, and achieved at various levels of the translated text, micro-, meso-, and/or macro-structural. Defining the nature of these correspondences and minimum and maximum degree of fit should be a major task of the translation theorist" (Holmes, 1988, p. 101).

production of translation texts. Similar to the rule "No context, no text," there would be no successful translation without a translational eco-environment. Therefore, it is necessary to maintain the harmony and balance of the translational ecosystem as a whole. Otherwise, there would be no sound development of Eco-translatology nor an academic Eco-translatological mission to maintain lingual variety and cultural diversity. On the other hand, regarding the inner mechanism of Eco-translatology, the translational eco-balance can also be illustrated not only as the compromise and tolerance between various elements in the translational ecosystem but also as the coordination and balance of various elements, including the author, the translator, the source text, and the target text, and between the ecology of the translator-community and the textual ecology, by which the translator can have equal dialogues with the source-text author, overcoming the temporal–spatial limitations and various obstacles so that the actual requirements and adaptation abilities can be fully realized; the point of balance between the author and the readers can be sought; the visions of the author, the translator and the readers can be mingled together in sympathy; and the mutually beneficial and soundly ordered eco-cycle can continue.

As indicated by the statement above, from the perspective of Eco-translatology, Eco-translation, so to speak, is the art used to coordinate the textual ecology, translator-community ecology, and translation-environment ecology.

Discussions on textual ecology and the balance of textual ecology can also be found in this work. The discussion on *textual ecology* can be found in Sect. 3.7.1, the discussion on *balanced harmony* in Sect. 3.7.1, and the discussion on *translatability* and *style translation* in Sect. 3.7.1.

5.3.4 *Translation as Textual Transplants*

It has been mentioned in the preface of this work that Eco-translation is a general term used to perceive and describe translation from the perspective of ecology. It is an integral conception with profound connotations. For instance, it refers not only to the overall vision of translation from the perspective of ecology but also to the translational ecology in the metaphorical sense of natural ecology, and it can also refer to the selection of translated texts with ecological adaptation, to the constraints on the translator-community imposed by eco-ethics, and, of course, to the selection of Eco-translatological texts and the translation of the natural ecological world according to ecological ideas.

However, viewed solely from the perspective of text, Eco-translation can also refer to "textual transplants" on the basis of the ecology of the source-text language to the target language. In this narrow sense, translation is the textual transplants from one lingual ecosystem to another.[10]

[10] According to the Eco-translation idea, translation is "the textual transplantation from one lingual eco-system to another." Eco-translation has been well-explicated in an article titled "On the Exemplary Introduction of *Eco-translation*" (Jiang, 2013). As is held in the article, each language (including each dialect) possesses its own unique lingual ecology and cultural ecology. Each text

For "transplants," various explanations have emerged. Rosanna Warren says in the preface of *The Art of Translation: Voices from the Field*, edited by her, "It (translation) is a cognitive and living mode. When a literary work is transplanted from one language into another, it is like moving a plant or an animal from one place to another place. They must live like an individual or a race who should adapt and grow up, as only the adapter to the new environment can survive" (Warren, 1989, p. 6). Similar remarks have also been made by Susan Bussnett and other scholars. The preface of *Constructing Cultures*, compiled by Gentzler for Bassnett and Lefevere, states, "Bassnett suggests that we think in terms of transplanting a *seed*" and "The seed, once *transplanted*, flourishes" (Bassnett & Lefevere, 1990, p. xviii). Recent years have seen continuous research on "cultural transplantation" (such as Fang, 1996; Ma, 1999; Tan, 2001; Zheng, 2010, etc.).[11] However, the *textual transplants* mentioned in Eco-translatology have even more extensive connotations and also conforms to the translational eco-environment. Thus, it is not only a special term but also a core idea in Eco-translatology.

The so-called "textual transplants" is related to the understanding of the nature of translation according to Eco-translatology. The internal "transplantability" of the source text is highlighted. Before the implementation of the transplants, the translator focuses on the transplantability of the source-text eco-structure, which results in the selection of the text to be translated; in the process of transplantation, the transplants of the lingual ecology, cultural ecology, and communicative ecology of the text, together with the *restructuring* of the translational eco-environment and *reestablishment* of the translational ecology, should be highlighted. After the completion of transplantation, the living conditions of the transplanted text in the eco-environment of the target-text language should be highlighted, together with the

lives in the unique lingual ecology and cultural ecology. The more the lingual ecology and cultural ecology are embodied in the text, the more difficult it is to translate that text. Eighty years ago, when Mr. Wen Yiduo saw Li Bai's poetic lines "人烟寒橘柚, 秋色老梧桐" "correctly" translated as "The smoke from the cottage curls/Up around the citron trees, And the hues of late autumn are/On the green paulownias" by Obata, he thought that the version was dull and flavorless. Then he asked a well-known and interesting question: "Why is the English version so dull, while the original poem is of great charm?" (Wen, 1926). Then, he explained, "There is nothing wrong with the translator's translating skills, and it is wrong for him to intend to translate such naturally wonderful poetic lines which are too beautiful and too delicate to translate. You will ruin the poetic flavor if you want to translate it!"(Wen, 1926). Wen Yiduo was talking about the problem of Eco-translation. Typical Eco-translation should either strive for the lingual-cultural ecology of the original language or that of the translated language. The former is actually related to the high degree of foreignizing translation and the latter to the high degree of domesticating translation. With many translation examples, Hu Gengshen's Eco-translation idea has been discussed in relation to the four layers of pronunciation, vocabulary, syntax, and culture in the paper.

[11] See "Various Problems in Cultural Transplantation" by Fang Zao, *Foreign Language Journal* 1996 (1); "On Cultural Transplantation in English-Chinese Translation" by Ma Aixiang, *Journal of Lanzhou Medical College,* 1999 (4); "Compromise and Compensation: Cultural Transplantation in Translation" by Tan Hailing, *Journal of Zhongnan Industry University* (Social Science Edition), 2001(1); and "On Cultural Transplantation in Cross-cultural Translation" by Zheng Lijun, *Information of Science and Technology,* 2010 (26), etc.

cultivation of the eco-environment of the target-text language so that the transplanted text can survive and last.

The survival of the translated text embodies the ultimate result of translation activity. Hence, the survival of the translated text must be one of the focuses of Eco-translatological research. The living conditions research refers to the survival of the transplanted text; the living conditions of the translated text and the relationship of the audience to the acceptance and distribution of the target text in the eco-environment of the translated language; the relationship of the target text and the source text in the living conditions; the relationship of the target text and the eco-environment of the target-text language; the relevance of the target text and the timing elements of the living conditions; the alienation of the target text from the eco-environment of the target-text language; and all the living conditions of transplantation, survival, and the long duration of the text in other translational eco-environments. In this sense, not only "inward-translation" but also "outward-translation" is stressed in Eco-translatology. In other words, Eco-translatology is related not only to translation with Chinese as the target-text language but also to translation with Chinese as the source language.

The connotations of different lingual ecosystems have been mentioned in the section titled "Textual Ecology": The textual ecosystem of the source-text language refers to the lingual ecology, cultural ecology, and communicative ecology of the source-text language system; the textual ecosystem of the target-text language refers to the lingual ecology, cultural ecology, and communicative ecology of the target-text language system (See Sect. 3.4.1). Hence, when the text is transplanted in Eco-translation, the work to be translated should be selected in accordance with the internal eco-structure of the source-text work, and during the process of translation, the eco-structure should be represented in the target-text language in accordance with the source-text eco-structure. In other words, in the text transplants of the Eco-translation, the translator should be responsible for the maintenance of the source-text ecology and the target-text ecology, be aware of the coordination between the source-text ecology and the target-text ecology, strive for the balance of the source-text ecology and the target-text ecology, and play a role in establishing (restructuring) the source-text eco-environment in the ecosystem of the target-text language.

For the source-text ecology and the target-text ecology, "maintenance" is required initially, and "coordination" is needed after that; coordination aims for "balance," without which the textual transplants and translation transformation will be needed. Then, the establishment and "restructuring" will be needed, namely, an eco-environment that conforms to the source-text ecology needs to be created in the target-text language system, and through the translator's effort or by means of technology, the world of the source-text language system in which the source-text work has been created is revived so that the target text can survive and last in the new eco-environment of the target-text language.

5.3.5 Translation as Adaptation and Selection

Adaptation is vital for survival, and selection is indispensable for evolution (development). Translation has been perceived from the perspective of An Approach to Translation as Adaptation and Selection: the translation process is related to the translator's selection activity, which, whether conscious or unconscious, is influenced by the elements of the translational eco-environment. Such selections will occur in various aspects of translation activity, will exist in various stages of the translation process, and will emerge in various layers of the translation transformation. The mechanism and purpose underlying the selection are the doctrines of "survival of the fittest" and "the elimination of the weak in favor of the strong" (Hu, 2004a, 2004b, 2004c, 2004d).

In other words, in "Translation as Ecological Balance" (see Sect. 5.3.3) and "Translation as Textual Transplants" (see Sect. 5.3.4), the translator's selective adaption or adaptive selection is ultimately relied on, namely, the translator's adaptation and selection. The reason is that eco-balance, textual transplants, and translator selection are progressive, causal, and interactive relationships. For instance, when translation is perceived as the adaptive-selective transformation of the lingual dimension, the lingual conformity of the source-text language and the target-text language is the basis for the translator's lingual transformation in the translation process, while the lingual alienation will enable the translator to make adaptive selections according to the translational eco-environment. If the source-text language is transplanted to the target text without any alteration, objections will result because of its ambiguity or informal lingual expression. At the same time, the coordination and balance between the source-text language and the target-text language should be maintained in the aspects of vocabulary, syntax, passage genre, pragmatic function, and even rhythm and tone. To realize such coordination and balance, the translator needs to conduct selective adaptation or adaptive selection according to the different translational eco-environments of the source-text language and the target-text language.

Therefore, the so-called "Translation as Adaptation and Selection" refers to the translator's selective adaptation and adaptive selection. On the one hand, when one cannot obtain the perfect stage of conformity of "faithfulness (*xin*), expressiveness (*da*), and elegance (*ya*)," "spiritual similarity and formal similarity," and "the beauty in meaning, form and sound," the translator should be relied on to make an independent judgment and make an adaptive selection on the basis of the selective adaptation of certain translational eco-environments.

On the other hand, the understanding and implementation of translation activity, including comprehension, interpretation, operation, and application, should be judged by the translator for independent adaptive selection, no matter whether it is "cultural adaptation," "purposed activity," "poly-system," or the dominance of ideology. For instance, the reason that Mo Yan (a noted Chinese writer) received the Nobel Prize for Literature is that Howard Goldblatt translated freely and creatively but not literally (Xie, 2013, p. 4). Here, the overly free translation is an example of the translator's adaptation and selection according to the translational eco-environment.

The above discussion is the truth of *Translation as Adaptation and Selection*, and the translation process possesses the nature of "the translator's selective activity in the adaptation of the translational eco-environment and in the selection of the transplantation of the text." The translator's adaptation and selection are discussed in my book titled *An Approach to Translation as Adaptation and Selection,* and there is no need for further discussion.

In the final analysis, the understanding and comprehension of Eco-translation can be concluded as follows:

The so-called "Eco-translation" is a general term used to perceive and describe translation from the perspective of ecology. It is an integral conception with profound connotations. Specifically, Eco-translation refers not only to the overall vision of translation from the perspective of ecology but also to the translational ecology with metaphorical reference to natural ecology; to the maintenance of the diversity of the translated language and translated culture and the protection of eco-environments and development of eco-civilization by means of translation; to the selection of the translated text by ecological adaptation; and to the regulation of "translator-community" according to eco-ethics. In addition, of course, it can include the selection of the Eco-translational text and the translation of the ecological natural world according to ecological ideas. From the perspective of the text, Eco-translation can also refer to the "textual transplants" based on the source-text ecology and the target-text ecology.

5.4 "Who Translates?" from the Perspective of Eco-Translatology

When discussing "Who translates?" we obviously focus on the translator. The translator is highlighted in Eco-translatology, not merely because Eco-translatology has been developed from *An Approach to Translation as Adaptation and Selection* but also because the idea of the "translator as the center" of the translation process was initiated in the approach, which is related to the meso-level and micro-level research on Eco-translatology. What is more important is that the exploration of the relationship between the translator of the translational eco-environment is one of the research objects of Eco-translatology. The issue of the translator's survival and ability development is one of the "three eco-themes" of Eco-translatology, namely, text ecology, translator-community ecology, and translation-environment ecology.

Thus, this section is related to the progress of research from "translator-centeredness" in the translation process to "translator's responsibility" in translation ethics, which serves as the concentrated statement of the viewpoint on the translator in Eco-translatology.

5.4.1 "Translator-Centeredness" in the Translation Process

The idea of "translator-centeredness" has been given a comparatively thorough discussion in my work titled *An Approach to Translation as Adaptation and Selection* (2004), and further studies have been done in my articles, "From 'Translator-subjectivity' to 'Translator-enteredness'" (2004) and "My Response to Questions on 'Translator-centeredness'" (2011). However, disapproval of the idea of "translator-centeredness" has emerged. Based on my advocacy of "translator-centeredness," a brief reflection on and re-explanation of the position, orientation, and definition of "translator-centeredness" are given in this section.

First, seen from the perspective of a "translation link," my proposal of "translator-centeredness" refers not to the pre-translating stage or to the post-translating effect but to during-translating behavior. (The term "translation link" can be seen in the section of terms of Preliminaries: The Name and Nature of Eco-translatology in this book, and it will also be mentioned in the first item in the "re-explanation" section).

Second, because the idea of "translator-centeredness" refers mainly to the during-translating stage, "translator-centeredness," from the perspective of translation layers, is oriented toward the "translation process" and "translation operation"; in other words, it refers to "the translator's actions of selective adaptation and adaptive selection". In terms of the preconditions, when talking about the translation activity itself and when discussing the target text created through the translator's behavior, we confess that the translator is the very "center" of the process and plays the "dominant" role; otherwise, it is not translation activity, which is another topic.

Third, the translator's "central" status and "dominant" role have been studied without mentioning the translator's "subjectivity." It should be considered that, because of disagreement on the definition of the translator's subjectivity and because the translator cannot have a "central" status and "dominant" role even if he/she possesses subjectivity (see the first section, "Translator's Centered Status," in Chapter 4, titled "Translator as Dominator," of *An Approach to Translation as Adaptation and Selection*) and because there is no "intersubjectivity" in the "during-translating" action stage, only in the discussion of the "pre-translating" preparation stage and "post-translating" effect stage can the issues of "creating a subject" and "audience subject" be addressed. Obviously, once the issue of "intersubjectivity" is involved, what is discussed actually refers not to the "during-translating" action stage but to the issues of "literary creation" or "adopted culture." It is held that the translation subject is none other than the translator; the dispute on the issue of "intersubjectivity" is but a "misleading section" that arises before the translating stages are divided. (Further discussion can be found in the first item in the following section titled "Re-explanation.")

Fourth, regarding puzzles and worries, I wish to do more to explain or solve them, which can be regarded as "rethinking" or "re-explanation."

1. The idea of "translator-centeredness" and research in related realms

There is a worry that if the translator is regarded as the center, the translator will become a "transgressor" so as to play the central role in other research fields related to translation.

There are many research fields related to translation, such as culture, literature, society, and communication, as well as influences upon readers, etc., which are indispensable in studies of the relationship between translation and the target-text language.

I am of the opinion that research in such fields as "culture," "literature," and "society," which are related to translation, refers to the post-translating "effect," "communication," and "influence" by passing over or even excluding the scope of "translation itself" or "during-translating" stage research. The reason is that the focus of such research has turned to studies on "culture," "literature," "communication" or "society" (while studies on culture, literature, communication, and society belong to independent academic disciplines). With the themes changing to the realms of "cultural studies," "literary studies," "communicative studies," or "social studies," the evaluation criteria have correspondingly changed to those for cultural studies, literary studies, communicative studies, or social studies rather than those for translation or translation criticism. Based on such preconditions, when studies have focused on "post-translating" explorations of "accepted culture," "translational literature," "communication issues," or "social problems," the scholars have perceived the translation, analyzed the target text, and criticized the translator from the perspectives and criteria of cultural studies, literature, science of communication and sociology. The translator in such research realms is of course not the "center"; therefore, the translator cannot play a "dominant" role in the related research areas.

Other research fields related to translation are research on the source text, the author of the source text, the reading of the source text, etc.

Similarly, the source text is dominated by the source-text author before publication in that the author is the subject, and the work is the product created by the author; thus, the research focuses on the problems that arise during the "creation" process. Although creation here is relevant to translation, yet it is not translation behavior or translation activity; thus, I refer to it as a "pre-translating" topic. Obviously, "translator-centeredness" is apt not to refer to the "pre-translating" stage.

It is evident that "translator-centeredness" refers to the "during-translating" stage if it does not refer to the "pre-translating" stage or to the "post-translating" stage. Therefore, the "dominant" role is reasonable, and "translator-centeredness" cannot be denied as a condition in which the problems of the translation process, translation behavior, and translation operation are discussed. Otherwise, it has nothing to do with translation but rather is related to research in other fields. Regarding other studies relevant to translation, when the research focus shifts to "culture," "society," "creation," or "reading," different scopes and focuses will arise in the related research fields. Though they are associated to a certain degree with translation behavior and the translation process, research on translation behavior and ontological studies on translation are no longer focused on such issues. Therefore, except in the translation process, translation behavior, translation operation, and translation activity, "translator-centeredness" does not apply, nor does it mean that "translator-centeredness" is universally applicable in all fields associated with translation.

2. The idea of "translator-centeredness" and constraints on the translator

There is a concern that, with the idea of "translator-centeredness," the translator will lose control and behave in an unbridled manner in the translation field. There is no reason to worry about such behavior, and we can reasonably believe that "translator-centeredness" will not drive the translator to lose control, as a translator-confining "mechanism" is at work: the "survival of the fittest" and "post-event penalty." (See Sect. 3 titled "Translator's Adaptive Mechanism and Confining Mechanism," of Chapter Four of *An Approach to Translation as Adaptation and Selection.*) The relationship between the translator's translation and the translator's survival is stressed here without illustrating the details of the thought.

To survive and develop, the translator should perform well in translation. The basis for a sound translation is the translator's ability. The interrelationship between the translator's translation and the translator's survival is illustrated as follows:

Translator's Survival → Translator's Development → Translator's Ability → Translation Success

That is, only when the translator survives can he/she develop for the better → The translator should possess the ability for his/her development → Only an able translator can be successful in translation, and vice versa:

Translation Success → Translator's Ability → Translator's Development → Translator's Survival

The significance of the above relationship is that successful translation depends on the translator's ability → the translator's ability relies on the translator's development → the translator's development is based on the translator's survival. The doctrines of "survival of the fittest" and "post-event penalty" naturally serve as constraints on the translator's translation behavior.

Whether or not the translational eco-environment serves as the arbitration and final judgment of the translator's selection is not only highly relevant but also very important for the implementation of the idea of "translator-centeredness" because after the idea of "translator-centeredness" has been initiated, the translator's intelligent work and subjective creation can be acknowledged, and the translator's "self-dominating power" will be increased; correspondingly, the translator's "responsibility" (Venuti, 1995, p. 290) will increase as well. However, the translator needs not only "self-esteem" but also "self-discipline" and "other discipline." It is evident that the corresponding "confining" mechanism that constrains the translator is highly necessary. As pointed out by Thomas S. Kuhn, a famous scientist and scholar, "In order to accomplish his assignment, the scientist must be controlled by a series of complicated confinements, ideological and operative" (Kuhn, 1962; Fu, 2002, p. 403). Obviously, the "post-event penalty" is highlighted in the confining mechanism to the translator in *An Approach to Translation as Adaptation and Selection* and can be commonly seen in translation activities (Hu, 2004a, 2004b, 2004c, 2004d, pp. 111–113). The above mentioned relationships between the translator's behavior and the translator's survival conditions are illustrated as follows (Fig. 5.1):

Figure 5.1 serves as a brief illustration of the selection and confinement of the translational eco-environment encountered by the translator. It is obvious that the doctrine of "survival of the fittest" controls the translator's translation behavior. In other words, the translator's survival conditions, namely, the "translational eco-environment," are selected and constrained by the doctrine of "survival of the fittest"; this concept is seemingly lacking in other systems of translation theory.

3. The idea of "translator-centeredness" and text research

There is an opinion that the emphasis on "translator-centeredness" will cause translators to fail to be concerned about the "text." In response, I examined this concept metaphorically. It is similar to the binary-opposites "cold war" way of thought, with the "translator" being set in opposition to the "text." Therefore, focusing on the "translator" will certainly lead to ignoring the "text." There is a small

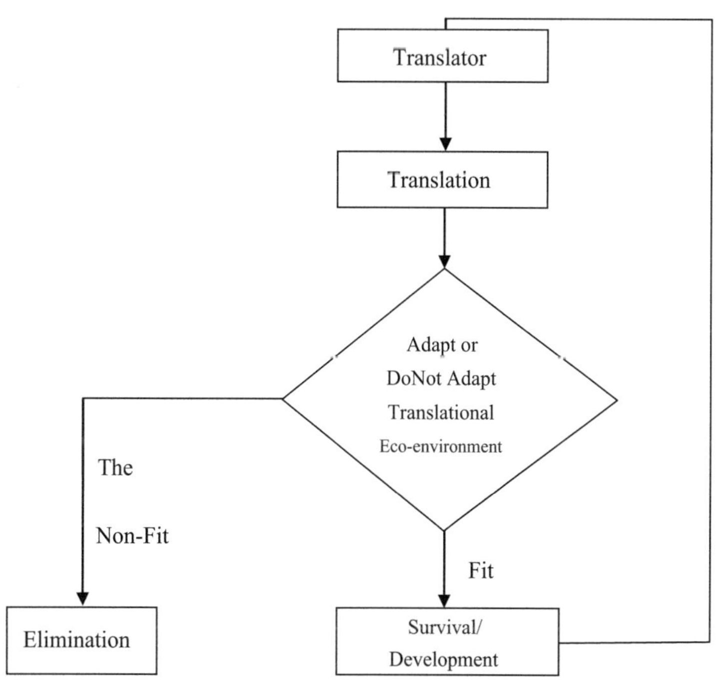

Fig. 5.1 Illustration of interrelations between translator's translation and translator's survival

misunderstanding that the emphasis on the translator actually refers to an emphasis on the translator's ability to translate the text, as the translator is highlighted on the precondition that the enhancement of the entire target-text quality should be highlighted. In fact, the development of the idea of "translator-centeredness" refers to the emphasis on the fact that the improvement of the text quality must depend on the translator's ability and his/her creative work, and only when the translator relies on his/her successful textual translation can he/she be worthy of and enjoy his/her "central" status. Thus, the aim of the idea of "translator-centeredness" is to focus on the translator's development, to fulfill the translator's potential, and to support the translator's creative work. Only when the translator is truly respected, improved, and cultivated will there be an increasing number of successful translators.

In China in the past hundred years, we have seen that "faithfulness (*xin*), expressiveness (*da*), and elegance (*ya*)" refer to the "text" rather than the "person," and over scores of years, we have seen that "similarity in spirit," "similarity in form," and "natural transformation" (*hua jing*) also refer to the "text" rather than the "person." In the ideas of translation theory or in the definitions of translation, the ignorance and obstruction of the translator are obvious. This situation can exist in translation theory both domestically and abroad. In reflecting on the research and development of "source-text centeredness" and "target-text centeredness," "translator-centeredness" is actually an improvement in general over the experience and reflections of "source-text centeredness"/"target-text centeredness" and is a revision of the partial-vision studies on translation theory in which the "text" is highlighted rather than the "person." In the long run, the emphasis on the static "text" rather than the dynamic "person" inevitably leads to the degradation of the translator's status, as with the "text" superior to the "person," the translator inevitably descends to the status of "servant" or "slave" and has to "dance in shackles and chains." In fact, when the translator creates the translated target text, it is produced in accordance with the idea of "translator-centeredness." Although the translator is not the host of the source text, yet, he/she is the "host" in his/her translation work. It is the translator who hosts the translation activity! Although think that we should not be influenced by what others say, yet we who are learners, teachers, workers, and "managers" of translation should at least keep to ourselves rather than belittle ourselves!

The focuses of translation theory research have been developing, passing from the "two ends" of the source text/target text to "translator-centeredness," from "static" text to "dynamic" translator, from "how to translate" to "who translates." However, there would be no "centeredness" without the "two ends"; there would be no embodiment of the "dynamic" translator without the "static" text; and without "how to translate," there would be no significance of "who translates."

The idea of "translator-centeredness" actually refers to higher requirements for the translator. In a sense, it refers to a movement from a focus on concrete language and wording to a focus on the development of the translator's overall ability, which makes it possible for the translator to perceive his/her value and capacity. To improve the target-text quality and to cultivate translation professionals, translators' knowledge, intelligence, skills, and creativity should be developed, and more efforts should be made to enhance the translator's techniques and abilities and to develop the translator's potential translation capacity. That is the true emancipation of the translator and is the basis for the development of translatology.

From this perspective, therefore, the idea of "translator-centeredness," which we have developed in relation to the translation process and translation behavior, refers not to the negation or belittlement of the "text" but rather to the fundamental route by which the text is examined and, finally, improved from the perspective of the "person," or "translator."

4. The idea of "translator-centeredness" and translation criticism

If the "translator" is the "center" of translation, should the translator or the target text be evaluated in translation criticism? Since the idea of "translator-centeredness" has been developed, the corresponding issue of translation evaluation criteria has not been specifically researched except for certain reflections and superficial visions of it. In our studies, the evaluation and assessment of the "holistic degree of adaptation and selection" of the target text have been generalized into "three referring indexes," namely, the degree of multidimensional transformation (with a focus on the textual quality), the reader's response (including the direction of translation criticism), and the translator's quality (including other factors related to the "person"). Among these, "translator quality" is uniquely considered one of the referring indexes of translation criticism, which indicates the concrete embodiment of the idea of "translator-centeredness."

Translation practice has shown us that the success of translation and the quality of the translated target text at the core depend upon the quality of the translator him-/herself. As said by Wolfram Wilss, whether the translator succeeds depends on his/her understanding of the source text and his/her ability to express this understanding in the target text. The two kinds of abilities are based on a collection of various elements, including mental allocation, personal background, translation experience, intelligence, motivation, flexibility, endurance, acceptance capacity, individual attitude, habitual response, depth and breadth of thought, similarity of meaning and style in the source text, matching (or non-matching) degree of target-text difficulty and translator capacity, degree of difference between the source text and target text in various aspects of vocabulary, idiom (habitual expression), syntax and social culture, as well as the constituting conditions of style in the source text and target text (Wilss, 1996, pp. 140–141).

Based on the idea of "translator-centeredness," the close relationship between the translated target text's "holistic degree of adaptation and selection" and the translator quality (including the translator's achievements, experience, abilities, credibility, and reputation) is highlighted. Generally, the translated target text's "holistic degree of adaptation and selection" should be directly proportional to the translator quality. That is, in normal conditions, the higher the translator quality is, the stronger his/her abilities to adapt and to select; the stronger his/her abilities to adapt and to select, the higher the translated target text's "holistic degree of adaptation and selection," and vice versa, of course.

There is a habitually used and persuasive example that to ensure the completion of an important translation task, the related side (or client) tends to choose the higher-quality translator (or interpreter) for security. There is an empirical presupposition that only when the translator (or interpreter) has been chosen can the quality of the translated target text or the holistic degree of adaptation and selection be predicted, even though the translator's (or interpreter's) translation is not yet available. Regarding the overall translation capacity, an experienced translator will prove to be more qualified and competitive than a newcomer.

It has always been stressed that the translated target text is actually the embodiment of the translator's translation capacity. In this sense, the assessment of the translated target text is also the assessment of the translation capacity of the translator, namely, the assessment of the translator. On this basis, regarding the assessment from the perspective of "translator-centeredness," there is always a combination of target-text assessment and translator assessment. Translator and translated target text are a "community" of profits, and the two should be considered to be in a state of divided unity, preconditioned by each other.

Antoine Berman once pointed out that the translator's translation motivation, translation aim, translation stand, translation scheme, translation methods, etc. make him/her the most active element in translation. Once his/her attitude, method, and standpoint have been chosen, he/she will select his/her destination. In addition, "each word he has translated becomes a pledge" (Berman, 1995, p. 75). "Undoubtedly, it is reasonable that the study of the translator is stressed, and *the translator is highlighted in translation criticism*" (Xu, 2003, p. 9) (emphasis is given by the author).

With the idea of "translator-centeredness" in the translation process, the "source text" and "target text" tendency stressed in the long run are avoided, and the translator's activity is examined from the perspective of the translator, who is at the center of the triple relationship of "source text—translator—target text," by which "a position has been found for the translator" (Hu, 2006, p. 39). Therefore, the translator's position and role have been securely guaranteed, and the theoretical basis for "doing things with translations" can also be identified. In this sense, compared to the traditional "two poles" of "source-text centeredness" and "target-text centeredness," "translator centeredness" can be taken as a sort of "deconstruction" and "balance." For a translator to improve the translated target-text quality requires inner motivation; in the research of translation theory, this is a recurring theme, while in the research on Eco-translatology, it is an important link used to balance and stabilize the ecology of translation theory research, as "translator-centeredness" should reasonably coexist with the other "two poles," namely, "source-text centeredness" and "target-text centeredness." Only by this means can "source text—translator—target text" be formed into an organic trinity with complementary effects so that the ecology of translation theory research around the triple relationship of "source text—translator—target text" can develop harmoniously.

5. The idea of "translator-centeredness" and Eco-translatological research

Is the translator truly "centered" in the translational eco-environment? It should be pointed out first that there is no statement that "the translator is 'centered' in the translational eco-environment." This is a mistaken expression.

"Anthropocentrism" in Eco-translatological research has been mentioned as evidence supporting the doubt of "translator-centeredness." However, the idea of "translator-centeredness" in the translation process has nothing to do with "anthropocentrism" in ecological research, as these are completely different concepts.

According to the so-called "anthropocentrism," the interests of human beings are regarded as the basis of the source-text value and the evaluation of virtue, and only human beings can be the subject of value judgments. The concepts of "anthropocentrism" arc as follows. (1) In the value relationship of humankind and nature, only human beings, who possess consciousness, are the subject, and nature is the object. The criteria of value judgment must always be grasped by the human beings, and the "value" mentioned at any time is of "human significance." (2) In the ethical relationship of humankind and nature, human beings are the destination. (3) All human activities are performed to meet the requirements for the survival and development of humankind, and an activity is of no significance if it fails to attain this goal; therefore, human interests should be regarded as the starting point and destination in any case. (4) Ecological "anthropocentrism" is the outcome of human beings reexamining their status in the universe and rejudging the human–nature relationship under the circumstances of the ecological crisis, which has become increasingly severe between the twentieth century and the present day. Its core viewpoint is the proposal that in the human–nature interaction, human interests should be regarded as the fundamental criteria by which to assess the relationship between human beings and their eco-environment.

The idea of "translator-centeredness" in the translation process, however, deals mainly with the demonstration of translator's "dominant" role in the translation process, which refers specifically to the fact that the translator should first "adapt" the translational eco-environment in the translation process and then "select" the target text according to the translational eco-environment. In other words, all the translation activities in the translation process, including the translator's "selective adaptation" and "adaptive selection," are judged and decided by the translator. That is the source-text core viewpoint of "translator-centeredness." It is obvious that the "translator-centeredness" in the translation process has nothing to do with the "anthropocentrism" of natural ecology, and the confusion of the two terms must be avoided.

Moreover, the backgrounds of the two terms are quite different. "Anthropocentrism" in ecology studies was proposed in the twentieth century, especially as ecological crises became increasingly serious, while "translator-centeredness" in the translation process was proposed against the background of the longtime dominance of "source-text centeredness" and "target-text centeredness" in the field of translation studies.

Over a long period, "source-text centeredness" and "target-text centeredness" were highlighted in translation studies. We proposed "translator-centeredness," on the one hand, to be true to the reality of the translation process and translation activity and, on the other hand, to serve as something of a balance, a deconstruction, or a counteraction of "target-text centeredness" and "source-text centeredness." The proposal of "translator-centeredness" broke away from the two extremes of "source-text centeredness" and "target-text centeredness" by means of the transition from "multi-centeredness" to "no-centeredness"; thus, a coexistent multi-eco-community of translation studies has been developed for the overall coordination and balance of the ecological environment of translation studies.

In these aspects, as a scholar pointed out in his speech at the First International Symposium on Eco-translatology, Eco-translatology has a certain function of deconstruction of traditional translation theories, namely, "source-text centeredness" has been deconstructed by "translator-centeredness": "That can be called the double deconstructivity possessed by Eco-translatology, namely, the deconstruction to 'translator-centeredness' and that to 'source-text superiority'" (Wang, 2011, p. 12). However, there are certain laws in translation, and there are also certain principles in translation studies. We are investigating the translation issues from the perspective of eco-reason because we cannot simply or mechanically transfer the natural/ecological contents into the realm of Eco-translational research.

The development of "translator-centeredness" and translator dominance will help to expand the scope of translator research in translation studies and to promote the theoretical layers. Meanwhile, it will stimulate improvement in translators' self-respect and self-discipline as well as enhancing their professional qualities. Because describing and explaining the translator's role in the translation process is one of the most fundamental problems in translation studies, issues such as the analysis of the features of "translator-centeredness," the proof of the notion of "translator-centeredness," the demonstration of the mechanism of "translator restriction," and the construction of the evaluation model of "translator quality" should all be meaningful research topics.

Overall, the translator is the overall sum of the contradictions in the translation process. The notions of "translator-centeredness" and "translator dominance" highlight the translator, who is motivating, sensual, and creative. Thus, translation theories can be established on the factual and concrete basis of the translator. The translator-oriented translation theory, which is related to the development of the translator's ability, will gradually show its superiority.

Inevitably speaking, the notions of "translator-centeredness" and translator dominance should be constantly enriched and improved in translation studies. However, as a point in the trinity of "source text—translator—target text," it could not be shaded any more. From the eco-ethical perspective of Eco-translatology, the basic principles of balance and harmoniousness in the eco-environment should be embodied. Therefore, "translator-centeredness" should coexist and co-develop with the other two points, namely, "source-text centeredness" and "target text centeredness." The reason is that only when "most of the forces are combined together can the new force energy be produced,"[12] while "the one into three"[13] can help to "get rid of the illusion of the dual contradiction and double polarization." Only by this means can the trinity of "ST-translator-TT" be formed into an organic community of triple effect and mutual complement so that Eco-translatological studies can develop evenly and harmoniously.[14]

5.4.2 *"Translator Responsibility" in Translation Ethics*

The above section shows that the notion of "translator-centeredness" in translation activity and the translation process is becoming ripe and rational and has been tested by time and practice. The main theoretical evidence lies in the fact that the development of the notion of "translator-centeredness" is illustrated in two layers: "translator dominance," which is part of the translation process, i.e., the operational micro-layer of Eco-translatological studies, and "translator responsibility," which is part of the translational ethics layer, namely, the macro-layer of reason.

The former layer has been described previously, and the first author has also explored the topic in recent years. Thus, no more discussion is needed in this section.

The latter layer, i.e., "translator responsibility," refers mainly to the responsibility of the translator in the overall translation process. In other words, the translator should, will, and must take responsibility for translation activities. The "translator responsibility" mentioned here is proposed as a notion related chiefly to the macro-layer, to the eco-rational layer, and, especially, to the layer of translational ethics.

From the perspective of Eco-translatology, although the translator, as a member of the translator-community, possesses an "equal dialogue"[15] relationship with other members of the community, yet, the other members do not participate directly in the translation process, nor do they undertake translation activity. Thus, only the translator, as the representative of the translator-community, can take concrete responsibility for the coordination of the interrelationships between the "translational environment," "translational texts" and other members of the "translation-

[12] See *Complete Works of Marx and Engels*, Vol. 25, Beijing: The People's Pressing House, 1972 edition, p. 120.

[13] Pang Pu proposed a premise of *one as three* in an article titled *On One as Three*, which was published in March 2003 by Shanghai Classics Publishing House.

[14] Recently, I read an article in which "translator-centeredness" is interpreted from a philosophical perspective. On the basis of the explanation in *An Approach to Translation as Adaptation and Selection,* the author highlights the core idea of "translator-centeredness" in her exploration, and, combined with the philosophical dialectical unified viewpoint, the unity of the translator with multiple roles, the unity of part and entirety, and the unity of the absolute center and relative center in translating activity are further explored (Li, 2010, p. 2).

[15] "Equal dialogue" refers to an attitude and a manner, while "translator's responsibility" serves as an implementation and a result. In translation, there are many people who need to be consulted in dialogue and even more who will be interfered with or dominated, but there is only one who will take responsibility: that is, the translator.

community." Only by means of "translator responsibility" can translational eco-unity be embodied, with the "translational environment," "translational texts" and "translator-community" interacting in balance and harmony. In this sense, "translator-centeredness" and "translator dominance" in the translation process are merely an embodiment of "translator responsibility" as an ethical principle of Eco-translation in the translation process and translation activity.

The reason is that only by or through the translator can eco-holism be established, and only by or through the translator can eco-reason be highlighted. From the perspective of translational ethics, however, "translator-centeredness" and "translator dominance" in the layers of the translation process and translation activity will be illustrated mostly as "translator responsibility" in the macro-layer of Eco-translatological research or of translational ethics (see also Sect. 3.7.4).

In the ecosystem of the "translator-community," the translator should take responsibility for coordinating the relationships between various sectors, for ensuring eco-reason, and for maintaining the eco-balance and eco-harmony. In other words, it is through the ethical principle of "translator responsibility" that the translator can embody the unified view of Eco-translation, which refers to the balance and coordination of various ecosystems in the general eco-environment of translation. Only by means of the responsibility taken by the translator for all the other units, including the translation texts, "translator-community" and translational eco-environments, and only when the translator examines the interrelationships between him-/herself and all the others from the perspective of eco-holism and eco-reason can he/she incorporate a larger sense of responsibility into his/her translation activity.

Entirety and relevance are emphasized in Eco-translatology in that importance is attached to the balance and harmony of translational ecology. Fundamentally speaking, however, who is supposed to implement, to undertake, or to maintain this translation activity? Only the "translator-community," represented by the translator, can implement it. Only the translator can undertake it. Only the translator's communication and coordination with "the others" can maintain it. That is the translator's responsibility! In this sense, all the rational perceptions, far-reaching statements, and expectations related to translation and translation studies are meaningful if they can be transferred to the translator's sense, are effective if they can be turned into the translator's ability, are implemented if they are regarded as the translator's obligation, and are realized if they become the translator's duty.[16] Without all this, there would be great losses, or even vacancies.

[16] In fact, the multiple responsibilities of the translator are stressed in various statements such as "ethics of representation," "ethics of service," "ethics of communication," "norm-based ethics,' and "ethics of 'commitment'" by Andrew Chesterman and "the reproduction of the original," "accomplishment of the requirement of the commissioners," "the social and cultural regulations conforming to the target language," "the requirements met by the target language reader," and "abiding by professional morals" by Sun Zhili. (See *On Translator's Responsibility* by Sun Zhili, published in *Chinese Translators Journal,* 4th issue of Sun, 2007, pp. 14–18.)

In a sense, Eco-translation ethics is actually a new ethics of 'translator responsibility.' It is on this basis that in Eco-translatology, 'translator responsibility' is defined as an important ethical principle.[17] It is safe to say that the development of the ethical principle of translation is on a solid theoretical basis.

5.4.3 *The Translator's Adaptation and Construction of the Translational Eco-Environments*

5.4.3.1 Eco-Environment

The relationship between the translator and the translational eco-environment is concentrated mainly in the ecological role of the translational eco-environment, the translator's adaptation and selection of the translational eco-environment, and the translator's reform and reconstruction of the translational eco-environment.

First, the translator's survival and development are closely related to the translational eco-environment in which he/she lives. The translator's survival and development are unsustainable at any moment without a translational eco-environment in which he/she lives. The translator will obtain what he/she wants from the translational eco-environment, including "the source text and the world illustrated in the source-text language and target-text language, namely, the mutually connecting and interacting unity of language, communication, culture, and society, as well as the author, the reader and the client" (Hu, 2003, p. 288). This can be considered the so-called "ecological effect" of the translational eco-environment on the translator.

In addition, the change in the translational eco-environment will certainly affect the translator, who is closely related to it. Specifically, changes in the translational eco-environment will be reflected in different layers, such as vocabulary, syntax, discourse, pragmatics, style, culture, and communication, during the translator's creation of the target text. In other words, changes in the translational eco-environment will certainly affect and confine the selection of translation strategies; in other words, the translation strategies will be selected in response to the changes in the translational eco-environment. The acceptance of effect and confinement is again called the translator's "ecological adaptation."

Moreover, the translator's reformation and restructuring of the translational eco-environment are tremendous. A successful translator will highlight "ecological effects," accept "ecological adaptation," and then, in different translational eco-environments, actively select different translation strategies and criteria to realize his/her translation purpose. Meanwhile, in the adaptation and selection of the translational eco-environment, the translator will create various translation strategies and skills; shape rich and colorful ideas; and then actively adjust, dominate, construct, and promote changes in the translational eco-environment. In other words, all the

[17] See *Foreign Languages in China,* 6th issue of 2011, p. 98.

information, energy, time, and space utilized for the translator's survival and development can be regarded as the translator's translational eco-resources. In this sense, the nature of translation is the process of the translator's utilization, evaluation, distribution, preservation, processing, dominance, and rebirth of the translational eco-resources, with the purpose of maintaining coordination and balance between the source-text ecology and the target-text ecology and the aim of realizing "doing things with translations" through the translator's efforts and the function of the translated target text.

Eco-translatology, which was developed on the basis of *An Approach to Translation as Adaptation and Selection*, aims to study the translator's powers of adaptation, selection, dominance, and adjustment to the translational eco-environment as well as the translator's ability, translation behavior, and translation effect under the influence of a certain translational eco-environment.

5.4.4 Mechanism of "Post-event Penalty"

The history of translation studies shows that in translation theory, there are two kinds of situations of translator confinement: the "pre-event precaution" and "post-event penalty," both which can be regarded as metaphors used to distinguish two situations for the sake of convenience.

The so-called "pre-event precaution"[18] refers to the "regulations," "restrictions," and "criteria" that serve as precautions for translation in translation theory (Arrojo, 1998, p. 26). The "regulations," "restrictions," and "criteria" are usually expressed in prescriptive language, with imperative terms (such as "must," "shall" "never will," or "may not") and with many strict rules, directions, or corrections; to put it vividly, the "pre-event precaution" is similar to putting the translator under firm control, or putting the translator into a straitjacket (Gentzler, 1993, p. 144). From this perspective, translation should be undertaken cautiously without breaking any rules. "The translator should seek equivalence timidly" (Fan, 2000, p. 14). Researchers are always concerned with "what translation should be, or what translation can be, and they are discussing some ideal criteria without analyzing and researching seriously what most translators are doing, how on earth to translate, or what the translation process should be like" (Liao, 2001, p. 389). Affected by the traditional view of value, "the translation criteria would be worked out beforehand, with which to evaluate the research object" (Zhang, 2002, p. 1).

The so-called "post-event penalty" is also a metaphor, referring to the judgment and disposition of the target text after the translation activity. It refers not to the beforehand "direction," "warning," or "confinement" of the translation activity but to the translator's active "dominance/management/handling/rewriting" of each

[18] It refers not to the concrete translation methods or deeds but to the requirement of translating or directing thoughts.

stage of the translation activity; that is, all is left to the translator for his/her adaptive selection. After the translator's selection at each stage, the final selection and adaptation will be made by the translator according to the principles of "survival of the fittest" and "elimination of the weak" applied in the translational eco-environment, and the translator as "the center," "the dominator," or "the ruler" will fully embody the translation activity.

Therefore, the final selection and arbitration of the translator in the translational eco-environment are closely related, and very important, to the implementation of the translation views of "translator-centeredness" and "translator dominance." The reason is that from the perspective of "translator-centeredness" and "translator dominance," the translator's intelligent work and subjective creativity are accepted, and the translator's "power of dominance" becomes greater, and, correspondingly, his/her "responsibility" (Venuti, 1995, p. 290) becomes greater as well. Thus, the translator needs not only "self-esteem" but also "self-discipline" and especially "the confinement of others." Obviously, in the implementation of the doctrine of the translator as "the center," "the dominator," and "the ruler," the corresponding mechanism of "confinement" is of great necessity to the translator. As Thomas Kuhn points out, "To accomplish his/her task, the scientist must be given a series of restrictions ideologically and operatively" (Kuhn, 1962; Fu, 2002, p. 403). Therefore, the "post-event penalty" serves as a "confinement of others" or "restriction" of the translator's selection.

For instance, "although we have spent large quantities of human labor and materials—organizing first-class translators such as Yang Xianyi and Gladys Yang and foreign experts, providing the best printing facilities and paper, and, meanwhile, investing large sums, the translated target texts are far behind the expectations in their communication, influence and acceptance: *The Dreams of the Red Mansions* translated by the Yangs is ignored in the English-speaking world, while the target text by David Hawks was mostly accepted by Western readers; fewer and fewer readers wish to read our *Panda Series*, which aims to introduce modern Chinese literature in foreign-language editions; and *The Journal of Chinese Literature* in the English edition and French edition, which had been existed for half a century or more, had to stop printing, sadly, for various reasons" (Xie, 2013, p. 6).

Obviously, the confining mechanism stressed in *An Approach to Translation as Adaptation and Selection* highlights not the "pre event precaution" but the "post-event penalty," which can be universally seen in translation activities (See *An Approach to Translation as Adaptation and Selection*, pp. 111–117).

5.4.5 *The Translator's Survival Conditions and Ability Development*

Survival and development are eternal themes for human beings. The historical impact and significance of human existence can be demonstrated and realized in development. The true significance of human life is the pursuit of development.

The exploration of the translator's survival conditions from the philosophical perspective and against the global background will not only show the alteration of the translator's survival conditions but also reflect the development of Eco-translatology and rich historical connotations in the translator's survival conditions and ability development.

The translator's survival conditions are not only the source of the translator's individual esteem, admiration, and protection but also the source of the tension, pressure, conflict, and frustration that directly affect the construction of the translator's individual spiritual paradise and, at any time, affect and confine the selection and implementation of the translator's individual behavior.

In turning from the traditional to the modern and from localization to globalization, modern translators are independent social, historical subjects who are self-governed, self-sufficient, self-supporting, and self-disciplining. Translators rely on their own ability and struggle for their survival and development, to extend their living space and to realize their own value. Their views of values and aesthetic interests are changing, and their modes of thought are changing from the isolated to the open, from the single to the multiple, and from the static to the dynamic. Together with the establishment of translatology as an independent discipline, the subjective sense of the "translator-community" is obviously stronger, and the communication stage is wider.

One of the research interests of Eco-translatology is the relationship between the translator and the translational eco-environment, including the overall theme of translation from the perspective of the translator. In such research, we should study not only how the translator should comprehend "the source text" and express "the target text" in the translation process but also the translator's exertion, ability stimulation, achievements and contributions, survival conditions, emancipation, development, freedom, rights, values, individuality, and subjectivity as well as the survival experience and human revelation tendency of translation theory, aesthetic interests, how the translator becomes him-/herself in the creative work of translation activity, how to solve the contradictions and problems arising in the process of the translator's development, etc. All of these research outcomes are the specific embodiment of the translator's subjectivity. The translators seek survival, development, and their maximum capacity of freedom of "selection," so that they can maintain their "central" status in translation activity as well as the translation process.

There are many changes in the development of the translator's ability. First, the enhancement of the translator's ability is focused on improving the translation ability and professional ability of the translator. Undoubtedly, such improvement is necessary and important. However, it is not sufficient according to the changes in the situation and the requirements of the actual life. The development of the translator's ability includes the improvement of the translator's learning ability, job-hunting ability, work-shifting ability, and creative ability as well as self-governing ability, the ability to use tools and to interact with the outside world, and the abilities exerted in different social groups. Meanwhile, regular education, irregular education, and whole-life education should be relied onto enhance the translator's ability.

5.4.6 The Translator, "Translator-Community" and "Translator-Community" Ecology

We know that the "translator-community" refers to the collection of various factors that affect and interact with the occurrence, development, operation, outcome, function, and effect of certain translation activities and is wholly related to translation activities. It is a human-related collection in which the source-text author, the target-text reader, the target-text critics, the target-text inspector, the target-text publisher, the target-text seller, and the translation donor and clientage included. The translator is naturally the representative of the group.

The translator-community is an important component of the translational ecosystem. In ecological terms, the translational eco-community is composed of producers, consumers, and dissolvers. The producer is the translator, who, as the creator of the translation product, is in the most central position in the translator-community; the consumer, including the translator and the target-text reader, is the user of the translation product; the dissolvers the translation researcher. Therefore, the factors related to translation, including the source-text author, the target-text reader, the target-text critics, the target-text inspector, the target-text publisher, the target-text seller, and the translation donor and client, are interassociated and interconfined in a "person-oriented" ecosystem that is part of the overall translational ecosystem, namely, the "translator-community" ecosystem.

In the "translator-community" ecosystem, the translator, as the representative of the "translator-community," has the responsibility to coordinate the relationships among the various parts (including the interpersonal relationships between various elements of the "translator-community"), to put eco-reason into practice, and to maintain ecological balance and ecological harmony. In addition, through the ethical doctrine of the "translator's responsibility," the translator embodies the integral view of translation ecology as the balanced harmony of various translational ecosystems. Only in this way can the translator be responsible for all the "others," including the translational ecology, textual ecology, and ecology of the "translator-community," perceive the relationship between him-/herself and all the "others" from the perspectives of eco-holism and eco-reason, and incorporate a larger sense of responsibility into his/her translation activities (See Sect. 3.6.4).

5.5 "How to Translate?" from the Perspective of Eco-Translatology

"How to translate" is not only a problem of translation practice but also an inevitable problem in research on translation theory. "How to translate" is a rather complex problem that is related to various elements such as the translator's experience, textual type, translation purpose, target-text reader, client's requirements, and degree of adaptation and selection to translational eco-environments. The issue of "how to translate" from the perspective of Eco-translatology will be comprehensively discussed in Chap. 6. In this section, the problems of "how to translate" will be explained and described from the perspective of Eco-translatology, including the translation process from the perspective of Eco-translatology, the translation doctrine of "multidimensional transformations," the translation method of "tri-dimensional transformations," and the "holistic degree of adaptation and selection" in the translation effect evaluation.

5.5.1 *The Illustration of the "Translation Process"*

Eco-translatology has developed on the basis of the earlier theory of Translation as Adaptation and Selection. The description of the "translation process" in *An Approach to Translation as Adaptation and Selection* is indebted to Darwin's thoughts on "adaptation and selection" (namely, "natural selection") transformed and interpreted in the context of translation practice. The nature of the doctrine of "adaptation and selection" is that any living being possesses the ability to adapt to the natural environment, and only when adapting to the natural environment can the living being survive and reproduce; in other words, any living being's survival and reproduction are but the outcome of the action of natural selection and adaptation to the natural environment. The basic law of the living being's adaptation to the natural environment is survival of the fittest.

The above thoughts and principles are applied to translation in the notion that the translator (or translated target text) should adapt to the translational eco-environment and be confined by the translational eco-environment. On this basis, the translation process in which the translator creates the translated target text can be illustrated as follows.

The target-text creating process can be roughly divided into two stages: the selection of the translator by the "translational eco-environment" (the so-called "selection by nature") and the selection of the target text by the adapted "translational eco-environment" (the so-called "human selection").

According to the basic principle of "natural selection," the selection of the translator by the translational eco-environment with the source text as the typically important component is highlighted in the first stage of selection. This stage can also be regarded as that of the translator's adaptation to the translational eco-environment, namely, the reader's adaptation.

In the second stage of selection, the translated target text is selected by the adapted translational eco-environment with the translator as the typically important component since the translator is highlighted in the first stage of selection by the "translational eco-environment." In other words, the translator has become able to implement selections at this stage in the capacity of the translational eco-environment, and the accumulation of the selected outcomes results in the produced target text. (See Fig. 3.2 above in this book).

In Fig. 3.2, the dotted-line square refers to the translational eco-environment. Roughly, the illustration of the translation process conveys a fundamental meaning, that is, the *translation process is dominated by the translator; the translation process is the translator's "adaptation" to the translational eco-environment with the source text as the typically important component; and the target text is "selected" by the adapted translational eco-environment with the translator as the typically important component.* This can be taken as the source-text meaning of the approach to Translation as Adaptation and Selection.

The small square in the dotted-line square on the left side of Fig. 3.2 is "the source text." The thin arrow beneath the dotted-line square is pointing to the translator, demonstrating that the translation operation at this stage is the selection of the translator by the translational eco-environment with the source text as the typically important component (namely, the selection of the translator). The arrow pointing to the translator beneath the dotted-line square is a thin line, illustrating that it refers to the selection of the translator by the translational eco-environment with the source text as the typically important component, which, furthermore, by contrast with the thick line above, highlights the fact that the operational stage is the translator's adaptation of the translational eco-environment with the source text as the typically important component. The small square on the right side of Fig. 3.2 is "the target text." The arrow pointing at the target text from the dotted-line square around the translator is a thick line, illustrating that this is the stage of the selection of the target text by the translational eco-environment with the translator as the typically important component (namely, the selection of the translator).

In the above explanation of the translation process, the translator is worth highlighting. As the translator should be selected in the first stage by the translational eco-environment, the translator is not included in the translational eco-environment of the first stage. In the second stage, however, as changes in the status of the translator have taken place, the translator is no longer "***a** translator*" in a general sense but has become "*the translator*" who has "accepted the selection of the translational eco-environment" or has "adapted to the confinement of the translational eco-environment." Hence, the translator at this time not only is an ordinary member of the translational eco-environment who can create the target text but also has become a "typically important component" in the translational eco-environment who can implement the selection of the target text as the embodiment and representative of the translational eco-environment. The translator here possesses a dynamic "double identity": On the one hand, he/she accepts the selection and confinement of the translational eco-environment, and on the other hand, he/she implements the selection and dominance of the target text as the representative of the adapted translational eco-environment.

For instance, the question "Who translates?" is proposed by Professor Xie (2013) in an article titled *To Look at Translation from Another Perspective—My Reflection on Mo Yan Winning the Nobel Prize for Literature*. I think that it is feasible to consider the question from the perspective of Eco-translatology. If the source text in the translational eco-environment is Mo Yan's novels, then what will be selected in the translational eco-environment with the source text as a typically important component is the translator who should be an expert in translating Chinese novels. Meanwhile, if Mo Yan's novels in the translational eco-environment are intended to be translated into English for readers in English-speaking countries, then what the translational eco-environment with the source text as the typically important component selects should be the translator who is good at grasping the nuanced speaking habits, unique wording preferences, and particular aesthetic flavor of readers in English-speaking countries. Otherwise, if the translator fails to understand the language, culture, and society of English-speaking countries, fails to understand the Chinese novel, or fails to be good at translation, then, according to the doctrines of "survival of the fittest" and "elimination of the weak," the translator will have difficulty being selected by the translational eco-environment with the source text as the typically important component, or the translator will be "naturally" eliminated at this point (that is, he/she will fail to be universally selected). In fact, the translator of Mo Yan's novels is Howard Goldblatt, who, with the title of *Chief Translator of Modern and Contemporary Literature* and with English as his native language, is second to none in his advantage as the translator. Therefore, Howard Goldblatt was selected by the translational eco-environment with Mo Yan's novels as the typically important component, or was "universally selected." In other words, Howard adapted to the translational eco-environment with Mo Yan's novels as the typically important component better than other translators did. This is the first stage of the selection of the translator in the "translational eco-environment."

In the second stage, namely, the stage of the selection of the target text by the translational eco-environment with the translator as the typically important component, since Howard Goldblatt, in the first stage, "adapted" himself to the translational eco-environment with Mo Yan's works as the typically important component, he could, with the "identity" of the translational eco-environment, implement "selection" (human selection) before the target texts are created with the accumulation of the selection results. The above description is the explanation of the translation process as "selection by nature" to "human selection." Howard Goldblatt's translation is an example in that Mo Yan's works have been successfully transplanted from the eco-environment of Chinese literature to that of Western literature.

Briefly speaking, the translation process, which is interpreted from the perspective of "adaptation" and "selection," is the adaptation and selection of the translator. On this basis, in Eco-translatology, translation is defined as "the activity of selection in which the translation adapts the translational eco-environment to transplant the text with the translator as the dominator, with the text as the basis, and with the transformation of cross-cultural information as the doctrine." Here, what the translator "adapts" to is the "world" (namely, the translational eco-environment) demonstrated by the source text, the source-text language, and the target-text language, and what the translator "selects" is the target text of "multidimensional selections" (at the linguistic, cultural, communicative, aesthetic, pragmatic, and other levels) in the target-text eco-environment.

The definition is expressed by the following formula: the translation process = the translator's adaptation (the "world" demonstrated by the source text, the source-text language, and the target-text language, namely, "adaptation" to the translational eco-environment) + the translator's selection (the selection of the degree of adaptation to the translational eco-environment and the selection of the final wording of the target text).

Another formula is expressed more simply: the translation process = the translator's adaptation + the translator's selection.

The most simplified formula is as follows: Translation = Adaptation + Selection, i.e., (T=A+S).[19]

5.5.2 *The Theoretical Basis for Three-Dimensional Transformation*

The translation method of Eco-translatology can be called "multidimensional" transformation, with a focus on "three-dimensional" transformation. In other words, in the principle of "multidimensional adaptation and multidimensional selection," the transformation of adaptive selection is relatively focused on the lingual dimension, cultural dimension and communicative dimension. In terms of translation as "textual transformation," what is concerned in textual transformation is the textual transformation in the "three-dimensional" (the linguistic dimension, cultural dimension, and communicative dimension) eco-environment in the system of the source-text language and the target-text language.

It is theory-oriented and practice-based to briefly generalize one of the translation methods of Eco-translatology as the "three-dimensional" transformation:

1. From the theoretical perspective, the linguistic, cultural, and communicative translation approaches are based on systematic studies on translation practice, while language, culture, and communication have always been highlighted by translation theorists. For instance, from the perspective of functional linguistics, the linguistic dimension refers to the concern for the lingual expression of the translated target text → the cultural dimension, for the contextual effect of trans-

[19] When Prof. Song Zhiping comments on *An Approach to Translation as Adaptation and Selection*, he holds that although the complex translation activity has been expressed as T = A + S, yet the translator's central role and dominant status have not been highlighted. After all, the S (selection) looks the same as "selection" in the theory of evolution, and the author keeps re-stating, "In the process of translating operation, all the activities of adaptation and selection should be decided and implemented by the translator" (2004, p. 100); therefore, if the notion of "the translator as the center" is illustrated in the formula of T = tA + tS, the dominating "human" element is added to the formula of natural science, which can not only illustrate the features of research on the humanities and social sciences but also conform to the author's core idea of "the translation process = the translator's adaptation + the translator's selection" (2004, p. 180) (Song, 2009). However, I have a different opinion because (1) "adaptation" and "selection" in the translation process are implemented by the translator, and all the discussions in this book are related to the translator's adaptation and selection; therefore, the word translator need not be added to the formula; and (2) since T is used for translation, there will be two Ts (T/t) in the brief formula if another T is confusingly used for translator. Therefore, the original formula has been retained here.

lation → the communicative dimension, and for the interpersonal intention of translation. From this, we can see that there are, to a certain degree, relevance and connection to A. Halliday's ideational theory, interpersonal theory, and textual theory as well as other theoretical scopes of the lingual field, lingual meaning, and lingual patterns.
2. From the practical perspective, language, culture, and communication have universally been acknowledged by the translation world as the focus of translation studies and are related to the perspectives from which the key transformations are needed in the translation process; the translator is apt to conduct the transformation of adaptive selection according to the different stages or orders of language, culture, and communication.
3. From the logical perspective, translation is the lingual transformation, language is the carrier of the culture, and culture is the accumulation of communication. Hence, language, culture, and communication have internal logical relevance, which is embodied as the basic content of the translation transformation.
4. From the perspective of the maintenance of "textual" ecology, the translator, by means of "selective adaptation" and "adaptive selection," should bear the responsibility for maintaining and transforming the best of the lingual ecology, cultural ecology, and communicative ecology of the source text; meanwhile, through "selective adaptation" and "adaptive selection," he/she should also take the responsibility for making the best of the transformed lingual ecology, cultural ecology, and communicative ecology survive and last in translational eco-environment of the target-text language. The maintenance of the coordination and balance of the lingual ecology, cultural ecology, and communicative ecology of the source-text language and the target-text language, in turn, contrasts with the "three-dimensional transformation" of the translation operation and finally realizes the "balance" and "harmony" of the source-text language and the target-text language in the lingual, cultural, and communicative ecologies.

The feasibility of undertaking the transformation of adaptive selection from the perspective of the lingual dimension is not to be mentioned here (Hu, 2004a, 2004b, 2004c, 2004d, pp. 134–135). Although translation is "the activity of selection in which the translation adapts the translational eco-environment to transplant the text with the translator as the dominator, with the text as the basis, and with the transformation of cross-cultural information as the doctrine," and the translational eco-environment is composed of multiple dimensions and various elements, this does not mean that other dimensions and other elements cannot play a role in the translation process when a certain dimension or element is stressed for the convenience of description. In fact, in the process of concrete translation, the lingual, cultural, and communicative elements usually interlace, interrelate, and interact, and it is sometimes rather difficult to distinguish them clearly. This conforms not only to the transformation of the lingual dimension in this section but also to that of the cultural dimension, the communicative dimension, or other conditions of adaptive selection.

As the ecology of the source-text culture and that of the target-text culture are different in nature and content, to avoid misunderstanding the source text from the starting point of the target-text culture, the translator needs not only to pay attention to the transformation of the source-text language but also to adapt the entire cultural system, namely, the cultural ecology, to which the language belongs and, in the translation process, to pay attention to the transmission of the connotation of bilingual culture, namely, the so-called "transformation of adaptive selection of the cultural dimension" (Hu, 2004a, 2004b, 2004c, 2004d, pp. 135–137). Overall, the so-called transformation of the adaptive selection of the cultural dimension means that the translator should have a sense of culture in the translation process, realizing that translation is a cross-language and cross-cultural exchange, noticing the obstacles caused by the cultural differences, and striving for balance and harmony between the cultural ecologies of the two languages to guarantee the smooth realization of information exchange.

The adaptive selection transformation of the communicative purpose in the translation process means that the translator, in addition to the transformation of lingual information and transmission of cultural connotations, should focus on the selection of the transformation on the communication layer in that he/she should ensure not only that the author's entire communicative purpose in the system of the source-text language has been embodied in the system of the target-text language or been transmitted to the target-text reader but also that the communicative purpose, including the source-text language/cultural forms and language/cultural connotations in the system of the source-text language, has been conveyed to the reader. Essentially, the adaptive selection transformation of the communicative purpose in the translation process is related to the pursuit of the situation in which the communicative ecology of the source text language and the target-text language will earn the best maintenance.

Although we have emphasized the "three-dimensional" transformation in language, culture, and communication, there are many aspects of adaptive selection in the translation process, with many aspects of adaptive elements. For instance, Dr. Zhou Zhaoxiang once listed more than ten elements: "the author, the author's purpose, information, the source text, truth, society, the communicative channel, the target-text language, the client (the employer, the translation office and the customer), the translator and the target text, etc." (Li, 1996, pp. 9–11). He also pointed out, "Before translation, careful consideration should be given to the various aspects of culture, reader, passage, market, aim, and chance, etc., and after that, one can decide if it will be translated, or if the translation strategies should be taken" (Zhou, 1996, p. 33). In addition, "to adapt the accepting degree, etc., the translation methods should be disposed differently. This is an objective reality not only for nonliterary works but also for literary ones, which need corresponding adjustment according to the ideological patterns and reading habits, etc." (Wang, 2000, pp. 24–39).

From the perspective of the actual situation of translation, statements on translation, from Nida's statement on biblical translation (Nida, 1982) to Daniel Gile's remarks on interpretation (Gile, 1994) and Xu Yuanchong's comments on literary translation (Xu, 1998); from "selective translation," "incomplete translation," "partial translation," "editorial translation," and "revised translation,"[20] which have been repeatedly emphasized by various figures, to "adaptation," "dominance," and "reflection," which have been constantly discussed in recent years, and the target text's "*yi sishen si*"(faithful in content and equivalent in effect), "*yi li shen si*" (unfaithful in content and equivalent in effect), "*yi si shen li*" (faithful in content and nonequivalent in effect), and "*yi li shen li*" (unfaithful in content and nonequivalent in effect), are the cumulative results of the translator's selection of adaptive strategies in order to adapt to the translation eco-environment.

Finally, two points should be stressed: first, the "three-dimensional" transformations of the "lingual dimension," "cultural dimension," and "communicative dimension" discussed above, as stated at the beginning of the section, have been proposed mainly based on translation practice, the development of translation theory, and the inner logic of the three dimensions. There are more transformations in the translation process than in the "three dimensions," but the "three dimensions" should be the most fundamental. Second, this section shows that there are not only multidimensional, multilayered, and multiaspect adaptations and selections in translation but also various contents that interlace, interconnect, and interact in each dimension, each layer, and each aspect. Together with other aspects of the translational eco-environment, especially the interpersonal elements of the "translation-community," such as the author, the reader, the donor, the publisher, and the translation critic, and with the statement that "the issues related to the 'human' elements are always multivariant and complex,"[21] it can be concluded that translation is certainly a complex activity of adaptation/selection and a "mechanism of complexity" (Docherty, 1993, p. 30) and that the translator will never be excessive in his/her efforts to seek the best "multidimensional" selective adaptation and adaptive selection in the translation process.

5.5.3 *The Definition of the "Holistic Degree of Adaptation and Selection"*

The "holistic degree of adaptation and selection" refers to the criteria of evaluation and measurement in translation. It is well-known that the translation criteria most often mentioned are actually related to the translation method. For instance, "*xin*" (faithfulness), "*da*" (expressiveness), and "*ya*" (elegance) serve as the translation criteria in the target-text evaluation and as translation methods in translation process. Similarly, the "theory of multiplicity and complementation" by Gu Zhengkun can be regarded either as the translation criteria or as the translation method; that is, in the translation process, the translator can absorb various translation styles, adopt

[20] See Huang Zhonglian: *Studies on Translation Variants,* Beijing, China Foreign Translation Publishing Company, Huang, 2001.

[21] See Hu (1991, pp. 28–32).

various translation methods, and develop various aspects of his/her translation abilities.

As for the evaluation and measurement of the "holistic degree of adaptation and selection," we refer to the totality of the degree of "selective adaptation" of the lingual dimension, cultural dimension, and communicative dimension when the translator is producing the target text, and the "adaptive selection" with which other elements of the translational eco-environment are concerned.

Under normal circumstances, if the degrees of "selective adaptation" and "adaptive selection" of a certain translated target text are high, the "holistic degree of adaptation and selection" will consequently be high. From the perspective of Eco-translatology, the best translation should be the translation with the highest "holistic degree of adaptation and selection."

The evaluation and measurement of the target text's "holistic degree of adaptation and selection" can be generalized into "three reference indicators": namely, the degree of multidimensional transformation, reader feedback, and translator quality.

The so-called "degree of multidimensional transformation" means the target text's "holistic degree of adaptation and selection," whether or not the translator has multidimensionally adapted to the specific translational eco-environment. In other words, first, whether the source-text ecology has been maintained, at best, or whether the source-text ecology has been destroyed, at a minimum, should be determined; and second, whether the target-text ecology has been maintained, at best, or whether the target-text ecology has been destroyed, at a minimum, should be determined. Since the source-text ecology and the target-text ecology are embodied mainly in the aspects of lingual ecology, cultural ecology, and communicative ecology (see Sects. 3.4.1–3.4.3), the "holistic degree of adaptation and selection" is embodied in whether the translator has realized the "three-dimensional" (lingual dimension, cultural dimension, and communicative dimension) transformation. Only when the translator truly realizes the "three-dimensional" adaptation, or at least the "three-dimensional" selective transformation, in the translation process can he/she create a properly translated product (the target text). Comparatively, the more dimensions of adaptive transformation there are, the greater the selective adaptivity is, and the relatively higher the holistic degree of adaptation and selection of the translated target text possibly is. The reference index shows that the proper target text addresses not only the lingual transformation but also the lingual connotations and communicative purpose.

The so-called "reader feedback" refers to the feedback and criticism of the translated target text by, for example, the general reader, expert reader, translation client, target-text publisher, and target-text critic. In a sense, the reference index is actually the evaluation through "market feedback" of the translated work. In the term of Eco-translatology, the "market feedback" here is only the evaluation of the translated work's survival degree in the target-text language. There are many variants of market feedback as well as many elements. For instance, from the perspective of the purpose of translation criticism, there are elements of "target-text proofreading, appreciation, introduction, assessment scoring, teaching demonstration," etc. From the perspective of the focus of translation criticism, there are the elements of "the author, the author's purpose, the information, the source text, the truth, society, the communication channel, the target-text language, the client, the translator, the target-text user, the target-text," etc. From the perspective of the operation of translation criticism, something should be done to "examine and assess the translator's purpose; to point out the highlighted focus of the translation task according to the purpose; and to analyze and check the target text to see whether the purpose of highlighting the focus has been realized" (Zhou, 1996, pp. 5–16). Generally, the better the "market feedback" is, the higher the target-text's "survival degree" in the target-text language is, and, as has been illustrated, the higher the target text's "holistic degree of adaptation and selection" can be.

The so-called "translator's quality," from the perspective of Eco-translatology regarding research on the translator, briefly includes the previous achievements, experience, abilities, integrity, and reputation of the translator. The translator's quality is embodied in the translator's cross-cultural acuteness, familiarity with the theme, judgment of the translational eco-environment, understanding of the market, background knowledge, translation experience, and working attitude. The significance of the reference index lies in the fact that it can make the criteria of the translation criticism predictable, namely, not only the "text" but also the "translator," and the "context" can considered—namely, the consideration and judgment can be based on the integral translational eco-environment. This discussion illustrates that the criteria of translation criticism in Eco-translatology are multidimensionally complementary and integrally unified, and therefore, should approach the truth of the translation effect.

From the perspective of Eco-translatology, two points should be stressed here. First, in assessing the translator's quality, attention should be paid to the translator's fundamental quality in his/her adaptation to certain translation assignments. Second, in measuring the translator's quality, attention should be paid to the fundamental quality shown in his/her previous translation work in adapting to similar translation assignments.

5.5.4 *Textual Transplants and Ecological Balance*

"Translation as Eco-balance" and "Translation as Textual transplants" (see Sects. 5.3.3 and 5.3.4) can be regarded as the generalization of the nature of translation from the perspective of translation ideas and as the strategy and method of translation from the perspectives of translation behavior and translation operation. Why, then, can the realization of the nature of translation be regarded as a translation strategy or translation skill? The reason is not that the realization of the nature of translation is closely related to the selection of translation strategies and translation skills—a translator will undertake translation in the way he/she views it—but that there are some differences in the perspectives from which the problems can be considered. For instance, "*xin*" (faithfulness), "*da*" (expressiveness), and "*ya*" (elegance), which were proposed by Yan Fu, one of the most famous Chinese translators, can be regarded as translation criteria or criteria for translation criticism[22]; from the perspective of the translation method, however, "*xin*" (faithfulness), "*da*" (expressiveness), and "*ya*" (elegance) can also be referred to and implemented as translation methods in the translation operation. Therefore, the translation ideas of 'Translation as Eco-balance" and "Translation as Textual transplants" in Eco-translatology can also be applied as translation strategies or translation methods—namely, the source text will be fully transplanted into the target-text language with ecological equivalence and balance between the source-text language and the target-text language, including in the lingual ecology, cultural ecology, and communicative ecology.

To maintain and balance the "genes" and "blood" in the source text and the target text, or to keep the genes and blood of the source text alive in the target text, the translator can adopt a "text-retargeting" Eco-translation strategy in the disposition of the text (See Sect. 6.5). The so-called "text-retargeting" translation strategy, from the perspective of Eco-translatology, means that the translator, in the translation process, adapts and reverts to the best of the eco-environment of the source-text language (on the basis of the source-text ecology of the source-text language) to choose the target text or adapts and reverts to the best of the eco-environment of the target-text language (on the basis of the source-text ecology of the target-text language) to choose the target text.

For an instance, to maintain, coordinate, balance, and reconstruct the eco-environment adapted to the source-text language, the translator can, during the translation process, transform or even extract the best of the source-text "ecology" in his/her brain (see Sect. 6.3) to transplant the new eco-environment that is adaptive to the source-text ecology into the target-text language.

Furthermore, to balance the source-text language ecology and the target-text language ecology, the translator should use "selective adaptation" and "adaptive selection" to compensate for the insufficiencies in the target-text language ecology so as to "translate by addition," "illustrate with notes," "supplement with information," "omit for conciseness," or "add for improvement," etc. These translation activities, in Eco-translatological terms, exert the function of "balance" in the target-text language ecology by constructing, repairing, and adjusting in the target-text language the eco-environment for the target text to survive and last (see Sect. 6.4).

The above ideas illustrate that the translator's adaptation and selection in the translation process mean that the translator starts with the inner ecostructure of the source text, conducts the selection of the would-be translated text, and, in the translation process, represents the innate ecostructure of the source text in the other lingual system.

[22] "The translation essence and translation criteria are two interdependent questions, and, so to speak, are two facets of the same question" (Yang, 2012, p. 1).

5.5.5 Eco-Translational Strategy/Optimizing the Selection of Methods

The translator optimizing his/her selection through the Eco-translational strategy and method is mainly illustrated by the translator, on the precondition of adapting to the changes of the translational eco-environment, optimizing alternative translation strategies and skills to obtain a higher "holistic degree of adaptation and selection."

Generally, literal translation, semantic translation, and foreignizing translation are common in their closeness to the source text, while free translation, communicative translation, and domesticating translation are common in their closeness to the target-text language or the target-text reader. They are interwoven with certain differences.

From the perspective of Eco-translatology, the translator, constrained by the source-text language ecology and the target-text language ecology, or on the precondition of adaptation to the translational eco-environment, actively selects to obey or break away from the constraints of the source text or the dominance of the target-text culture—from complete obedience or subjection to partial obedience or subjection—which can be viewed as the selection of different translation strategies by the translator in order to adapt to the translational eco-environment. Therefore, from the perspective of Eco-translatology, no matter whether the selection is foreignization or domestication, literal translation or free translation, semantic translation or communicative translation, formal equivalence or functional equivalence, it can be regarded as the selection of the translation strategy by the translator for adaptation to the translational eco-environment. In Eco-translatology, translation is defined not from the perspective of the author/the source text or from the perspective of the target text/the reader but from the angle of the translator; that is, translation is "the activity of selection in which the translation adapts to the translational eco-environment to transplant the text with the translator as the dominator, with the text as the basis, and with the transformation of cross-cultural information as the doctrine." Therefore, regarding questions such as "Which is better? Foreignization or domestication?" or "Should we translate literally or freely?" we can explain that we should translate for the better—that is, the translator should select the better option for survival. It is simple to work out the answer to the question from the perspective of the adaptation and selection of the translator. As translation is defined as "the activity of selection in which the translator adapts to the translational eco-environment to transplant the text, with the translator as the dominator, with the text as the basis, and with the transformation of cross-cultural information as the doctrine," while the translational eco-environment, including the society, culture, and various elements, is in constant alteration, it is natural for the translator to adapt to the dynamic and ever-changing translational eco-environment and to make the selection between domestication and foreignization or between literal translation and free translation that corresponds to the translational eco-environment.

To answer the question of "how to translate" more concretely, or for the text transformation at the microlevel from the perspective of Eco-translatology, references have been given in various sections of Chapter Five on the concrete illustration of Eco-translational strategies and methods, including "transformation," "emptiness," "supplementary construction," "text retargeting," "bionics," and "interference."

5.6 "Why to Translate?" from the Perspective of Eco-Translatology

In my paper titled *On Initial Exploration of An Approach to Translation as Adaptation and Selection*, presented at the "The Third FIT Asian Translators' Symposium" held in Hong Kong in 2001, the phrase "doing things with translations" was used for the first time.[23] This proposition appears in my doctoral dissertation, titled *On Exploration of An Approach to Translation as Adaptation and Selection with the Translator as the Center* (p. 28), which was completed in September 2003 at Hong Kong Baptist University. The formulation of "doing things with translations" was specially noted in my monograph titled *An Approach to Translation as Adaptation and Selection*, published by Hubei Education Press in June 2004 (Hu, 2004a, 2004b, 2004c, 2004d, pp. 10, 98) and was listed as a special key term in Appendix V of that book in the form of English-Chinese contrast (ibid, 218–219).

Although "doing things with translations" has been quoted by other scholars and used in related published papers,[24] however, the systematic demonstration and

[23] The title (*How to Do Things with Words*) of Longfellow J. Austin's academic work was translated by Xu Guozhang into "论言有所为"(*lun yan you suo wei*). As translation is creative work, and translations possess the function of changing one's ideas and moving civilization forward, which "can not only expand the discourse of translation theory for the renewal of the theories, but (by means of translational practice) also expand the culture of the accepted side for a better social reform" (Gentzler, 1993, p. 197), I, in imitation of the above English pattern of the term, used the term *yi you suo wei* for the first time in 2001 by translating it as "doing things with translations" (Hu, 2001, 2003).

[24] To generalize and further describe the connotations of "doing things with translations," I once searched for the term with the *Baidu* search engine and found hundreds of pieces of information related to the term. However, after a careful examination of the relevant articles, I found that except for very few articles in which the term of "doing things with translations" developed by me was quoted or similar expressions (such as "to do something" or "not to do something") were used, most such terms are related to a translated work titled "译有所为—功能翻译理论阐释" in Chinese. By contrast and analysis, "译有所为" in the title of the work is quite different from the term developed by me (see the footnotes at p. 246) in that, first, "译有所为—功能翻译理论阐释" is a translated work; its original author is Christiane Nord, a German scholar, and its chief translators are Prof. Zhang Meiifang and Prof. Wang Kefei. The original work is titled *Translating as a Purposeful Activity: Functionalist Approaches Explained* ("翻译即一种有目的的活动:功能主义翻译途径阐释"; "译有所为" is merely the expression used by the two translators to understand

description of the proposition are still insufficient. To compensate for the incomplete aspects, the issue of "doing things with translations" will be further discussed in this section in relation to such aspects as the subjective motivation of the translator, the objective effect of the target text, and the spaces for future development.

Generally, "doing things with translations," as described in *An Approach to Translation as Adaptation and Selection*, is illustrated in two main aspects: first, the translator has a certain motivation (focus on subjective motivation) in undertaking translation activity; second, something can be done with the translated target text (focus on objective effect).

5.6.1 Translator's Subjective Motivation in "Doing Things with Translations"

The translator undertakes translation activity for fame and reputation, for love of ethical virtue, for reasons rooted in religious faith, for a thirst for universal truth and to meet his/her daily requirements. Therefore, from the perspective of the translator's subjective motivation, the reason for "doing things with translation" can be (1) "survival," (2) the "realization" of ideals, (3) the "pursuit of the interesting," (4) the "transfer" of moods; and (5) "competition".

1. "Doing things" for "survival"

For the translator, "survival" refers to the engagement in translation as a means of earning his/her living. For instance, Zhu Shenghao, the master translator, "had to translate to earn a certain amount of fees to support the life of his family with five members, as he, apart from a small sum of rent, hardly had any other income" (Zhu & Wu, 1988, p. 32). "Translators are human beings with flesh-and-blood bodies in that they, too, want to live, to take their family and children into consideration, to strive for the better by wandering about between the cities and different cultures" (Pym, 1998; Ke, 2002, p. 31). "Translators are human beings, too, in that they should live and pursue entertainment" (Gu, 2003, p. 359). It was said in ancient China that "the senses of glory and humiliation depend on sufficiency of clothing and food". It is evident that "survival for all living creatures is the primary demand" (Qian, 2001, p. 18). These remarks illustrate that "all that people strive for is related to their interests".[25]

2. "Doing things" for the "realization" of ideals

and translate the words "translating as a purposeful activity" in the title of the original work. Second, the main content of the translated work is the introduction of the functionalist theory of the German school of translation studies. Third, the translated work was published in October 2005.

[25] See *Complete Works of Marx and Engels* (Vol. I), Beijing: people's Publishing House, 1956, p. 82.

The "realization" of ideals refers to conducted translation based on the ideals of the translator. For instance, over a few years at the turn of the nineteenth and twentieth centuries, Gu Hongming translated *The Analects* and *The Doctrine of the Mean* into English; these works were published and then printed abroad, resulting in large book sales and widespread influence. Gu Hongming was undoubtedly an upholder of Chinese culture and a promoter of Chinese civilization. Yan Fu, who lived in the same period as Gu Hongming, realized his far-reaching political aspirations by means of his translation of "the Western classics." He accomplished his translation of *Evolution and Ethics,* by which he developed his viewpoint of survival for the better and saving China from extinction according to the evolutionary theory of "selection by natural competition" and "survival of the fittest" and promoted gathering the people's power, developing the people's intelligence, renewing the people's morality, strengthening the nation independently, and saving China from extinction. As another example, Fu Lei, too, had very clear initial motivations in his pursuit of translation, namely, to help his people seek light in the darkness; to enhance the courage and confidence of youth; to stimulate people's love for the world, human life and all beautiful things, and to enhance the intelligence of the Chinese people. In summary, he intended to broaden the vision of his Chinese readers and to revitalize the Chinese people.

3. "Doing things" for the "pursuit of the interesting"

The "pursuit of the interesting" refers to how the translator's interests or hobbies affect his/her translation choices. For instance, Fang Zhong, one of the China's famous translators, "began translating Chaucer out of interest," He said, "Chaucer began to draw my interest because of his narrative poems, novels and stories, especially *The Canterbury Tales*. I was more interested in a narrative poem he created in his early years" (Fang, 1989, p. 112). Yang Yi, another famous translator, said, "At first, I must be interested in a book, an article or a poem before I have love for it and have a thirst for translating it to the readers" (Yang, 1989, p. 312). Xu (1998), a famous translator in China, proposes that a good translator should have a love for translation and feel pleasure in translation. Many translators do enter the realm of translation because of their interest in the source-text author, the source-text works, or translation itself.

4. "Doing things" for the "transfer of moods"

The "transfer of moods" refers to translation for the purpose of transferring a mood (usually sadness). At the age of eighteen, Lin Shu, "the king of translation" in early modern China, married Liu Qiongzi. They proved to be a model couple with deep love for each other. When Liu Qiongzi died from a disease at the age of 46, Lin Shu was preoccupied with great sorrow each day. Then, his friend Wang Shouchang returned from France with a novel titled *The Lady of the Camellias* by Alexandre Dumas fils. Wang invited Lin Shu to translate the novel to alleviate his painful sorrow in mourning for his wife, and Lin Shu agreed. Another instance is Mu Dan, one of the great translators in China. After the 1950s, he stopped his poetic creation of translation for the rest of his life because of the constraints of the political conditions of the time. In 1958, he was labeled a *preliberation counter-revolutionary*, for which he was confined and criticized, and lived as a prisoner for more than 10 years. It was the physical torture and mental pain, however, that made him strive for the translation of Russian and English poems. Mu Dan, so to speak, "translated" his great sorrow into his painful and great target texts.

5. "Doing things" for "competition"

"Competition" refers to translation for the better (with the intention of competing or the ambition to defeat other translators in translation). For instance, Xu (1982) thinks that literary translation is a competition between the two languages and even the two cultures for which the translator should give full play to the advantages of the target-text language by using the best expressions of the target-text language. He regards the truth-beauty relationship in literary translation as a dynamic adjustment and balance; it is by means of "competition" that the target text can converge with the expectations of potential readers. Such "competition" by "advantageous" means could be seen in ancient times. For instance, Horatius (65-8 BC), a literary critic and translation theorist in ancient Rome, upheld "the superiority of the target text," and Quintilianus (35-97 AD), a rhetorician and translation theorist in ancient Rome declared for the first time that the target text should be superior to the source text as a result of competition (Tan, 1991).

Translators engage in translation with other various motivations and aims; for example, some translators translate for "education,"[26] for "examination," or for "promotion," and there are many cases of a "combination" of different purposes. However, generally, the connotation of "doing things with translations" is explained in this way from the perspective of the subjective motivation of competition.

5.6.2 *The Textual Objective Function of "Doing Things with Translations"*

The translation that is discussed here includes various types of translation, such as oral translation (interpreting), written translation, and machine translation. Undoubtedly, the tremendous functions and great achievements of translation are undeniable. However, from the perspective of "doing things with translations," what can translation truly "do"?

1. Translation is "done" to promote communication and exchange

Human translation activities have existed since different human groups speaking different languages first connected to one another. The relationship between two primitive tribes, in hostility or friendship, depended on translation for the

[26] For instance, some persons want to learn a foreign language and engage in translation work in order to teach or guide their own children.

communication of languages and thoughts to achieve mutual understanding. Naturally, the translation activity performed at that time was merely oral translation. Along with the continuous development of human history and society, human groups and nations speaking different languages began to develop written languages; hence, not only oral translation but also written translation was needed for mutual communication.

According to reliable historical records, Chinese written translation started with *Ode to the Yue People* (*yue ren ge*) during the Spring and Autumn period approximately 2500 years ago. During the long historical period since then, written translation in China, in terms of scale and size, experienced four consecutive stages: (1) the translation of Buddhist scriptures from the East Han Dynasty to the Sui and Tang Dynasties; (2) the translation of science and technology during the Ming and Qing Dynasties; (3) the translation of foreign literature and social science during the period of the May Fourth Movement (1919); and (4) the surge of all-around translation from 1949 (the founding of the People's Republic of China) to the present day. In these four periods, many translators accumulated abundant translation experience and translated many Buddhist scriptures as well as numerous works of literature, politics, science and technology; China produced large quantities of excellent translators who were great indebted to the communication, understanding and exchanges of Sino-Western civilizations.

Translation in the West can also be divided into four great epochs. The first great epoch refers to translation during the ancient Greek and Rome period. Generally, our knowledge of Roman culture, to a great degree, is attributable to translation. The second great epoch took place from the eleventh century to the sixteenth century. The third great epoch was mainly the period of Enlightenment and Romanticism (mainly the seventeenth and eighteenth centuries). The development of Western culture at the time benefited from translation, which was continually expanding. The fourth great epoch has emerged in modern times, mainly in terms of translation activities since the end of World War II. In this period, mainly owing to interest in various subjects such as economics, politics, and military affairs, relevant works could be introduced by means of translation. Therefore, the translation scale is larger than that of translation in any previous period.

Translation has made it possible for human beings to exchange ideas and communicate in language, thought, and behavior. In a broad vision, we can perceive the human history of translation as lingual and cultural contexts exerting influence upon one another. In the history of translation, both in China and abroad, translation activities have become increasingly larger in scale, abundant in content, and wider in coverage, communicating, exchanging, and mingling continuously to enable the coexistence of human beings. Thus, translation has certainly “done” things that it should do.

2. Translation is “done” to stimulate lingual innovation

From the source-text language system to the target-text language system and by means of translation, there are countless cases of innovation in terms of words, sentences, passages, genres, and meaning. Buddhist translation in the early period in China directly influenced at least three aspects of the Chinese language.

First, the Buddhist translation influenced the Chinese vocabulary. According to the historical records, just from the Han Dynasty to the Tang Dynasty, as many as 35,000 new Chinese characters were created because of translation. In other words, just by means of Buddhist translation in the early period, the Chinese vocabulary permanently increased by 35,000 or more words. Many expressions in colloquial Chinese come from the Buddhist scriptures, but we fail to sense this association because of our habitual use. The expansion of vocabulary means the expansion of the content of thoughts and of our ability to realize things.

Second, Buddhist translation influenced the Chinese style. For instance, many typical uses of Chinese conjunctions or prepositions, such as "*zhi,*" "*hu,*" "*zhe,*" and "*ye,*" originated in the classic writings of ancient China, along with the usage of parallel style. This kind of Chinese syntax was affected by the target texts of the Buddhist scriptures simply because Buddhism, which originated abroad, did not have the same sentence structure as the regular Chinese lingual pattern and was translated in accordance with the source-text syntax style. Consequently, this style was gradually accepted through translation, and the source-text Chinese style gradually changed. This shows that translation played a great role in promoting the modernization of the Chinese language. In addition, the Buddhist scriptures were finely categorized into chapters and sections, as the writings in the languages of the Indo-European school featured detailed structures that are illustrated in the translation of Buddhist scriptures. This not only made it possible for the Chinese people to learn the new pattern of writing but also enriched the Chinese structure of thought.

Third, Buddhist translation influenced the themes of Chinese literature (such as literary theory). When Chinese scholars discussed literary theory, they adopted some terms and expressions from the Buddhist scriptures. For instance, the literary term "*hua jing*" (natural transformation), which is popular in the field of translation theory, was proposed by Qian Zhongshu, a noted Chinese scholar, and originated in Buddhism. The doctrines of "flavor" and "spirituality" are also related to Buddhist and Zen theory. Meanwhile, the themes of many literary works, such as *A Journey to the West*, one of the Four Classic Works in China, come from the Buddhist scriptures.

The same is true in the West. The period of the Renaissance, especially in the sixteenth century, witnessed the peak of translation activity. Martin Luther, the German religious reformer and translator, translated the first people's Bible. Before he translated the Bible into the common language, it was written in Latin, which was difficult for the common people to read. The target text by Luther brought the German language into a new era (Gu, 2003, pp. 358–484). A translated target text should reform the language of a country! From the above examples, we can see the role of translation

3. Translation is "done" to stimulate cultural evolution

As a bridge to connect different cultures, translation can undoubtedly stimulate the evolution, progress, and combination of culture. Chinese culture has been absorbing nutrition from foreign cultures in the development process to fulfill and enrich itself and to directly or indirectly transport new cultural achievements to near or distant countries or regions; this process contributes to the cultures of all human beings. Taking the successful translation of the Buddhist scriptures in Chinese history as an example, the translation of Buddhist scripture in China began with An Shigao, who began translating the Buddhist scriptures in 148 AD during the reign of Jianhe of the East Han Dynasty and ended in 1287 (the 24th year of the reign of Zhiyuan of the Yuan Dynasty), when *Yuan Zhi Yuan*, the Tibetan-Chinese compiled bibliography of Buddhist scriptures, was compiled. The translation of Buddhist scriptures lasted for more than one thousand years, resulting in a great variety and large quantity of translated target texts. Hence, the translation of Buddhist scriptures serves as a translation action with a long history. With so many translators and translated target texts, China's translation of Buddhist scriptures constitutes a unique achievement in the world history of translation, although no exact statistics are available. "In the 32 volumes of *Taisho Tripitaka*, which was compiled from 1912 to 1925 in Japan, 1658 of the Buddhist Scriptures (in 6100 volumes or more) were collected with a large number of Buddhist Scriptures uncollected" (Sheng & Wang, 1984, pp. i–ii). "The translators in ancient China played important roles in the forums on the translation of the Buddhist Scriptures. They not only gave profound speeches on the Buddhist Mottos, but also translated a lot of Buddhist Scriptures into Chinese" (Baker, 1998, p. xvi). The translation activity evidently influenced Chinese culture in that the introduction of Buddhism changed the cultural balance of China, in which Confucianism and Taoism had formerly been dominant, by creating a cultural trinity of Confucianism, Taoism, and Buddhism. Overall, it was translation that led China to the new cultural trinity; thus, the impact of translation cannot be denied!

Prof. Ji Xianlin, a modern Chinese master of cultural studies who recently passed away, once made a wonderful statement on Macau's status and role in Sino-Western cultural communication: "In five thousand years of the Chinese history, there were several climaxes of cultural communication. The last and the most important one was the introduction of the Western culture, which began at Macau at the end of the Ming Dynasty and the beginning of the Qing Dynasty" (Ji, 1999). Historical studies have demonstrated that Macau was the bridge for the most important Sino-Western cultural communication in 287 years (from 1553 to 1840). The important means of Sino-Western cultural communication in this period was translation activities by Macau translators, who for the most part were preachers.

Since the May Fourth Movement, Chinese history has seen three peaks of foreign literature translation. The first peak occurred during the period of the May Fourth Movement, the second in the early years after the founding of the People's Republic of China, and the third during the period of reform and opening up. During the period of the May Fourth Movement, many foreign literary works were translated into Chinese. The May Fourth Movement was an anti-imperialist and anti-feudalist thought liberation movement in that two great banners promoting the new morality and new literature were held aloft. In the formation of the new thoughts and new morality, the translation and introduction of foreign literature played an important role. Since the foundation of the People's Republic of China, the translation and introduction of foreign literature have been thrived. Since the beginning of the reform and opening up period, the categories and volume of translated target texts of foreign literature have greatly increased, and the quality of the target texts has also increased. The large quantities of translated and introduced foreign literature have played an important role in enriching the culture, promoting aesthetic standards, increasing international understanding, widening vision and modes of thought, improving writing techniques, acquiring information about nutrition, etc.

Overall, it is through translation activity that new thoughts, new concepts, new technology, new artistic schools, and new expression skills have been introduced, that the target-text language has been enriched, that the academic atmosphere has been revitalized, and that cultural communication and progress have been promoted.

4. Translation is "done" to stimulate social reform

There are countless cases of translated target texts changing people's concepts and behavior and promoting social evolution with translation as the medium. Dr. Simon Chau held that translation work is "the vanguard of human civilization," and he generalized the process of the transfer of the human civilization as a chain: new information → new viewpoint → change of decision → individual behavior → social reform (in interaction with system transformation) → transformation of civilization. He pointed out that translators in the twenty-first century "tend to become the key leader in the fore of social reform by introducing the latest materials, theories, thoughts, and customs from other places by publishing them by the most effective means to be accepted by various layers of the society, which can indirectly bring about an incalculable effect of civilization transfer." Dr. Chau also generalized translators' contributions as FIRST (facilitate, interflow, reconcile, serve, and transform) (Zhou, 1998, pp. 39–40).

From the macro perspective, translation can reform a country's customs, reshape a nation's beliefs, and change the cultural vision of values. In the early period of the Middle Ages, for instance, the development of Western science was "as poor as a church mouse." However, from the end of the eleventh century to the end of the thirteenth century, a large-scale translation movement emerged in the Western world. By means of translation, Westerners accepted the baton of the scientific culture of the Mediterranean world from the hands of the Arab people and, through constant combination and innovation, enabled Western civilization to step out of darkness into enlightenment. By means of translation, Westerners redeveloped algebra, geometry, the Indo-Arabian digital system, astronomy, dynamics, optics, magnetism, hydraulics, mechanical engineering, internal medicine, surgery, pharmaceutics, chemistry, and natural history (including zoology, botany, and mineralogy) (Xu, 2002, pp. 46, 78). The translation movement in this period also resulted in the prosperity of education and the proliferation of universities, thereby greatly enhancing the civilization of the Western world, reviving the long-decayed ancient Greek civilization, and ending the "dark" Middle Ages with the great "Renaissance." Without translation, modern Western culture would not exist, or at least there would have been no "Renaissance."

From the micro perspective, translation has the same function in triggering social reform. In the early years of the twentieth century, Yan Fu's translated works were popular in China, which attracted many readers, from scholars to young students, to talk about evolution in newspapers and to discuss evolution in schools until "the thought of natural selection was well-known to the Chinese people, and the cultural atmosphere in China did change". Thus, a translated work exerted an enormous huge influence on Chinese society. This influence was certainly related to China's social environment at that time, which can also be attributed to the great achievements of Yan Fu's particular contributions in translation. After Yan Fu's translation endeavors, a translated work promoting the theory of evolution became the alarm clock for the slumbering Chinese nation as well as a weapon of thought for reform and innovation to the point that it drove wise people to seek reform, triggered an upsurge in youth patriotism, and enlightened and educated several generations of Chinese people. What the translator sought is, so to speak, "doing something with translation" (Hu, 2009).

In 1985, as another example, a book by Harvey and Marilyn Diamond titled *Fit for Life* was published by Bantam Publishing House, Toronto, Canada, and later published in the USA. According to the introduction, "the publication of the English source text in the USA caused a sensation, and for many years, the book has been listed No. 1 of the best sellers of the *New York Times* with several editions. By 1991, more than three million volumes of the book had been sold, with the target texts of various foreign languages sold throughout the world, by which an unprecedented food revolution has been caused. In 1992, the Chinese edition of this book, titled "Fit yi-shi (壹世)," which was translated by Yuan Yuan, was published in Hong Kong by One Publishing Co., Ltd., creating a sensation in Hong Kong, too, and drawing many people's attention" (Li, 1996, p. 177).

5. Translation is "done" to promote eco-civilization

It is an important responsibility for Eco-translatology to promote the protection of the eco-environment and the development of eco-civilization as an academic mission. This promotion can include the following aspects. First, the translator can, by means of translation, reform and construct the translational eco-environment to promote the coordination and balance of translation ecology and Eco-translation. Second, the translator can promote a variety of languages, cultures, and knowledge by means of selecting the translated text, the translation strategy, and the translation time. Third, by means of translation, the translator can promote environmental protection, and "the green revolution" (for example, by translating works of eco-literature, eco-culture, eco-environmental protection, eco-scientific knowledge, and eco-civilization). Fourth, through the demonstration of "translational ecology" and "Eco-translation" and through the exploration of the eco-reason and eco-significance of translation activity, Eco-translatology helps to study and exchange the eco-thoughts that affect the translator and the target text, thus playing a constructive role in the promotion of the development of eco-civilization. The issues listed above show that further study and further translation practice should be undertaken.

6. Translation is "done" to shape the reputation of the country

In the diplomatic promotion of a country, the translator's task is to translate his/her mother tongue into different foreign languages and to publish and spread the viewpoints of his/her country by means of various media such as books, journals, newspapers, radios, televisions, and the Internet. Published translation is the embodiment of a country's level of foreign exchange and the construction of a humanitarian environment. As pointed out by Zhang Jian, a Chinese scholar, "Published translation is related to the image of a country." He adds, "In recent years, there existed a severe 'deficiency' and huge 'cultural deficit' in China's exchange and dissemination to the foreign countries, and the problems in translation proved to be the bottleneck that restricted Chinese culture from spreading to the world, so that China's international image on the international stage is far from that which is acknowledged in the world."[27]

Studies on shaping the national image have been noted by translators from various countries (Valdeón, 2012; Kumar, 2013). For instance, Roberto A. Valdeòn, a Spanish scholar, has conducted a study on "Translation Studies and National Image Reconstruction" together with his colleagues.[28] Tejaswini Niranjana, an Indian scholar, has also unearthed the problematic relationship between nation and translation in her *Siting Translation: History, Post-Structuralism, and the Colonial Context* (Hyderabad: Orient Longman. 1992).

Therefore, it is of inestimable actual and academic significance to study published translation against the larger background of cultural globalization to promote China's cultural communication, establish a sound international image, and enrich the cultural content of globalization, which will demonstrate the important role of the translator.

7. Translation is "done" to promote the development of translation studies

Over thousands of years during which the human beings communicated with each other, accumulated culture and civilization, and promoted social progress by means of translation, generations of translators exchanged their thoughts and ideas, created books, researched translatology, and promoted the construction and development of translatology as they provided society with their translated works.

In the historical river of translation, Western translators' views on translation represent a high level of generalization and categorization, from the distinction between "literal translation and free translation" by Cicero (106-43 BC) to the optimizing idea of "the translator as dominator" by Horatius (65-8 BC), from "popularity" in translation and "the Seven Principles" of translation by Martin Luther to the 'three divisions" of translation (literal translation, free translation and imitation) by

[27] See "Preface I" of *On Process of Publicity Translation: Translator's Adaptation and Selection*, Beijing: the People's Publishing House, 2010 (11), p. 1.

[28] Roberto A. Valdeón. Recycling in the Translation Ecosystem: the Case of Stable and Unstable Sources in News Production [P]. Paper presented at the 3rd *International Symposium on Eco-Translatology*. Chongqing, November 24, 2012.

John Dryden (1631–1700) and to "functional equivalence" and "functional translation" by Eugeue A. Nida (1914–2011). Similarly, in China, translators have approached translation studies in various ways, from the distinction of "*wen*" and "*zhi*" and "faithful to the source-text" by Zhi Qian to "five losses and three non-translations" by Shi Dao'an; from the "eight musts" and "ten doctrines" introduced by Yan Zong to the idea of "faithful and plain" and "five nontranslations" by Xuan Zang (600-664 BC); and from the "good translation" by Ma Jianzhong (1845–1900) to the famous "xin," "da," and "ya" by Yanfu. The "masters" of translation inherited the idea of translation, expanded the vision of translation theory, renewed the discourse on translation, erected monumental markers and became the basis for the development of translatology.

From translation practice to translation theory and from the translation discipline to the translation industry, the translation movement, ancient or modern, domestically or abroad, is the initial motivation for the development of translatology, while the translators, both ancient and modern, domestically and abroad, are the motivating factor that has enabled translatology to obtain a secure academic position.

5.6.3 Further Studies on "Doing Things with Translations"

We have studied the issue of "doing things with translations" from the above two perspectives of the translator's subjective purpose and the objective effect. However, studies of "doing things with translations" can always be performed in different ways and from different dimensions.

On the one hand, "doing things with translations" is not only part of translation studies but also serves to demonstrate the target text's cultural value and the existence of translatology. It is indispensable for raising the status of translation and the social status of the translator.

On the other hand, the study of "doing things with translations" has great potential. For instance:

Divided by subject, "doing things with translations" can be studied from the various perspectives of literature, science and technology, social science, and environmental protection (refer to Sect. 5.6.2).

Divided by region, "doing things with translations" can be studied internationally, nationally and regionally.

Divided by time, "doing things with translations" can be studied in ancient times, modern times, and contemporary times.

Moreover, other topics of translation studies, such as "why to translate," "what to translate," and "how to translate," can be studied.

Overall, "Translation is greatly useful," and "The benefits of translation are great." In the present context of globalization, the role and function of translation have become increasingly important because globalization has accelerated the pace of cultural variety and development toward ecologization. Therefore, "doing things with translations" will be highlighted in research.

Translation activities and translation theories are abundant in every place and at every time with oceans of translation material. Regarding "doing things with translations," there is a great variety with vast coverage. In this section, "doing things with translations" has been introduced as a topic for further study.

5.7 Summary

The sections above discuss topics such as "What is theory?" "What is translation theory?" "What is systematic theory?" and "What is universal theory?" On this basis, systematic translation theory and ontological theoretical questions such as "what to translate," "who translates," "how to translate," and "why to translate" from the perspective of Eco-translatology have been considered. The research in this section can be regarded as a meso-level study of Eco-translatology. From the perspective of function, meso-level studies focus not only on Eco-translatological understanding but also on rational description; meanwhile, they also serve as a link to macro-level studies (Chap. 4) and micro-level research (Chap. 6).

In the entirety of Eco-translatological research, the theory of Translation as Adaptation and Selection approach is considered the backbone of Eco-translatology.

References

Arrojo, R. (1998). The revision of the traditional gap between theory and practice and the empowerment of translation in postmodern times. *The Translator, 4*(1), 25–48.

Baker, M. (1998). *Routledge encyclopedia of translation studies* (p. xvi). London: Routledge.

Baker, M. (2010). *Critical readings in translation studies*. London: Routledge.

Baldwin, J. M. (1909). *Darwin and the humanities*. Baltimore: Review Publishing Co.

Bassnett, S., & Lefevere, A. (Eds.). (1990). *Translation, history and culture*. London: Pinter. Bassnett, 1998: xviii.

Beeby, A., Ensinger, D., & Presas, M. (1998). *Investigating translation: Selected papers from the 4th international congress on translation. Barcelona*. Amsterdam: John Benjamins Publishing.

Berman, A. (1995). *Pour une critique des traductions: John Donne*. Paris: Editions Gallimard.

Blackmore, S. (1990). *The meme machine*. New York: Oxford University Press.

Casagrande, J. B. (1954). The ends of translation. *International Journal of American Linguistics, 20*, 335–340.

Catford, C. (1965). *A linguistic theory of translation*. London: Oxford University Press.

Docherty, T. (1993). Theory and difficulty. In R. Bradford (Ed.), *The state of theory* (pp. 20–34). London: Routledge.

Fan, W. (2000). *Rethinking of translation: Translated or not translated*. Taibei: Bookman Books Press.

Fang, M. Z. (2010, September 9/10). *Construction on eco-translatology from the perspective of core terminologies*. Paper Presented on the First International Eco-Translatology Symposium, Macao.

Fang, Z. (1989). Translation memories. In *Discussion on contemporary literature translation* (pp. 74–81). Beijing: Peking University Press.

Feng, W. (2008). An existential study of translation and its embedded value. *Foreign Language Education*, (4), 82–85.
Fish, S. (1989). *Doing what comes naturally: Change, rhetoric, and the practice of theory in literary and legal studies*. Durham: Duke University Press.
Fu, Y. L. (2002). Paradigms in translation studies: Shift, development and innovation. In Z. J. Yang (Ed.), *New Explorations in translation studies*. Qingdao: Qingdao Press.
Gentzler, E. (1993). *Contemporary translation theories*. New York: Routledge Inc.
Gile, D. (1994). Opening up in interpretation studies. In M. Snell-Hornby, F. Pochhacker, & K. Kaindl (Eds.), *Translation studies: An interdiscipline*. Amsterdam: John Benjamines B.V.
Goethe, J. W. (1827). The importance of translation. As cited in Jin Di's article: The objectives of translation. *Journal of Tian Jin Foreign Studies University, 2*, 1–10.
Gu, Z. K. (2003). *Comparative appreciation of sino-western poems and translation theory studies. Beijing: Tsinghua*. University Press.
Gutt, E.-A. (2000). *Translation and relevance: Cognition and context*. Manchester: St. Jerome Publishing.
Hatim, A. B. (2001). *Teaching and researching translation*. London: Taylor and Francis Ltd.
Hermans, T. (2000). *Norms of translation*. Oxford: Oxford University Press.
Holmes, J. S. (1988). *Translated! Papers on literary translation and translation studies*. Amsterdam: Rodopi.
Hu, G. S. (1991). A discipline needed to work hard at it. *External Communication and Exchange*, (8), 28–32.
Hu, G. S. (2001, December 6/8). *Study on translation as adaptation and selection*. Paper presented on the third Asian translators symposium of international federation of translators. Hong Kong.
Hu, G. S. (2003). Translation as adaptation and selection. *Perspectives: Studies in Translatology, 11*(4), 283–291.
Hu, G. S. (2004a). From "Translator's Subjectivity" to "Translator-centeredness". *Shanghai Journal of Translators*, (3), 9–15.
Hu, G. S. (2004b). Theoretical props of the approach to translation as adaptation and selection. *Shanghai Journal of Translators, 4*, 1–7.
Hu, G. S. (2004c). *Translation as adaptation and selection*. Wuhan: Hubei Education Press.
Hu, G. S. (2004d). Translator-centredness. *Perspective: Studies in Translatology, 2*, 106–117.
Hu, G. S. (2009). Eco-translatology: Trans-disciplinary integration on translation studies. *Shanghai Journal of Translators, 2*, 3–8.
Hu, G. Z. (2006). A theory of "Translator's Turn": A review of "Translation as Adaptation and Selection". *Chinese Book Review Monthly*, (2).
Huang, Z. (2001). *The theory of translation variation*. Beijing: China Translation & Publishing Corporation.
Hunt, S. D. (1991). *Modern marketing theory: Critical issues in the philosophy of marketing science*. Cincinnati: South-Western Publishing Company.
Jakoboson, R. (1959). On linguistic aspects of translation. In R. A. Brower (Ed.), *On translation* (pp. 232–239). Boston: Harvard University Press.
Ji, X. (1999, December 4). The portrait of Macau culture. *Canton Evening News*.
Jiang, X. H. (2013). "Eco-translation" exemplified. *Journal of Eco-Translatology, 1-2*, 39–49.
Jiao, W. (2010). An analysis of "Evolution and Ethics" (Yan Fu from "the Sequence Chain", Trans.). *Shanghai Journal of Translators*, (4).
Ke, F. (2002). Study of translation history should be on basis of people. *Chinese Translators Journal*, (3), 31–32.
Kelly, L. (1979). *The true interpreter: A history of translation theory and practice in the West*. Oxford: Basil Blackwell.
Kuhn, T. S. (1962). *The scientific revolutions*. Chicago: The University of Chicago Press.
Kumar, R. (Ed.). (2013). *Role of translation in Nation Building*. New Delhi: Modlingua.
Lawendowski, B. (1978). On semiotic aspects of translation. In T. A. Sebeok (Ed.), *Sight, sound and sense*. Bloomington: Indiana University Press.

Li, C. (1996). *Translation appreciation*. Hong Kong: The Commercial Press.
Li, D. (2010). Philosophical reflections on the translators in translation as adaption and selection. *Xue Li Lun*, (2), 17–19.
Liao, Q. Y. (2001). *Comtemporary translation studies in U.K.* Wuhan: Hubei Education Press.
Liu, J. Z. (2005). Tradition and development of translation theory studies (Preface). In J. Zhang & K. Chen (Eds.), *Famous expert-famous speech-famous translation*. Shanghai: Fudan University Press.
Liu, M. Q. (1998). Orientation of Chinese translation studies. In *Asia Translation Tradition and Modern Tendency Symposium Pre-Proceedings*. Taibei: Translation Studies Institute of Fu Jen Catholic University. 7-8-9.
Luo, X. Z. (2009). Innovation is a must to development. *Shanghai Journal of Translators*, (4).
Lv, J. (1997). Translatology—A special field of communication theory. *Journal of Foreign Language, 2*, 39–44.
Meng, F. (2002). The influence of the evolving Chinese cultural pattern on translation studies in China. *Chinese Translators Journal*, (2), 13–17.
Mou, Z. S. (1997). *Nineteen lectures of Chinese philosophy*. Shanghai: Shanghai Guji Press.
Neubert, A. (1998). Theory and practice of translation studies revisited: 25 years of translators training in Europe. In A. Beeby, D. Ensinger, & M. Presas (Eds.), *Investigating translation* (pp. 13–26). Amsterdam: John Benjamins Publishing Company.
Newmark, P. (1988). *A textbook of translation*. London: Prentice-Hall.
Newmark, P. (1997). Why translation theory? What translation theory? *Journal of Translation Studies, 1*, 94–101.
Nida, E. (1982). *Translating meaning*. California: English Language Institute.
Nida, E. (1997). Theories of translation. *Journal of Translation Studies, 1*, 102–108.
Nida, E., & Taber, C. (1969). *The theory and practice of translation*. Leiden: E. J. Brill.
Pym, A. (1998). *Method in translation history*. Manchester: St. Jerome Publisher.
Qian, J. (2001). *The selection of existence*. Beijing: Chinese Social Science Press.
Sheng, Y., & Wang, W. (1984). *The textual study of the Chinese target-texts of the Buddhist classics* (pp. i–ii). Taibei: Tianhua Publishing Company Ltd.
Snell-Hornby, M. (2006). *The turns of translation studies: New paradigms or shifting viewpoints*. Amsterdam: John Benjamins Publishing Company.
Song, Z. (2009). A nonlinear approach to translation selection. In G. S. Hu (Ed.), *Translation and cross-cultural communication: Integration and innovation* (pp. 111–120). Shanghai: Shanghai Foreign Language Education Press.
Tan, Z. X. (1988). On translation studies. *Foreign Language Journal*, (3), 22–28.
Tan, Z. X. (1991). *Western translation history*. Beijing: Commercial Press.
Tan, Z. X. (1997). Reflections on the science of translation. *Babel, 43*(4), 331–352.
Valdeón, R. A. (2012). Tears of the Indies and the power of translation: John Phillips' target-text of Brevisima relacion de la destruccion de las Indias. *Bulletin of Spanish Studies, 89*(6), 839–858.
Venuti, L. (1995). *The translator's invisibility: A history of translation*. London: Routledge.
Wang, K. F. (1997). *Translation culture*. Shanghai: Shanghai Foreign Language Education Press.
Wang, N. (2011). Eco-literature and eco-translatology: Deconstruction and reconstruction. *Chinese Translators Journal, 2*, 10–16.
Wang, X. Y. (2000). Revision of literary translation. *Translation Quarterly*, 24–39.
Wang, Z. H. (2012). *Yan Fu's translation of evolution and ethics*. Unpublished MA Thesis, Changsha University of Science and Technology.
Warren, R. (Ed.). (1989). *The art of translation: Voices from the field*. Boston: Northeastern University Press.
Wen, Y. D. (1926). English translation of Li-Tai-Bai's poem. In the Editorial Board of Chinese Compilation and Translation (Ed.), *The art of translating poems* (p. 1987). Beijing: China National Translation and Publishing Corporation.
Wilson, E. O. (2002). *The future of life*. New York: Alfred A. Knopf.
Wilss, W. (1992). *Űbersetzungsfertigkeit*. Tübingen: Günter Narr Verlag.

Wilss, W. (1996). *Knowledge and skills in translator behavior*. Amsterdam: John Benjamins Publishing Company.
Xie, T. Z. (2001). On the misconceptions in translation studies and theories in China. In *The third Asian translators' forum of FTI: "Program. Abstract"* (p. 135). Hong Kong: Translation Society.
Xie, T. Z. (2007). *Introduction of medio-translatology*. Beijing: Peking University Press.
Xie, T. Z. (2012). New breakthrough in contexts of Chinese translation studies in modern times. *Chinese Translators Journal*, (1), 13–15.
Xie, T. Z. (2013). A study on translation from another perspective—Thinking of Mo Yan's Nobel Prize for literature. *East Translation Journal*, (1), 4–8.
Xu, J. (2003). *On translation*. Wuhan: Hubei Education Press.
Xu, J., & Mu, L. (2009). Translation studies in China in the past three decades (1978–2007). *Foreign Language Journal, 32*(1), 77–87.
Xu, S. W. (2002). *The dissemination of oriental learning and the revival of the Western culture*. Shanghai: Shanghai Peoples Press.
Xu, Y. C. (1982). Faithfulness and smoothness. *Foreign Language Teaching, 1*, 19–25.
Xu, Y. C. (1998). On how to take full advantage of language priority in translation. *Foreign Languages and Their Teaching*, (6), 35–38.
Yang, X. R. (2004). Why we don't have theoretical schools in our translation studies. *Foreign Languages and Their Teaching*, (4), 39–42.
Yang, X. R. (2012). *Binary, pluralism, integration—Translation essence and standard* (p. 1). Shanghai: Shanghai Foreign Language Education Press.
Yang, Y. (1989). A Tart fruit: 40 years in translation. In W. Shoulan (Ed.), *On contemporary literary translation* (pp. 310–315). Beijing: Peking University Press.
Yuan, S. P. (2011). Yan Fu's translation from the perspective of eco-translatology. *Times Literature, 10*, 108–109.
Zhang, N. (2002). *Studies on objectivity and neutrality of translation*. Tsinghua University, "Evolution and Theoretical Study of Translation of in the New Century" Advanced Translation Forum Speech, Beijing.
Zhengkun, G. (2002). Metatranslatology. *Chinese Translators Journal*, (4, 5).
Zhili, S. (2007). The responsibilities of translators. *Chinese Translators Journal*, (4).
Zhou, Z. X. (1996). Comments: Theories and practice. In C. Li (Ed.), *Comments on translation* (pp. 1–19). Hong Kong: Commercial Press.
Zhou, Z. X. (1998). *Translation and life*. Beijing: China Translation & Publishing Corporation.
Zhu, C. S. (2000). Out of misunderstanding and into the World - Chinese translation studies: Reflections and prospects. *Chinese Translators Journal*, (1), 2–19.
Zhu, C. S. (2001). *Studies on translation: Language, text, poetry*. Taibei: Bookman Books.
Zhu, H., & Wu, J. (1988). Zhu Shenghao's translation chronicle. *Chinese Translators Journal*, (6), 29–32.

Chapter 6
The Microlevel: Eco-Translation Operational Illustrations

6.1 Introduction

China has an academic tendency to prioritize practice that has descended from the tradition of "unity of knowledge and practice." There is an old Chinese saying in *Xunzi*:

> Not having heard something is not as good as having heard it; having heard it is not as good as having seen it; having seen it is not as good as knowing it; knowing it is not as good as putting it into practice.

Another saying, by Zhu Xi, a Song Dynasty Confucian scholar, is "With respect to order, knowledge comes first, and with respect to importance, action is more important" because "action produces the effect of knowledge, while knowledge does not produce the effect of action." For Wang Yangming, a philosopher during the Ming Dynasty, it is only through simultaneous action that one can obtain knowledge; knowledge necessarily/automatically leads to action; knowledge means knowing how to respond to a given situation, and action is responding to a given situation. Based on the traditional Chinese eco-wisdom, Eco-translatology naturally seeks the *academic unity of theory and practice*, which is the primary reason for one of its three paradigmatic features—*practicality* (see Sect. 8.2.3).

Since "Translatology" is a study of the law and the art of translation, the author approves of the English proverb: *Any science begins with philosophy and ends in art.* The author also thinks that what Peter Newmark says is true:

> I feel uneasy if I write more than a dozen lines about translation without producing an example, partly to explain, illustrate and support my "theory", partly to invite discussion reaction in favour or against, and in the hope of finding a rapprochement, a conciliation if not an agreement (Newmark, 1991, p. 5).

In addition, translation practice can reflect upon concepts and theories regarding translation. What translation strategies and methods the translator selects can also reflect upon his/her cognition and understanding of the essence and criteria of

(Hugs) G. S. Hu, *Eco-Translatology*, New Frontiers in Translation Studies,
https://doi.org/10.1007/978-981-15-2260-4_6

translation. Therefore, Eco-translation strategies and methods, as well as illustrations and demonstrations, can thoroughly embody the theoretical deliberations of Eco-translatology.

Based on the above, this chapter attempts to focus on Eco-translatology at the microlevel, namely, to comment on, analyze and interpret with actual examples the cognitive, directive, interpretive, and descriptive functions of the basic concepts of Eco-translatology in translated text generation and translation phenomenon interpretation. Naturally, some operational translation strategies and methods are included.

6.2 "Transplanting" Through Multidimensional Adaptation and Selection

6.2.1 "The Last One Was Delicious; Bring Me Another One!"

"Multidimensional transplantation" is one of the significant ecological translation concepts and methods of Eco-translatology. In the process of text translation, "multidimensional transplantation" usually manifests itself as the translator changing a single-dimensional translation treatment in the wake of an overall consideration of a whole sentence, paragraph, passage, and text. That is, the translator should not only execute language transformation but also take care of such multiple dimensions as cultural connotations, communicative intentions, psychological hints, and aesthetic pursuits, especially the balance between the source language ecology and the target-language ecology and the holistic and interrelated factors of the micro-, meso-, and macro-translational eco-environments.

Now let us consider, for example, precise translations of a public sign concerning a crocodile.

The original language of the sign is ***The last one was delicious; bring me another one!***

General context: The sign is posted at the crocodile pool of a zoo in a country in North America. The signboard beside the pool reads as shown above. The original is speaking in the voice of the crocodile, showing a particular cultural expression and sense of humor. To "warn" and "attract" visitors (especially Chinese tourists who are not proficient in English), the zoo manager plans to set up another sign in Chinese.

Here is a brief analysis of the different translated Chinese versions of the original English sign.

- Version 1: 上一个好吃, 再带来一个. (shang yi ge hao chi, zai dai lai yi ge)

Version 1, translated by a sophomore English major, is a word-for-word translation from the linguistic perspective. The first and second parts each consist of five Chinese characters, which appear to be symmetric. However, it is evident that the communicative intention of the original is not achieved, nor are the "Western" cultural expression and sense of humor adequately reflected on the Chinese sign. Even the language itself seems to be nothing special without the "brief" and "vivid" characteristics of the original language. From the perspectives of adaptation to the translational eco-environment and "multidimensional transplantation," the translator has neither carried out multidimensional adaptation to the translational eco-environment nor performed selective transformation to adapt to the translational eco-environment. Utilizing the terminology of ecological translation, we can say that version 1 has apparently "damaged" the original ecology, including the humorous expression, language function, communicative intention, "narrator" identity transformation, etc.; thus, the translator's "intervention" in the production of this translated version is unsuccessful.

- Version 2: 鳄鱼伤人, 禁止入水!(er yu chang ren, jin zhi ru shui!)

Version 2 is taken from the monograph *Translation Studies and Practice* by Prof. Fang Mengzhi of Shanghai University, which was published in 1999 by Qingdao Publishing House (see Fang, 1999, p. 109).

Version 2 is neat in its use of words with antithesis, and, more importantly, the translation, from the communicative perspective, better realizes the "sign" function of the original. In this sense, version 2 is indeed a good translation.

However, after careful analysis, we find that version 2 does not demonstrate that the sign is speaking in the voice of the crocodile. In contrast, it seems to be ponderously issuing, with a straight face, an "order" in the tone of a "person." In this case, the "identity" of the "speaker" in the original has been changed, and the Western cultural expression and sense of humor are gone. In addition, this translated version, in terms of the "sign" function fulfillment, appears to "warn" rather than "attract." However, the "attraction" function of the original sign is distinct. We can imagine that if the original author (the owner) aimed only to "warn," the original could be written as "Keep Clear!" or "No Approaching!" (jin zhi kao jin!), which would be more direct and concise. However, the original author (the zoo manager) instead tried to be unique and speak in the voice of the crocodile; therefore, it is easy to appreciate the pragmatic implications.

In view of version 2's translation effect of "warning" rather than "attracting," and because its expression is unduly direct and the pragmatic implications, especially the Western "cultural expression" and sense of humor, are gone, we can use the terminology of ecological translation to say that even though this preferable version retains the "communicative ecology" of the source language, the damage to the original "cultural ecology" and "linguistic ecology" is also apparent. In other words, version 2 shows the "imbalance" between the source language ecology and the target-language ecology caused by this sign translation.

- Version 3: (At the top of the Chinese signboard is painted a crocodile opening a big mouth stained with blood.)

人肉真香, 再来一口!(ren zou zhen xiang, zai lai yi kou!)
Here is the analysis of version 3:

1. The translator translated the word "delicious" in the original into "好吃," "可口," pointing out that human flesh is delicious for a crocodile.
2. From "the last" and "another one" in the original, we can infer that the crocodiles have already had the "human flesh." The word "再" (again) in the translation can express this meaning.
3. Considering that the original speaks in the voice of the crocodile, the translator has painted a crocodile opening a big mouth stained with blood above the Chinese version on the signboard, which is similar to the crocodile speaking, matching the original tone; thus, the "narrator" identity of the source language ecology is explicitly and vividly "reproduced." (Note, however, that the translation is making up, building, and reconstructing the "lacking" elements in the target eco-environment by utilizing auxiliary symbols.)
4. Furthermore, the ferocious crocodile opening its mouth stained with blood is very intuitive and vivid, the "warning" and "intimidating and prohibiting" effects are incisively and vividly produced as well.
5. Since the translation better keeps the Western cultural expression and sense of humor, it can play the role of attracting tourists to stop and enjoy the exhibition a way that can help the tourists (especially Chinese tourists) understand Western culture and can quickly become a kind of "conversation piece." Undoubtedly, it also fits the zoo manager's purpose of "attracting" tourists to enjoy the sight. With the help of the adaptive translation selection of symbolic language, the translator succeeds in rendering the translation accurate, expressive, vivid, and intuitive, thus better fulfilling the original communicative and "warning" functions.

Through the comparative analysis of the three actual translation examples mentioned above, we can see that in version 3, the original content and its communicative significance are transplanted together with a graphic from multiple dimensions, including linguistic, cultural, communicative, and aesthetic dimensions and the client's intention, the pragmatic scenario, and the readers' needs, which better reflects the translation principles of "selective adaptation" and "adaptive selection" of Translation as Adaptation and Selection, resulting from the translator's multidimensional adaptation and adaptive selection of the specific translational eco-environment. That is, in version 3, the translator better maintains the balance between the original ecology and the ecology of the translation and constructs an eco-environment closer to the source-text ecology, enabling the "survival" of the translation in the target-text ecology.

The above analysis not only illustrates the translation principles of "multidimensional selective adaptation and adaptive selection" of Eco-translatology but also demonstrates that in accordance with the "multidimensional transplantation" concept and method, a translation with a higher "degree of holistic adaptation and selection" can be produced. At the same time, the above illustrations and interpretations show how to reduce the "damage" to the original text ecology and the target-text ecology in the process of translation as much as possible and how to maintain the coordination and balance of the original text ecology and the target-text ecology in the process of translation as much as possible, in other words, how to create and reconstruct the target eco-environment for the "survival" of the translated version.

6.2.2 "Workers Unite to Protect the Factory Property!"

There are various factors of the translational eco-environment. The key to any translation, including Eco-translation, lies in how the translator follows an appropriate "selective adaptation" and "adaptive selection" process according to the specific translational eco-environment.

Many years ago, the author accompanied a group of American businessmen to visit a private enterprise on the mainland. Both the Chinese side and the American side were interested in investment at that time. The two groups hoped to have the opportunity to work together to develop some new product. However, on arriving at the gate of the factory, we caught sight of a slogan that read:

- 全厂动员，打击偷盗!(quan chang dong yuan, da ji tou dao!)

The American guests curiously asked the meaning of the sign. With no thought at all, the first author of this book instantly interpreted it as follows:

Workers of the factory should be mobilized to fight against thefts.

- (该厂的工人应该动员起来去打击偷盗!/quan chang de gong ren ying gai dong yuan qi lai qu da ji tou dao!)

Then, some guests shrugged their shoulders, some knit their brows, and others looked extremely puzzled.

The reason for this response is simple, as this English translation carries at least two implications: first, there must be many "thefts" in this place; therefore, it may be unsafe here; second, as the workers are mobilized to fight against thefts, the job of production may not be well done.

As a Chinese interpreter, the author thought then, regarding the "macro" translational eco-environment, that there was a definite intention of both the American side and the Chinese side to cooperate; that is, the Chinese side wanted to attract foreign investment, and the American side wanted to invest. Regarding the "meso" translational eco-environment, everybody arrived in high spirits, and the first impression was essential. Regarding the original intention of the slogan (i.e., the "micro" translational eco-environment), its real meaning might also have been to strengthen security measures and reinforce the factory's regulations and policies to make them more favorable for the production process.

Based on the above thinking and judgment, the author modified the translation slightly and suggested a supplementary explanation:

Workers get mobilized to protect factory property~!

- (工人们都被动员起来去保护工厂里的财产! gon ren men dou bei dong yuan qi lai qu bao hu gong chang li de cai chan!)

Compared with the former "blurted-out" translation, the revised "mental draft" of the translation resulting from the interpreter's consideration of the "macro," "meso," and "micro" translational eco-environments is a slight improvement. However, the words "get mobilized" suggest the passive meaning that "the workers' self-awareness is not very high," so they also need the management authorities to mobilize them.

Based on a comprehensive judgment and assessment of the various elements of the "macro," "meso," and "micro" translational eco-environments, the author finally corrected and supplemented the translation as follows:

Well, the slogan means—Workers unite to protect factory property!

- (哦, 对了, 刚才那条标语的意识是:工人们团结起来保护工厂的财产! ou dui la, gang ci na tiao biao yu de yi shi shi: gong ren men tuan jie qi lai bao hu gong chang de cai chan!)

Judging from the American guests' facial expressions, showing that "they caught on," the above "corrected and supplemented" translation achieved a good effect.

At this point, then, as an interpreter on the scene, the author had undergone the following processes of conception and expression.

Workers of the factory should be mobilized to fight against thefts.

↓

Workers get mobilized to protect factory property.

↓

Workers unite to protect factory property.

The above is a translation process of adaptive selection based on adaptation to the "translational eco-environment." This translation conception and selection process show that the production of a translated version is based upon the translational eco-environment. To make the source and target-text ecologies consistent with each other, the translator needs to follow a multidimensional "selective adaptation" and "adaptive selection" process.

The above case indicates that the translator's effort is actually to create, reconstruct, and improve the eco-environment of the target text to achieve coordination and balance between it and the source-text ecology and to ensure the "survival" of the translation.

6.3 “Emptying” the Translator’s Mind and Absorbing or Reshaping the New Eco-Environment Before Translating

Human memory function makes the human brain store much related information. For translation activity, much information concerning linguistic knowledge, cultural background, communicative strategies, and norms of conduct must also be stored in the brain of the translator. On the one hand, this information is vital for the translator to comprehend the original and express the translated version; on the other hand, from the perspective of Eco-translation, to adapt to the source-text ecology and construct a target-text ecology that conforms to the source-text ecology, the translator sometimes must “forget,” “transfer,” or “empty” this information. Here, the so-called emptying can be interpreted as an Eco-translation phenomenon but can also be used as an Eco-translation method.

6.3.1 Translating a Text of Mythical Experience

Hu K. T. (2011) of Fu Jen Catholic University in Taiwan explores the revelation of Eco-translatology for translating a text of mythical experience. He finds that during the process of translating a mystical text, one has the impression that the meaning of the words is minimal in terms of ecological conditions: the cultural eco-environment of the communication, the relationship between the translator and the author, the selective attitude of the translator towards the translational eco-environment, and the selection and adaptation process in forming the translational text. The translator, who is the center of the process, bringing a text to an audience in another cultural and psychological eco-environment, must pass through two crucial steps. First, he/she should “empty” him-/herself to gain insight into the spiritual experience of the author, which he/she, against his/her educational background, cannot explain logically. The second step is that he/she should select an adaptive strategy to produce the text.

For instance, Dr. Hu Kung-Tze translated the following text of mythical experience, the author of which explained how she described her own experience as well as her expected effect:

The Original Text

Ogni discorso, lo ripetevo poi alle pope[1] alla lettera e col calore che sentivo, per farle partecipare di ogni cosa e perché me sembrava che ciò che non è utile all’umanità o almeno agli altri, non ha valore. Lo comunicavo inoltre perché conservasse quella trasparenza divina e non s’aggiungessero elementi “umani” a guastare ogni cosa. (287)

---

[1] This is a word referring to girls in the original author’s dialect.

The Translated Text
每一次的谈话，我重复给我的同伴们听，重复我听到的每一个字，也用当时我所感觉到的热情，好让她们参与每一件事情，而且我觉得，假如我经历的事情，不能给人类带来益处，或者至少，不能为别人带来益处，就没有叙述的价值。除此之外，我讲这些事情，想原原本本地保留上天传达的讯息，不要加入"人为"的因素，而破坏任何一件事情. (The underlined parts are added by the translator.)

The translator believes that the original author's narrative is intended to allow readers to participate in her mystical experience; therefore, she is reluctant to add any aspect of the experience not given by God. In this paragraph, the translator's translation strategy for this text is also shown, i.e., to translate as much as possible according to the original without the translator's own interpretations.

However, let us determine how to achieve the purpose of allowing readers to participate in the translated text. In this text, the original author also proposes some requirements about the communication environment.

The Original Text
Alla sera, alla meditazione che durava circa mezz'ora, avevamo l'avvertenza di mettere tutte l'anima nella più assoluta passività onde il Signore, volendolo, potesse comunicarsi. E le mie compagne facevano tacere tutto in loro, anche ciò che poteva esser ispirazione, affinché l'unità con me fosse perfetta. (289)

The Translated Text
傍晚时分，做默想的时候，我们小心地、尽量地把整个心灵放在被动状态，好让上主，假如祂愿意的话，可以和我们说话。而我的同伴们，将心灵放空，即使有来自上天的灵感也放到一边， 好让她们与我的合一达到完美的地步。(The underlined part is added by the translator.)

This is the original author's mental preparation before communicating with her peers. Under this circumstance, even the original author's so-called Questa vision will become more clear.

The Original Text
Questa vision—diciamo così—lo ricordo bene, mi è stata Chiara solo quando anche le "pope" fecero fare, sul loro nulla, lo stesso patto a Gesù Eucaristia onde unirsi con noi. (288)

The Translated Text
这个神视:— 让我们暂时这么称呼它 — 我记得很清楚，只有当我的同伴们在她们放空的心灵上，也和耶稣圣体做同样的盟约，好让祂与我们结合为一，我才能够看清楚(我的神视)。

Thus, emptying the mind in combination with the sacred body of Jesus is the condition in which the original author can see Questa vision. Given that Dr. Hu Kung-Tze's translation practice of "emptying" the translator's mind involves a text of mystical experience about mental processes, that the original language is German, and that the analysis is relatively complex, no more specific citations will be given here. Interested readers can refer to the paper "On Applying Eco-translatology: A Case Study" (published in the *Journal of Eco-translatology*, 2012(2):28-32)

However, the translation case below is relatively simple and is also illustrative.

6.3.2 *"Classified Instruction" or "Dispatching Commander"?*

The Original English Version
The Alteration Management System (AMS)

The AMS is an automated accounting procedure designed to assist the Type Commander (TYCOM) in fulfilling his responsibility toward monitoring and controlling the configuration of attached units and in satisfying the requirements of the Fleet Modernization Program (FMP). Using a combination of the Maintenance Data System (MDS) and AMS administrative procedures, the TYCOM can direct the accomplishment of alterations, throughout his force. In support of the FMP, the AMS computer programs produce reports of the alteration in convenient formats so that the TYCOM may adequately advise the Chief of Naval Operations concerning the structure of the FMP.

The Translated Chinese Version
备用处理系统

- 备用处理系统是一种自动计数程序，用于协助分类调配指令系统，监控所附子系统的结构并满足"舰队现代化程序"的要求。综合利用"维护数据系统"与"备用处理系统"两个管理程序，分类调配指令系统可直接实现备用方案的处理，发出不同优先级别的指令，并监控整个备用方案处理的情况。
- 在支持"舰队现代化程序"方面，该"备用处理系统"的计算机程序可以简易的格式打出备用方案的各种报告，以便分类调配指令系统能恰当地向海军作战行动首长显示"舰队现代化程序"的结构情况。

Significantly, the person engaged in marine engineering technology produced the following translation for the same text:

调动管理系统

- 调动管理系统是一种自动描述过程，目的是为了协助分类调度司令官完成他监控所属部队的结构情况，以及满足"舰队现代化计划"的各种要求。利用"维护数据系统"和"管理系统"两个管理程序的组合，分类调配司令官可以在他职权范围内指挥调变，确定优化方案、监控调度进展。为了支持"舰队现代化计划"，调动管理系统的计算程序可以用简易的格式打出调动情况的报告，以便使分类调动司令官能向海军军事行动首长适时建议"舰队现代化计划"的结构情况。

From the above translation, it can be seen that the translator comprehended and translated "Commander" as "司令官," "attached units" as "所属部队," "direct the accomplishment of alterations, throughout his force" as "司令官可以在他职权范围内指挥调变," "Naval Operations" as "海军军事行动," "to direct" as "指挥," etc. We can imagine that such "pre-established" or "existing" information in the translator's mind of the knowledge structure, cultural background, and linguistic habits enabled him/her to completely map out a schema of the "dispatching management system of naval operations." This translation case shows that because the translator often starts from the world with which he/she is familiar and understands the original text with his/her own "pre-knowledge," "pre-view," and "pre-grasp" in order to maintain, coordinate, balance, and construct an eco-environment that conforms to the original language ecology, the translator often must "transform" or even "empty," as much as possible, the existing "ecology" in his/her mind. Only in this way can the translator create a new eco-environment that conforms to the original language ecology.

The concept of "emptying the mind" is often referred to in the Chinese literary tradition as well. For example, in *Anthology of Four Poets of the Song Dynasty* 《宋四家词选》, it is stated that when learning to write a poem, "the poet can draw inspiration only by emptying himself first" (quoted from Hu K. T., 2012, p. 32).

There is no lack of such examples in the Western translation field. For instance, the renowned French translator Charlotte Mandell points out that "before starting to translate, a translator tries to empty him/herself out, absorb the text, and then speak the text anew in his/her own language."[2]

6.4 "Making Up" for Elements Lacking in the Target Eco-Environment

6.4.1 Additional Explanations in the Target Text

Consider *Tian Yan Lun* (*Evolution and Ethics*), translated by Yan Fu, for example. *Tian Yan Lun* (*Evolution and Ethics*) is a translated work whose basis is the traditional Chinese culture with a combination of Western and Chinese elements. In it, Yan Fu teases out the Western culture and the traditional Chinese culture to integrate and contrast them with each other in a "natural evolution." To achieve this end, Yan Fu uses many translation means such as "prefacing," "example changing," and "footnoting." In terms of Eco-translatology, he employs these means to "make up" and "reconstruct" the translational eco-environment of the target text so that the translated versions can adapt to it and "survive" in it.

Following are a few examples.

1. Prefacing

   In the Preface to Prolegomena XIII, Yan Fu wrote: "不能爱则不能群, 不能群则不胜物, 不胜物则养不足。群而不足, 争心将作."

   Yan Fu cites the words of Ban Gu, a famous historian in the Eastern Han Dynasty of China, as evidence that the ethical starting point of "being good at being cooperative in the society" (善群) advocated by the scholars of ancient

[2] See BretaAart (2012).

China is "love," which is analogous to Huxley's "inborn sympathy." Huxley holds that the development of the universal sympathy of humankind brought about the ethical evolution of humankind (Tian & Hu 2005, p. 60). Huxley states that "the fittest" in "the survival of the fittest" has the meaning of "the best," and "the best" bears a kind of ethical implication. "The survival of the fittest" lies not in making the fittest survive but in making more people suited for survival. Therefore, "ethically, the best would survive" (Huxley, 1932, p. 57).

Yan Fu also compares Herbert Spencer's thought that "the human mind should be subject to natural laws in governing the society" (任天为治) with Daoism: "斯宾塞氏之言治也, 大旨存于任天, 而人事为之辅, 犹黄老之明自然, 而不忘在宥是已" (in the Preface to Prolegomena V).

Here, "Letting Be, and Exercising Forbearance" is a well-known writing of Chuang Tzu in which Chuang Tzu expresses his concepts of action less governance and letting things develop naturally. Yan Fu quotes Chuang Tzu's "闻在宥天下, 不闻治天下也" (I have heard of letting the world be, and exercising forbearance; I have not heard of governing the world) as a foil to Spenser's concept "the human mind should be subject to natural laws in governing the society" (任天为治).

It is thus clear that Yan Fu in the Preface to *Tian Yan Lun* as much as possible interrelates "natural evolution" (Tian Yan) Chinese traditional cultural classics to produce strong readability, the final purpose of which is to make *Tian Yan Lun* "survive" in the translational eco-environment of the target text.

2. Example Changing

In translating, Yan Fu often transforms the original examples into examples with which Chinese people are familiar. For instance, in Prolegomena IV, Yan Fu changes Huxley's Cretaceous example into the example of a flooded botanical garden to illustrate that human control of nature is limited; thus, Chinese readers can better understand Huxley owing to the many rivers in China.

For another example, in Prolegomena VIII, Huxley uses pigeons as a metaphor to satirize the ridiculousness and impossibility of the selection of men by men: "The pigeons, in short, are to be their own Sir John Sebright." Sir John Sebright, the nineteenth century British agronomist, was famous for improving poultry and livestock and was especially good at raising pigeons. Yan Fu transforms the illustration into Chinese allusions as follows: "今乃以人择人, 此何异上林之羊, 欲自为卜式, 汧、渭之马, 欲自为其伯翳, 多见其不自量也已。" Yan Fu uses Pushi (卜式) and Boyi (伯翳) to replace Sir John Sebright to achieve the same effect among readers because Pushi (卜式) was a minister in the West Han Dynasty who became rich by raising sheep, and Boyi (伯翳) was the ancestor of the Ying (嬴) family in Chinese history. The original would have been hard for Chinese readers at that time to understand, as they likely would have had no idea what Huxley was referring to, so Yan Fu turns them into familiar, easily understandable Chinese allusions.

For the third example, in Prolegomena VIII, Yan Fu uses an English proverb, "粪在田则为肥, 在衣则为不洁," and adds "然则不洁者, 乃肥失其所者也" to express the original sentence "Dirt is riches in the wrong place, and that sound aphorism has moral applications." His addition enhances the meaning.

3. Footnoting

In Prolegomena VI, Yan Fu adds his explanation of Utopia: "夫如是之群, 古今之世所未有也, 故称之曰乌托邦。乌托邦者, 犹言无是国也, 仅为涉想所存而已。" There is no doubt that officials are necessary to explain.

In Section 9 of Lecture IV, Yan Fu adds: "二者本为同物" (originally, both versions are the same), which is necessary for readers who did not know the common sense of *Brahman*.[3] In Section 16, in translating "Karma,"[4] "羯磨," Yan Fu adds "used for iron absorption," "吸铁之用," in consideration of the fact that knowledge of natural science was not widespread in China at the time.

In Section 4 of the body part (《论五·天刑》), Huxley mentions the hero, Hamlet, of Shakespeare's "Hamlet," which is a household name in Britain, as Li Kui is to the Chinese people. For fear of people familiar only with ancient Chinese books not understanding, Yan Fu further explains, "罕木勒特, 孝子也。乃以父仇之故, 不得不杀其继父, 辱其亲母, 而自刃于胸。"

"The translated work *Tian Yan Lun* (*Evolution and Ethics*) was published in Guowen Daily in 1897, the singles of which was cut in blocks for printing in 1898, and the lead print of which was published in 1905 (Chen, 1989, p. 49). Due to its adaptation to the translational eco-environment of the day, it became a best-seller and was successively reprinted within a short period of time 'to the 20th edition in 1921' (Han, 1969, p. 11)" (see Hu, 2004, p. 148).

As previously mentioned, Yan Fu uses many translation means, such as "prefacing," "example changing," and "footnoting," to "create" and "reconstruct" the translational eco-environment of the target text so that the translated texts can be adapted to it and "survive" or "last" in it.

In fact, in exploring the translator's actions of translation treatment in the process of translation, it is clear that translators always adopt the translation strategies and means of "prefacing," "translator comment," "annotation," "footnote," "epilogue," "postscript," "bold type," "italic type," "capitalization," "compound word," "color transformation," "illustration," and so on to achieve the purpose of creating, remedying, and reconstructing a translational eco-environment for the survival of the translated text.

6.4.2 "Supplementation" in Consecutive Interpreting

"Supplementation" also refers to compensating for elements that are lacking in the target eco-environment. Consider the following example:

[3] See https://www.britannica.com/topic/brahman-Hindu-concept for etymological understanding of the Hindu concept.

[4] See https://www.britannica.com/topic/karma for further understanding of the Hindu term.

- *Scenario*: In a university lecture hall, in the course of an academic lecture given by the chief editor of an American professional journal, a Chinese interpreter is standing beside the speaker to make a simultaneous interpretation.
- *Source utterance*: Talking about copyright transfer, the journal editor states that if a submitted article is accepted for publication, the copyright of the article will then be transferred to the journal. Regarding the publication charges, they depend upon the journals themselves. For instance, no page charges are generally levied for journals published in England, while the printing costs of most American journals are usually high, for example, fifty dollars per page for some journals. *Well, here is a short price list for your reference.*
- *Real-life rendition*: 谈到关于版权转让的问题, 该刊物指出:如果所提交的文章已接受在本刊发表, 那么, 拟发表的这篇文章的版权便将转让给本刊。至于文章的发表费的问题, 主要取决于刊物本身。比如, 在英国刊登的文章, 一般不收取发表费, 而大多数的美国刊物的发表费就很昂贵。譬如, 有些刊物登载一页文章就要收费50美元。你看, 我这儿有一份简短的价格表。*[('视译'价格表)* 《物理评论快》报每页收费*110*美元, 《天文学杂志》每页收费*90*美元*;* 《科学仪器评论》每页收费*65*美元, 《勘误表》每页收费*10*美元*;* 特约发表的研究论文不收发表费。*]*
- *Comments*: The above interpretation is proper, but the 62 italicized Chinese characters do not exist in the source utterance but result from the sight interpretation of the interpreter. Why did the interpreter supplement the original remarks? It turned out that the addition was the speaker's original intention, for the source utterance contained the words "Well, here is a short price list for your reference," which showed that he originally intended to inform the audience. However, the hall was so large that it was impossible for the audience (especially people in the back rows) to see clearly, so the speaker handed the interpreter a list for sight interpretation.

Here, the "limitations" of the translational eco-environment of the interpretation scene or "deficiencies" of the translational eco-environment elements often bring about ecological "defects and insufficiencies" of the target language, which will further result in informational "defects and insufficiencies" for the audiences so that the "original intention" of the source utterance cannot be completely transmitted. Therefore, the interpreter not only must interpret the content of the source utterance but should also boldly employ the "supplementation" strategy to compensate for these deficiencies and reconstruct the target eco-environment, maintaining the relative balance of the source utterance and real-life rendition ecologies to achieve better cross-cultural and cross-linguistic communications (Hu, 1991, pp. 58–1991).

6.5 "Leaning on" Either the Source-Text Ecology or the Target-Text Ecology

The "leaning on" strategy refers not only to leaning on the source-text ecology but also to leaning on the target-text ecology. On the one hand, leaning on the source-text ecology mostly represents the highly foreignizing treatment of the source text in the translation process, which refers to the translator adapting him-/herself to the translational eco-environment of the source language. On the other hand, leaning on the target-text ecology mostly represents the highly domesticating treatment of the source text in the translation process, which refers to the translator adapting him-/herself to the translational eco-environment of the target language.

6.5.1 Leaning on the Source-Text Ecology

When the translator adapts him-/herself to the translational eco-environment of the source language and emphasizes maintaining the source-text ecology, leaning on the source-text ecology, generally speaking, will cause the translation to more accurately convey the meaning of the original.

Viewed only from the perspective of linguistic ecology, the source-text ecology involves the original linguistic form, the original syntactic structure, the original rhetorical need, the original author's ideas or thoughts and feelings, the original cultural background, etc.

Take, for example, the first sentence of *Pride and Prejudice*: "It is a truth universally acknowledged, that a single man in possession of a good fortune must be in want of a wife."

All of the previous translations published by Yilin Press, People's Literature Press, Shanghai Translation Publishing House, etc. inverted the order of the truth and the content of the truth, with the former behind the latter. Ma (2000/2003), Cheng (2003), and Zhang (2004) all think that this sentence should, in accordance with the original word order, be translated as "世间有这样一条公认的真理—凡财产丰厚的单身男人势必缺太太" (by Ma Hongjun), which is the best way to reproduce the original rhetorical device "bathos" and retain the original humorous charm of formal style followed by wit. This translation, to a certain extent, reflects the translation method of leaning on the source-text ecology.

For another example, consider the phrase "Looking for a needle in a haystack."

This idiom is generally translated as "大海捞针"; no one is likely to express it as "草垛里找针." However, consider the following sentence:

> The Peace Commission must, therefore, pack its bags and return to the United Kingdom rather than cause further resentment by looking for a needle in a haystack.

If the part "looking for a needle in a haystack" is translated as "大海捞针" without "leaning on" the source-text ecology, from the perspective of the original cultural ecology and communicative ecology, there will be doubts about the adequacy of the translation in terms of the source text, even though it is not bad in terms of meaning. The reason is that, in the above example, "the Peace Commission" was the public opinion investigation group sent to Rhodesia by the UK, but its investigation aroused strong opposition from the local people, and the African delegates compared the activities of the Peace Commission to "looking for a needle in a haystack" not only because the commissioners could not find the "public opinion" they were looking for but also because would turn the haystack into a mess. If it is translated as "大海捞针," it would not express the resentment of the local people.[5] As a result, to "lean on" the source-text ecology, namely, to adapt to the translational eco-environment of the source text, the translator expressed "looking for a needle in a haystack" as "草垛里找针," which seems to achieve an appropriate degree of holistic adaptation and selection.

6.5.2 *Leaning on the Target-Text Ecology*

When the translator adapts to the translational eco-environment of the target language and emphasizes maintaining the target-text ecology, leaning on the target-text ecology, generally speaking, will enable better care for the readers' reading habits and improve the translation's readability.

In this regard, the translations of Yan Fu, Lin Shu, etc. can be regarded as typical translation cases of "leaning on" the target-text ecology. As discussed previously, Yan Fu practiced "prefacing," "example changing," "footnoting," etc. in his translation work, which can be viewed as his choice to "lean on" the target-text ecology to adapt to the translational eco-environment of the day in the process of translation.

As a translation strategy or method, "leaning on" either the source-text ecology or the target-text ecology, as illustrated above, is a kind of "polarized" translation method. On the one hand, this method does not conform to the Eco-translation concept of "Translation as Eco-balance." In this sense, strictly speaking, it is inappropriate to consider such an approach "Eco-translation" (because "ecology" avoids "polarization," and extremes can bring about an "imbalance" between the source-text ecology and the target-text ecology). However, either the source-text ecology or the target-text ecology can be "leaned on." As a requirement of the translator's adaptation to a certain specific translational eco-environment or as a means of the translator's initial adaptive selection, this approach is sometimes applicable to the treatment of certain texts, especially nonstandard texts. Therefore, the author illustrates it with other translation strategies or methods, thus conforming to the "multiple symbiosis" principle of Eco-translation ethics.

The author has pointed out that if seen only from the text, Eco-translation can also refer to "textual transplants" based upon the source text and the target text. Moreover, in the operational process of "textual transplants," the translator can select varying degrees of "leaning on" either the source-text ecology or the target-text ecology. As a result, "leaning on" either the source-text ecology or the target-text ecology is, in the final analysis, still the translator's adaptation and selection based on the translational eco-environment. In other words, the translator adapts him-/herself to the translational eco-environment and makes the translation selection adaptive to the translational eco-environment. This kind of "selective adaptation" and "adaptive selection" is the most basic method of Eco-translation.

[5] See Chen (2005, pp. 98–93).

6.6 "Naturalizing" or "Greening" the Translation

"Naturalization," also called "ecologicization," has two meanings here: the first is to translate "natural" and "ecological" things and texts about "nature" and "ecology" (such as eco-literature, eco-criticism, eco-thought, eco-philosophy, and eco-civilization texts); the second is "shape-imitation" translation, namely, "naturalizing" the translation by a special translation treatment of specific texts to imitate their natural shapes and thus manifest the "natural" and "ecological" originality, image and style of the translated versions.

6.6.1 Translating the Natural World

Su (2010), CEO of the Taiwan Association of Translation and Interpreting, pointed out in the paper "'Ecological Translation': Some Common Mistakes in Translating Plant and Animal Names in Terms of Eco-translatology" that "translational eco-environment" and "ecological translation" are different terms with different interpretations. The former refers to the global and specific environments for translating, while the latter, in his paper, refers specifically to the translation of plants and animals in the natural eco-environment. Some plant and animal names have long been incorrectly rendered into Chinese yet have seldom been the subject of critical scrutiny. Most translators seldom question or doubt the accuracy of the equivalent plant and animal names that they find in English–Chinese dictionaries. Some classic mistakes have existed for so long that translators accept them as part of their vocabulary for example, "pu ti" 菩提 for "linden," "tu ying" 秃鹰 for "bald eagle," and "sha ren jing" 杀人鲸 for "killer whale." Some misleading translations are relatively new, such as "cha shu jing you" 茶树精油 for "Australian tea tree oil"; therefore, we had better correct them before they gain popular currency. In terms of Eco-translatology, Mr. Su in this paper advocates the use of scientific names as an effective and easy-to-follow method to assist translators in achieving accuracy when called upon to translate plant and animal names. He gives examples of some common mistakes in the rendition of plant and animal names and endeavors to provide translators with a reliable methodology for choosing target-language equivalents for plant and animal names commensurate with the context in which they work well.

It should be noted that the Eco-translatology naturally includes the "greening" translation in its true sense of plant, animal, mountain, water, soil and stone names and the related research, even though these are not necessarily the main study points.

6.6.2 Imitating Natural Shapes

Forming a translation based on "mountain," "tree," and "leaf" shapes is quite similar to the "creative spelling" game, which can also be seen as the "creative" treatment of the translation from the "natural" and "ecological" perspectives.

McNaught (2012) of London University provided some useful C-E and E-C instances in his speech at the 3rd International Symposium on Eco-translatology in Chongqing.

This kind of translation technique can create "natural" and "ecological" word images from certain "non-natural" and "non-ecological" English words. For example:

- (Italicizing): ***wind***ow, ***Mud***dled, ***stream***ing
- (Capitalizing): s***Land***er, **Mountain**ous, s***Tree***t
- (Hyphenating): **Bear**-er, f-**lake**, **sea**-mless
- (Combination of all three): ***Leaf***-let, fl-***Air***, t-***rain***, d-***River***, thr ***Ice***

The application of the above translation skills can also bring about the "natural" effect of the ecology of some dull texts. For instance:

The Original Text
Mentre stavo camminando per strada ho visto un volantino appeso in vetrina.…

The Translated Text
"While I was walking down the s**Treet,** I saw a **Leaf**-let in a shop **Wind**-ow.…

To show "environmental" consciousness, some people also utilize the "splitting" and "branching" methods to produce vivid "natural" and "ecological" effects. Consider the following two examples:

《鹿柴》

空山不见人，但闻人语响。

返景入森林，复照青苔上。

(in Yu 1980:234)

Deer Park

On

An empty

Mountain there is

no one to be seen, yet voices

of man can be heard. Sunlight pierces

the deep Forest illuminating the green Mosses.

(Mountain shape)

Jessica Smith / Greenaway

first leaves

a

s

k

l

d

i

w c

e

a

tt

i d

a

ch

st

(ä)

soft and green with

long

the

road

a

little specks of brown

ll

like bananas

eaves

some trees turn utterly yellow

more quickly than others

In addition, consider the following similar case.

The Original Text

A Christmas Tree

Star
If you are
A love compassionate
You will walk with us this year
We face a glacial distance, who are here
Huddld
At your feet.

(William Burford)

The above-cited pattern poetry is a dynamic integration of content and form: The poem title is "A Christmas Tree", and the lines are arranged in the shape of a Christmas tree with a glittering "star" hanging on the top of the tree, an umbrella-shaped "crown", and a thin "trunk" derived from the poet's deliberate omission of the letter "e" from the word "Huddled".

The Translated Text

圣诞树

星啊
你那爱中
如果含有怜悯
来年就和我们同行。
这里我们面对冰河距离
拥挤
在你脚底。

(Tree shape)

The above translation is a "natural" and "ecological" treatment of the source poem "A Christmas Tree," with "星星" for the "star" on the top of the tree, the shape of the "crown," and the sense of the strong "trunk" created by two bold characters, "拥挤," for "Huddld," which may be considered the best kind of "ecological" translation.

In addition, different images of "expansion," "growth," "down towards the earth," "up towards the sky," and "gradation" can be expressed with "color" to indicate the "natural" and "ecological" pattern of words, as in the following examples:

绿树Tree嫩芽Sprouts大地Earth根系**ROOTS**大山MOUNtaIN

6.7 "Intervening in" the Text Production and the Translation Activity

Eco-translatology at the microlevel focuses mainly on the restrictions and influences of the basic concepts of Eco-translatology on the translation text production, translation operations, and translation phenomenon interpretations, i.e., various "interventions" in translation texts and translation research. Specifically, in studying "interventions" in micro-texts and translation research, Eco-translatology emphasizes the restrictions and influences on the translation text production, translation operations, translation phenomena, or translation actions from different Eco-translatological perspectives, including the rational traits of the ecosystem, the holistic ecosystem, the interrelated translator community, and holistic adaptation and selection. However, the translations intervene only in the specific "translational eco-environments" and in the patronage of the "translator community," as explained and exemplified here.

6.7.1 Translation Interventions in Specific Translational Eco-Environments

The "translational eco-environment" is one of the key concepts of Eco-translatology and one of the major perspectives of the systemic theories of Eco-translatology used to interpret translation phenomena. The term "translational eco-environment" can be defined as the worlds of the ST and the source/target language, namely, the interrelated and interactive entirety comprising the linguistic, communicative, cultural and social aspects of translating, as well as the author, the client, and the readers (i.e., "translator community"). The translational eco-environment is essentially an aggregation of all the factors associated with the translator's optimal adaptations and selections. It is divisible to the macroenvironment, mesoenvironment, and microenvironment, including the intralingual environment and extralingual environment, material environment and spiritual environment, and, furthermore, subject environment (translators, authors, readers, publishers, editors, supervisors, etc.) and object environment (ST, TT, textual function, translation strategies, translation regularities, etc.). As discussed above, the social, cultural, and humanistic environments in which Yan Fu lived had a vast "intervention" impact on the production of his translations. Again, from the perspective of Eco-translatology, Xu Chi's first consideration of the selections (choices) of foreign works was to meet the needs of his psychological environment, which is one of the internal levels of the translational eco-environment referring to character, viewpoints, likes and dislikes, etc. (Hu, 2004, pp. 102–104, 2009, 2010). One prominent feature throughout Xu Chi's translation activity is that he attaches much importance to the psychological environment. The translator should select the original text to be adaptive to his/her psychological environment for translation; translation methods and translation criteria should be on a scale to fit the translator's and the reader's psychological environment. For the translator, "the ideal is to find the works of the writer whose disposition is close to his/her own; only in this way can he/she resonate with the writer and the writer's works and better reproduce the style of the original author." Eco-translatology holds that the translated texts with the highest degree of holistic adaptation and selection can achieve the "*survival of the fittest*." Many of Xu Chi's translations, including *Walden,* were widely accepted and still have strong vitality after more than a half-century, which can be referred to as "Fragrance survives forever" (Liu, 2011).

There are numerous such examples of the selections of foreign works influenced by translational eco-environment changes. A change in the social environment of economic development will bring a change in word connotations followed by a change in translation methods. For instance, the term "外向型经济" (wai xiang xing jing ji) was translated as "export-oriented economy," which was initially correct at the beginning of China's opening to the outside world. However, with further development, China encouraged and supported superior enterprises to invest in foreign countries, which are no longer limited to their export products, so the connotation of "外向型经济" (wai xiang xing jing ji) changed to "global-market-oriented economy," and the translation changed accordingly.

Similarly, the version "going out" for "走出去" is retranslated as "going global," the reason for which is straightforward: "going out" is to find a way to "economic globalization." Obviously, the production and the change of these translated versions are closely related to the holistic external translational eco-environment. From the above actual examples, it is evident that all aspects of the translational eco-environment in the systematic theories of Eco-translatology have an influence and restriction on translation activities and texts.

6.7.2 *Translation Intervention in the Patronage of the "Translator Community"*

The "translator community" has been defined as an aggregation of the "participants" involved in translating activities that interacts and interrelates with the generation, development, operation, results, function and effects of the translation activities, accommodating the author, target-text (TT) readers, translation critics, translation reviewers, publishers, marketers, patrons, consigners, etc., with the translator as their representative.

The "translator community" refers to all the "participants" related to a specific translation activity with a translator as its representative. That is, it refers to "humans." As we know that "the style is the man," it is indisputable that the "participants" in the "translator community" commonly have an influence and restriction on the translated texts.

As one of the participants in the "translator community," the patron may also intervene in the translated texts. For example, patrons of translation activities often play a crucial role in what works to translate/publish, how to translate/publish a work, etc.

The word "patron" was introduced by Lefevere in 1992. The "patron" manipulates the entire process of translation (Lefevere, 1992, p. 26). Generally, the translator has no choice but to show respect to the power holders, accept the texts prescribed by the patron, and try to make the translated texts satisfy the demands of the patron.

As patrons, classes, governmental departments, and religious groups, when they choose texts, first consider their ideological or educational meaning and do not pay much attention to the original artistic level and literary status. Compelled by the patron, the translator can only adapt him-/herself to the patron's requirements. One of the most striking features of the influence of classes, governmental departments, and religious groups on the translation of exotic ideology and cultural classics is that they can promote or hinder the course of translation and the dissemination of the translated texts. In other words, they, through their power, exert pressure on the translator in the choice of texts and the translation process to further influence and guide the dissemination and development of the target ideology (Gentzler, 1993).

The translation of Buddhist scriptures is the first significant period of China's translation history, which lasted more than one thousand years, from the late Han Dynasty to the early Song Dynasty. It was the first large-scale, systematic, and organized translation activity in the history of China. Before Buddhism was introduced into China, social conflicts were gradually intensifying, so the ruling class wanted to reinforce its rule with religion. As an exotic idealist theology, Buddhism fit in readily with the Chinese ruling class, who brought the idealist philosophy into full sway and took advantage of Buddhism as a tool of class and ethnic oppression. Therefore, the Buddhist scriptures were maximized. The patrons of Buddhist scriptures translations were feudal lords and other top rulers. When Buddhism was first introduced into China, it soon found its integration point with Chinese culture: the claim of the eternal soul. However, Indian Buddhism did not claim that humans have an eternal and unchangeable soul, which was incompatible with the traditional Chinese claim of the primordialqi soul, through which the introduction of the theory of *saṁsāra* and *karma* into China was accepted. Of course, to be conducive to the spread of Buddhism in China, the Buddhist Scriptures translations had to maximize the ruling-class ideological demands to win the rulers' sponsorship and support.

In contrast, if the demands of the patrons (the ruling classes) could not be satisfied, the translation activities would be hindered and even fail. In the early eighteenth century, the Vatican attempted to interfere with China's internal affairs, and many missionaries did not respect traditional Chinese liturgical rites and customs, which aroused the antipathy of the Qing government. In 1724, the Yongzheng Emperor of China banned Roman Catholicism and adopted the closed-door policy, the execution of which brought Chinese and foreign cultural exchanges and the cause of translation to a standstill. Certain translators even found translation activity to be fatal; for example, William Tyndale, a British religious translator, and Étienne Dolet, a French religious translator, were burned at the stake.[6]

The patron also manipulates the translator's selection of the original and target texts, which results in the translator's passive cooperation and adaptation. As a translator and thinker, Yan Fu had a definite political objective in the selection of translation materials. At the same time, he was an official appointed by the Qing government, which caused his selection of translation materials to be manipulated by the ideology of the patron. His translation of "*Missionaries in China*" (by A·Michie) in 1892 was carried out under the command of Li Hongzhang. Matteo Ricci was the first person in the Ming Dynasty to bridge Chinese and foreign cultures; his translation activities were manipulated by the monarch of the late Ming Dynasty and the top scholar-officials in feudal China. When the Chongzhen Emperor decided to adopt the Western calendar and started to modify the old calendar, the translation and introduction of *Hun Gai Tong Xian TU Shuo* and *Qian Kun TI YI* by Matteo Ricci thereupon came into being.[7]

Additionally, if we consider, for example, Lin Shu's translation activities, we will note that Lin Shu does not understand foreign languages, and his translated works are based mainly on his assistants' narrations and his own processing and transformation of the text.[8] Regardless of the original contents and forms, he translated Shakespeare's dramas *Henry IV* and *Julius Caesar* and Ibsen's drama *Ghosts* into novels.[9] What was the reason for the popularity of his translations? From the perspective of cultural and historical development, it was the extensive support of literary patrons at the time. In the era of Lin Shu, when the bourgeois reformist trend of thought proliferated, a large number of people such as Kang Youwei and Liang Qichao committed themselves to all kinds of literary works with political novels as their priority to achieve the goal of political reform. Lin Shu's translated works 《巴黎茶花女遗事》 (*The Lady of the Camellias*) and 《黑奴吁天录》 (*Uncle Tom's Cabin*) were created to adapt to this trend; thus, it went without saying that the top reformers of the day would support them. In discussing the Late Qing Dynasty translation compromise in content, Zheng (1936) points out in *The Influence of Translated Novels in the Late Qing Dynasty on the New Literature* that in the work of translation, the original was even amended in a compromise with China's old forces.

---

[6] Wang (2003, p. 16).

[7] Tu and Wang (2005).

[8] See Qian (1981).

[9] See Feng (1959).

A similar situation existed after the founding of the new China. In the 1950s and 1960s, the wide translation of Russian classical literature and literary works of the Soviet Union was inseparable from the strong support and guidance (even the necessary arrangement) of the relevant government departments and political parties of the day. In the early years after the liberation, China International Publishing Group focused mainly on the translation of works on Mao Tse-tung's and other national leaders' thought and the literary works of Russia and the Soviet Union; thus, the translation of Western modernist literature was pathetically meager.

With the development of modern translation and the commercialization of the translated texts, an increasing number of works were not directly presented on the basis of the translator's preferences. The production of the translated versions had to be integrated with many other factors that had already intervened at the pre-translating stage. The purpose of translation actions today is much more extensive than it formerly was; it may include propagating culture, establishing an image, the communication of ideology, or explaining prevalent business practices. Moreover, no matter what the purpose is, it will have an influence on the translator's practice and the final forms of the translated texts before the translator starts the translation process as well as the shaping of the translated texts. Obviously, in the era in which the translated works have been commercialized, the purpose of the patrons promoting the publication of the translated texts has increasingly involved manipulating the final forms of the translated works. The translator's respect for the original, the execution of the rewriting strategy, or the selection of foreignization or domestication have all become more dependent on the patron.

From the above, it can be seen that, at all times and in all places, the patron plays a vital role in translation activities as a member of the "translator community" and as one of the elements of the translational eco-environment.

6.8 Summary

Any science begins with philosophy and ends in art. Translation studies involve the interplay of theoretical models and case studies. This chapter illustrates Eco-translation at the microlevel. At the microlevel, the translation process is described as alternating cycles of translator adaptation and selection, translation strategies/methods generalized as "three-dimensional transformations" (linguistic, cultural, and communicative dimensions), and translation criteria are defined in terms of the degree of multidimensional transformation, specifically the holistic degree of adaptation and selection. Eco-translatological studies are thus characterized by holistic eco-vision at the macrolevel, epistemological adaptation/selection at the mesolevel, and operational applicability at the microlevel.

Different types of target texts can be produced following ecological principles. In this chapter, various text types covering literary, social, scientific, commercial and political/diplomatic translations are analyzed and interpreted in terms of textual ecologies.

Eco-translation strategies and methods are illustrated with 12 actual examples. These strategies and methods include (1) "transplanting" texts in multiple dimensions (linguistic, cultural, communicative, etc.); (2) "emptying" the translator's mind to absorb or reshape the new eco-environment before translating; (3) "compensating" for the elements lacking in the target eco-environment; (4) "leaning on" either the source-text ecology or the target-text ecology; (5) "naturalizing" or "greening" the translation; and (6) "intervening in" the text production and the translation activity by following the eco-concepts and principles of Eco-translatology.

Obviously, the above Eco-translation strategies and methods as well as the illustrations and demonstrations embody the theoretical deliberations of Eco-translatology.

Last but not least, the author would also like to express the following idea.

On the one hand, as this work focuses on the theoretical construction of a translation discourse system, it is reasonable not to cover micro-level translation operations. For example, some translation theories of the West, such as polysystem theory, Skopos theory, and cultural theory under deconstructionism, mostly lack monographic studies and descriptive studies regarding the specific aspects of micro-operation and translation method. The author believes that in a systematic translation theory, this should be regarded as a defect. On the other hand, if in a work of a theoretical nature, the author discussed certain translation methods and skills in depth or even proposed such "instructions" as "must be translated this way" and "cannot be translated that way," this theory might become trapped in the age-old "prescriptive" question of translation theory. At the same time, the prescriptive approach does not conform to the fundamental principles of eco-reason.

In view of the above, to balance content and form, the title of this chapter refers to "micro" research, but the content of this chapter puts a greater emphasis on the "explanation" of case studies from the perspective of Eco-translatology, which means it considers analytical interpretations of translation examples. Regarding how to translate in accordance with ecological translation principles, translators can follow the concepts of Eco-translatology in their translation process, translation actions, and text transplantation through their own judgment and choices according to factors such as people, causes, environments, and time.

In addition, although this chapter discusses different topics, with case studies and analysis of the translation process and interference in translation, from perspectives such as macro-level eco-reasons and entire translational ecosystems and subsystems, the interpersonal "translator-community," and the integration of adaption and selection, certainly, it can be said that there are still many perspectives, dimensions, and aspects that are not covered, and it is inevitable that these will emerge. At the same time, because the current stage is a period of the construction of the theoretical system of Eco-translatology, it is understandable that the author has no effort to spare, while it is not very necessary, to perform an in-depth systematic analysis of theoretical applications.

Nevertheless, it is obvious that the systematic theory of Eco-translatology can describe the phenomenon of translation, explain translation actions, and guide translation practices and is therefore proven workable. Even if some translation examples can be interpreted by other theories, in these cases, the concepts and methods of Eco-translatology are similar to those of the other theories in the interpretation of these translation phenomena. Therefore, from the perspective of the applicability and operability of translation theory, the theoretical system of Eco-translatology can describe and interpret the text production process and translation actions at the microlevel, which makes micro-level studies the foundation for the survival and development of Eco-translatology.

References

BretaAart. (2012). Translation as breathing new life in a new language: Conversing with Charlotte Mandell. *Emprise Review*, No. 10. http://emprisereview.com/interview/translation-as-breathing-new-life-in-a-new-language

Chen, Q. J. (2005). *Theory and practice of English-Chinese translation* (pp. 98–93). Wuhan: Wuhan University Press.

Chen, Y. G. (1989). *History of Chinese translated literature*. Beijing: China Translation & Publishing Corporation.

Cheng, Y. S. (2003). Descriptivism and communication-two themes in modern translation studies of China. *Journal of Anhui University, 1*, 89–94.

Fang, M. Z. (1999). *Translation research and practice*. Qingdao: Qingdao Publishing House.

Feng, Z. (1959). Translation and introduction of the literature of Russia and other European Countries during the May 4th Era. *Journal of Beijing University, 2*, 131–151.

Gentzler, E. (1993). *Contemporary translation theories*. New York: Routledge Inc.

Han, D. (1969). *History of modern translation*. Hong Kong: Chenheng Books Company.

https://www.britannica.com/topic/brahman-Hindu-concept for etymological understanding of the Hindu concept.

https://www.britannica.com/topic/karma for further understanding of the Hindu term.

Hu, G. S. (1991). A branch of learning with great efforts. *Out-bound Communication and Exchanges Journal, 8*, 28–32.

Hu, G. S. (2009). An "Ecological Turn" signified by the "Sequence Chain". In G. Hu (Ed.), *Translation and cross-cultural communication: Integration and innovation* (pp. 26–39). Shanghai: Shanghai Foreign Language Education Press.

Hu, G. S. (2010). Eco-translation vs. natural ecology: Relevance, similarity, isomorphism. *Shanghai Journal of Translators, 3*, 1–7.

Hu, G. Z. (2004). *Translation as adaptation and selection*. Wuhan: Hubei Education Press.

Hu, K. T. (2011). *Eco-translatological reflection on translating a mystic text*. Paper presented at the 2nd International Symposium on Eco-Translatology, November 11–14, Shanghai.

Hu, K. T. (2012). On applying Eco-translatology: A case study. *Journal of Eco-translatology, 2*, 28–32.

Huxley, A. (1932). *Brave new world*. London: Chatto & Windus.

Lefevere, A. (1992). *Translation, rewriting and the manipulation of literary fame* (pp. 22–26). London: Routledge.

Liu, A. (2011). Insight into "translational eco-environment". *Journal of Eco-Translatology, 1*, 80–92.

Ma, H. J. (2000/2003). *Translation criticism*. Beijing: China Translation & Publishing Corporation.

McNaught, D. (2012, November 23–25), *Ecotranslation theory: A new ecocritical approach to highlighting environmentalism in translation studies*. Presented at the 3rd International Symposium on Eco-Translation, Chongqing, China.

Newmark, P. (1991). *About translation*. Philadelphia: Multilingual Matters Ltd.

Qian, Z. S. (1981). *Lin Shu's translation*. Beijing: The Commercial Press.
Su, Z. L. (2010, November 9–10). *"Eco-translatology": Translation and misinterpretation of Flora and Fauna*. Paper Presented at the First International Eco-Translatology Symposium, Macau.
Tian, W., & Hu, W. X. (2005). On Yan Fu's ethical concept in "TIANYAN LUN" and Its influence on Chinese traditional concept of ethics. *Teaching and Research, 7*, 59–63.
Tu, G. Y., & Wang, F. H. (2005). On the translator's choices: A case study of Matteo Ricci. *Chinese Translators Journal, 2*, 20–25.
Wang, D. F. (2003). An invisible hand: Ideological manipulation in the practice of translation. *Chinese Translators Journal, 5*, 16–23.
Zhang, N. F. (2004). A critical introduction to aixelá's strategies for translating culture-specific items. *Chinese Translators Journal, 1*, 18–24.
Zheng, Z. D. (1936). The influence of late Qing Dynasty on new literature. In W. Kefei (Ed.), *Studies on translation culture* (pp. 157–162). Shanghai: Shanghai Foreign Language Education Press. 1997.

Chapter 7
Theoretical Applications and Applied Studies

7.1 Introduction

The source of theory is practice, and the value of theory is also reflected by practice. Just as "the value of theory lies in the degree of reality needed" (Karl Marx), "the establishment of a disciplinary system comes from the call of reality," and "the final value of a theoretical work lies in the fact that it can provide a brand-new way of thinking for people, put up a scaling ladder to the truth, and abstract some regular things."[1]

In the practice and development of Eco-translatology in recent years, researchers in different fields have developed colorful applied studies and theoretical applications to fill various "vacancies" of translation studies on the basis of understanding ecology, mastering translatology, and grasping Eco-translatology in combination with their personal research interests and the existing results.

In theoretical applications and applied studies, this chapter aims to answer or clarify the following problems. What theoretical viewpoints and academic standpoints of Eco-translatology can be applied? What are the relevant theoretical applications and applied studies in recent years? In particular, whether can theoretical applications of Eco-translatology provide "new interpretation" for "hot topics" (such as translatability/untranslatability, retranslation, translation strategy employment, "turns" of translation studies, etc.)? As a theory founder, what specific overviews and comments on the related theoretical applications and applied studies can the author make? This chapter will make the related discussions and interpretations centering on these problems.

[1] See Wang (1996, pp. 56, 59).

(Hugs) G. S. Hu, *Eco-Translatology*, New Frontiers in Translation Studies,
https://doi.org/10.1007/978-981-15-2260-4_7

7.2 Applied Cases of Practical Studies (1)

The life of theory lies more in application than in innovation. Based on the above-mentioned theoretical studies, projects of applied studies with an Eco-translatological nature have been undertaken in different areas.

7.2.1 Studies on the Translation of Literary Works

Zhai and Zhang (2005) of the School of Foreign Language Studies of Anhui Normal University discusses "Adaptation and selection: A study on Lin Yutang's translation strategies employed in *Six Chapters of a Floating Life (Fusheng Liuji)*." In this paper, the author chooses linguistic and cultural dimensions to analyze how Lin adapted and selected in the process of translating *Fusheng Liuji*, which is an expressive text. The study shows that in the linguistic dimension, adhering to the idea that "the beauty of the work lies not in its contents but in its form, which is the kernel of literature," Lin strove to retain the original form. He directly translated the four chapters of the original as "Wedded Bliss," "The Little Pleasures of Life," "Sorrow," and "The Joys of Travel," keeping the narrative first-person perspective and the author's inner narrative time order centered around bliss, pleasure, sorrow, and joy. Moreover, the division and arrangement of the original paragraphs in each chapter were retained in Lin's version. However, at the sentence level, due to the significant differences between classical Chinese and modern English, it was impossible to retain the same patterns in the translation. Conscious of his responsibility to the target-text readers, Lin did his best to conform to English syntactic norms. Regarding the cultural dimension, Lin Yutang, who had a high level of bilingual and bicultural proficiency, chose to employ both domestication and foreignization. Through domestication, Lin adopted a transparent, fluent style to minimize the strangeness of the source text for TL readers. At the same time, he also occasionally used the foreignization strategy to satisfy the curious TL readers' expectation of the distinct foreign cultural features of the original version. The different functions of the two translation strategies were employed properly and made mutually complementary in Lin's version.

Hemingway's short story *Cat in the Rain*, published in the 1920s, through a description of the heroine wanting to save a cat in the rain, implied the struggling and awakening of women's consciousness, revealed the barren and confused state of the post-war American spiritual world, and expressed the writer's worries about the mental condition of contemporary Americans. Wu and Li (2006) of the Foreign Language School of Ludong University perform a comparative analysis of the original text of Hemingway's *Cat in the Rain* and the translated version by Cao Yong from the perspective of Eco-translatology in which they point out the translator's central position and dominant function, describe the translator's adaptation and selection process, demonstrate the restriction mechanism for the translator, and explore the translator's role in literary translation.

"Hao-Liao-Ge" has been regarded as the keynote piece of Dream of the Red Chamber, which expresses the author's grief and indignation about Confucian ethics. Jiao (2008, pp. 124–131) offer a comparative analysis of two English versions, translated by Yang Xianyi and David Hawkes, from the perspective of the three-dimensional transformation method proposed by Hu Gengshen in his *Approach to Translation as Adaptation and Selection*, i.e., cultural dimensional transformation, communicative dimensional transformation, and linguistic dimensional transformation. The comparative study finds that Yang's version makes adaptive selections and transformations towards the source culture in terms of the cultural and communicative dimensions, while Hawkes' version has a tinge of Christianity in the target culture and replaces Confucianism and Buddhism in the source culture. The lesson that can be drawn from the "three-dimensional transformation" translation method is that the "three-dimensional transformation" is essential or indispensable and that the translator's selective adaptation and adaptive selection are practically useful when translations are performed.

Jiang (2009) of Macao Polytechnic Institute analyzes the three English versions of *The Peony Pavilion* by employing "multidimensional adaptation and adaptive selection" as the theoretical tool. Specifically, the characteristics, gains and losses of the three English versions of *The Peony Pavilion* are analyzed at the linguistic, cultural, and communicative levels. Through analysis and comparison, he concludes that according to the basic principle of "multidimensional adaptation and adaptive selection," the translator should in the process of translation consider the multidimensional adaptive selection transformations, which include the cultural, communicative, and linguistic dimensions; at the same time, he/she should pay close attention to the combination, balance and coordination of the three dimensions so that the successful translation can reflect the translation ecosystem of "holistic harmony" (Hu, 2008). Therefore, the translator needs to adapt to the translational eco-environment; the "participants" such as the author, the translator, and the reader, of the translational eco-environment need to maintain "harmony"; and the translated work and the original work also need to be "in harmony." All of this shows the translator-dominated "selective adaptation" and "adaptive selection."

7.2.2 Studies on the Translation of Philosophical and Sociological Literature

It has been more than 100 years since Yan Fu translated *Evolution and Ethics* and proposed the three difficulties of translation—"faithfulness, expressiveness and elegance"—in his *Introductory Remarks of Tian Yan Lun*. In the past 100 years, the topic of Yan Fu's translation and his translation criteria has been a constant presence in the translation field. Jiao (2010), with the "approach to Translation as Adaptation and Selection" as a translation tool, discusses and interprets Yan Fu's translation and his three difficulties of translation—"faithfulness, expressiveness and elegance"—from a new perspective by applying both the qualitative research method (including theoretical citation and logical inference) and the quantitative research method (including illustrations and case studies). The study shows that Yan Fu's translation of *Evolution and Ethics* is the corollary of Yan Fu's "adaptation" and "selection" of his translational eco-environment. Explicitly, Yan Fu's translation of *Evolution and Ethics* fully embodies the "three-dimensional" transformations (the adaptive selection transformations of the cultural, communicative, and linguistic dimensions). The study also shows that Yan Fu's "three difficulties" of translation (faithfulness, expressiveness, and elegance) and Hu Gengshen's "three-dimensional" transformations have intrinsic and logical corresponding relationships. Precisely, Yan's "faithfulness" corresponds directly to Hu's "linguistic dimensional" adaptive selection transformation; Yan's "expressiveness" corresponds to Hu's "communicative dimensional" adaptive selection transformation; and Yan's "elegance" corresponds to Hu's "cultural dimensional" adaptive selection transformation. This study surveys Yan Fu's translation and interprets the translation criteria of "faithfulness, expressiveness and elegance" from the perspective of the theory of Translation as Adaptation and Selection, which is believed to be conducive to broadening the research on Yan Fu's translation.

As one of the most authoritative classics of the Confucian school in Chinese history, *The Confucian Analects* has exerted a profound influence on both Chinese thought and culture and even on the development of civilization worldwide. In recent years, the deepening process of globalization and renewed popularity of Confucianism have led to the revitalization of both the study and the translation of Confucian works, while the English translation of *The Confucian Analects* has gained increased attention in the fields of Sinology and translation. In the thesis "An Eco-translatological Perspective of the Studies on the English Translation of *The Confucian Analects*," Sun (2004, 2010) of the School of Translation and Interpreting of Qufu Normal University conducts a preliminary study on the six representative English versions by James Legge (1861), Ku Hung-ming (1898), Arthur Waley (1938), Lin Yu-tang (1938), Ezra Pound (1951), and Roger Ames (1998) from an Eco-translatological perspective. With the multidimensional and multiparameter perspective on harmonious translation as the theoretical support, by putting the six versions in the corresponding macro-level and micro-level social environments of the translation, the researcher explores the relationship of those elements, such as historical context, cultural awareness, communication purpose, and language strategies, in the process of the translation.

In 2009, Associate Prof. Liu Ya-feng of the School of Foreign Languages of Huzhou Teachers College received her doctoral degree from Shanghai International Studies University with a dissertation entitled "Translator's Adaptation and Selection: Translation Process Research for China's Global Communication." She spoke on "Research on the Eco-environment of Translation for China's Global Communication" at the First International Symposium on Eco-translatology in Macao in November 2010. This paper studies in detail the eco-environment of translation for China's global communication. First, the significance of research on translation for China's global communication and its eco-environment from the perspective of Eco-translatology is expounded. Second, the eco-environment of translation for China's global communication is appropriately defined. Furthermore, this paper thoroughly discusses the function of the ecological factors of the eco-environment of translation for China's global communication and the interrelation of these factors. Finally, this paper proposes a viewpoint and suggestions on how to consciously conserve and maintain the eco-environment of translation for China's global communication and how to construct an eco-environment of translation for China's global communication of sustainable development.

Liu (2009) of the School of Foreign Languages of Zhongnan University of Economics and Law conducts a study on the translation of metaphorical idioms in news reports from the perspective of Translation as Adaptation and Selection. Such metaphorical idioms, the cream of language, account for a significant proportion of news reports. The translation of metaphorical idioms in news reports is not only the transformation of language but also a means of communication and exchange between Chinese and Western cultures, which is the difficulty of translation. As a new translation theory in the new century, the approach that "translation is the translator's adaptation and selection" advocated by Translation as Adaptation and Selection provides a new basis and perspective for translation practice. The study in this paper attempts to identify appropriate ways to translate metaphorical idioms from news reports from the perspective of Translation as Adaptation and Selection. The study shows that because Translation as Adaptation and Selection advocates the translator's adaptation to a particular eco-environment at "multidimensional" or at least "three-dimensional" (linguistic, cultural, and communicative) levels and the selection of an appropriate translation, the translator is required to start from the translation principles of Translation as Adaptation and Selection, consider the objective "eco-environment" in news reports, and adopt flexible translation methods to optimize and select the translation with the highest "degree of holistic adaptation and selection."

7.2.3 Studies on the Translation of Sci-tech/Commercial/Legal Documents

Disease names are essential terms in traditional Chinese medicine (TCM). Many TCM disease names originate from *Suwen of Huangdi Neijing* (*The Yellow Emperor's Classic of Internal Medicine*). Different translations of disease names have been given in different English versions of this book. Such chaos makes thorough research necessary. Li Y. Z. (2009) of Nanjing University of Chinese Medicine adopts the Translation as Adaptation and Selection approach as a theoretical guide; takes the translation of disease names in *Suwen of Huangdi Neijing* (*The Yellow Emperor's Classic of Internal Medicine*) as the study object; focuses on the similarities and differences between the translations of those disease names in the three English versions by Li Zhaoguo, Ni Maoshing, and Iliza Veith; and investigates the adaptive selection process of the different translators mainly from the three-dimensional transformation perspective of the linguistic, cultural, and communicative dimensions. The results of the research are as follows: First, it is the different translation ecological environments that have chosen the three different translators with different knowledge backgrounds. Second, different translators from different ages give different translations of each type of TCM disease name. Next, the proportions of the numbers of different types of TCM disease names are very different. Last, the Translation as Adaptation and Selection approach can explain the translation process and results.

Huang (2010) of Southwest University of Political Science and Law explores "Legal Texts Translation: An Eco-translatological Perspective," which focuses on legal text translation in "three dimensions": the linguistic dimension, the cultural dimension, and the communicative dimension. She holds that in legal translation, the translator should consider the unique characteristics of legal language, paying attention to the linguistic adaptive transformation; then, consider the cultural aspects of the source language and target language, such as history, tradition, religion, and legal systems, in making the cultural adaptive transformation; and finally, consider the communicative adaptive transformation by embodying the communicative intention of the source text in the translation. Therefore, under the guideline of the holism of Eco-translatology, translation can be accomplished appropriately.

Cao (2007) of the School of Foreign Languages, Changsha University of Science & Technology, explores several problems of C-E tourism translation in combination with actual examples under the guidance of Translation as Adaptation and Selection. She states that in the process of translation, the translator should make adaptive selection and transformation choices in the linguistic dimension, the cultural dimension, and the communicative dimension and make flexible use of translation strategies and methods, which is crucial in improving the quality of tourism translation and communicating Chinese culture.

Yingzhen (2010) of the Faculty of International Studies of Henan Normal University performs "A Textual and Strategic Analysis of Henan Tourism Translation from the Perspective of Eco-translatology." In this paper, she points out that Henan Province has abundant cultural heritage tourism resources. Henan has 96 state-protected historical sites and 1.3 million cultural relics, which constitutes one-eighth of the amount in the entire country. It is rich in not only humanistic tourism resources but also natural landscapes. There are many national scenic spots in Henan, such as Mount Song in Zhengzhou, Longmen Grottoes in Luoyang, Mount Yuntai in Jiaozuo, and Red-Flag Canal in Linzhou. The provincial government proposed the strategy of "Priority to Tourism Development" and the slogan "Rich Cultural Heritage and Magnificent Natural Landscapes" and targeted receiving 400 million domestic and foreign visitors (person-time) annually by 2015. However, the author's field survey indicated that significant issues exist in Henan tourism translation. She intends to conduct a textual and strategic analysis of Henan tourism translation from the perspective of Eco-translatology and holds that different translation strategies should be adopted to render different types of tourism versions to fit their environment.

Fan Min of the School of Translation and Interpretation of Qufu Normal University studies "The English Translation of Chinese Public Signs from an Adaptive and Selective Perspective." In this paper, she analyzes Chinese public signs and their translations from the perspective of Translation as Adaptation and Selection. With a view to the reader's response, the English version should be adaptive and selective according to the translation purpose. She points out that the translation of public signs is a selective activity that the translator applies to the translational eco-environment. The adaptation is achieved by the translator selecting the approach to the translation of public signs to conform to the target environment, and the selection is achieved by the translator selecting the approach to the translation of public signs to adapt to the target culture. The translation of public signs is a form of cross-cultural communication. In the process of translation, the selection process of retaining, substituting, or interpreting within the cultural differences depends on the translation purpose and the reader's acceptance. In addition, the adaptation strategy should be adopted to different degrees. The selection of the target version should be an adaptive dynamic process according to different contexts to maximize the needs of cross-cultural communication.

7.2.4 *Studies on Translation Teaching*

With the establishment and development of translation studies in recent years in China, the corresponding research on translation teaching and translation talent cultivation has become a top priority. Can the emerging Eco-translatology studies adapt to this change and contribute to it?

Song (2010) of the School of Foreign Languages of Shanghai Maritime University explores "Translation Pedagogy Ecosystem: An Eco-translatological Perspective." In this paper, he points out that as part of the microstudy of Eco-translatology, the translation pedagogy ecosystem is constituted by the study of disciplinary construction, textbook compilation, teaching methodology, testing and evaluation, etc. Guided by the view of holistic interrelationship, the kernel of the translational ecosystem, the study of translation pedagogy hopes to reveal the intrinsic nature of its constituents in terms of interrelationship, inter-restriction, and interaction. By making a preliminary study of the translation teaching subsystem of the translation education ecosystem at the level of the translation ecosystem and rethinking the present situation of translation teaching, he holds that the recommendations for translation teaching based on the findings can be summarized as follows: (1) to orient grammar-based teaching of translation techniques toward holistic interrelated practices; (2) to change the simple binary evaluation of a translated text to the appreciation of its aesthetic value; (3) to perfect the evaluation system, which is an important part of the Eco-translational teaching system; (4) to steer the grammar-based compilation of translation textbooks to the context of Eco-translatology; (5)

to adapt the give-and-take relation between teacher and learner to life interactivity in translational ecosystems, which is a natural expression of the balanced translation pedagogy system; and (6) to pay attention to the distinctive implications of Eco-translatology for the translation pedagogy ecosystem.

Mao (2010) of the School of Foreign Languages of Shanghai Maritime University explores "A Reflection on MTI Education from Eco-translatology." In this paper, he points out that in Eco-translatology, the Eco-translation system is composed mainly of four sub-ecosystems: translation management, translation marketing, translation education, and translation itself. The ecosystem of translation education can be roughly divided into three levels: first is the simple or complex educational ecosystem focusing on translation education, with the external translation, social and standard environments combined; second is reflecting the interrelations within the translational education system in a single school or educational institution; and third is a system studying the external environment, domestically and internationally, and centering on students'/trainees' personal development. MTI education can be regarded as a new element or minor subsystem of the ecosystem of translation education. It should have the basic characteristics of interaction, co-evolution, adaptation, balance and integral effect of the Eco-translation system. Therefore, the theory of the Eco-translation system of Eco-translatology can be the guideline for MTI curriculum design, textbook planning, proficiency testing, quality and management evaluation, and study of translation marketing. (1) The development of the social economy and the frequency of international exchanges in politics, economy, culture, science, and technology require high-level translation personnel, which creates a need for MTI education. The idea of MTI education should be adaptable to the external environment, emphasizing specialization and application, which distinguishes it from academic master of arts programs. (2) However, MTI education and traditional master of arts programs are mutually effective, as some basic courses in traditional master of arts program are necessary for MTI education, while some specialized courses together with the idea of MTI education can enrich traditional master of arts education. (3) MTI education is professionally oriented and should be modulated according to the needs of the ever-changing external translation education environment; the economic, cultural, and social development of the state; and the state's international competitive edge in economic globalization. (4) In its modulation process, MTI education should be equipped with the ability to modulate its own features, and these features constitute the stable educational eco-environment of different educational institutions, thus benefiting the healthy development of MTI education. (5) Though MTI and academic master of arts programs offer the same degree and are interconnected in terms of educational level and teaching modes, the difference is that the former is a professional degree program biased towards application and specialization.

Youlan (2010) of Fudan University investigates "Studies on Translation Textbooks in China: An Eco-translatological perspective." In this paper, she reviews the development of translation textbooks in Mainland China and interprets how translation textbooks are compiled, developed, distributed, and studied in the context of Chinese translation ecology from an Eco-translatological perspective, with a view to proposing that translation textbooks for translation majors should be ecologically designed in the new century. This proposal highlights the need to establish a strong ecological relationship between knowledge domains in such textbooks, for example, linking translation knowledge with life, with learners' personal experiences, and with other disciplines (language, literature, culture, and philosophy). The textbooks should be written in a descriptive way so that they can be dynamically interpreted. Specifically, four innovative concepts should be integrated into the new type of textbook: (1) redefinition of translation in a larger context: translation is a translator's way of living by adapting him–/herself continuously to the translation ecological environment; (2) emphasis on the adaptation and selection system: specific translation events are expected to be presented in the textbooks, with an emphasis on the translators' self-adaptation and selection; (3) foregrounding the status of translators: translators play a decisive role in the process of translation, keeping translation activities in harmony with the environment; and (4) dynamic assessment criteria: textbooks should establish criteria for evaluating translation by following the rule of "the survival of the fittest," emphasizing the role of readers' response, translators' competence, and the degree of multidimensional transformation.

Li (2012) of Xi-ya-si International Studies College of Zhengzhou University proposes his idea for constructing an Eco translatology-based textbook system for English translation majors. He states that first, the key concept "adaptation and selection" of Eco-translatology theory will be taken as the fundamental principle for both the macroscopic direction and for planning the textbook system construction and the microlevel embodiment in the organization and arrangement of the primary textbook content, such as translation methods and examples. Second, the system will focus on the construction of the types and system of translation textbooks according to the "multidimensional transformation" and "textual transplants" theories. Third, the system will use such translators' subjectivity theories as "translator-centeredness" and "doing things with translations" to construct the textbook system of translators' professional quality and ethics with the purpose of perfecting the textbook content for translators' cultivation. Fourth, the system will be based on the "degree of holistic adaptation and selection," "multidimensional transformation," and "reader feedback" to complete the assessment standard for translation text books and translators' education.

The author offers the following humble opinions on the translation textbooks study from the perspective of Eco-translatology:

From the perspective of Eco-translatology, the construction of a new ecological paradigm of the textbook system for translation majors could be guided by Translation as Adaptation and Selection, use eco-holism and interdisciplinary research as the basic methods, and be oriented towards the needs of the translation industry and the students' actual level and objective needs. The exploration of a new ecological paradigm of the textbook system for translation majors could follow the "three principles" in the general guiding ideology, develop the "four threads" in the overall mode of thought about compilation, and focus on the "five features" in the material selection of samples. Specifically, the "three principles" are as follows:

1. "the Sequence Chain," namely, the textbook reflects the expansion of translation nature cognition, translation knowledge and translation studies perspectives, and the relationships of knowledge ecologies, including aspects of the "knowledge chain" such as translation, language, culture, society/communication, and translational eco-holism.
2. "the eco-reason," that is, the textbook embodies emphasizing holism and relevance, seeking dynamics and balance, reflecting ecological aesthetics, identifying the "translator community," adhering to Eco-translation ethics and highlighting the unity in diversity.
3. "the eco-relationships," i.e., the textbook represents the relationships among the three ecologies: translation ecologies (studies on translational environments), textual ecologies (studies on source and target texts), and "translator community" ecologies (studies on translator actions).

The "four threads" to be developed are as follows:

1. "the thread of ontological translation theories" (longitudinal, hereinafter inclusive), which includes the translation definitions, translation process, translation principles, translation methods, translation criteria, translation phenomena explanations, translation criticism, etc. of Eco-translatology.
2. "the thread of translation ecologies," which is about translational environments, including international and domestic, different hierarchical, and differently oriented translation eco-environments.
3. "the thread of textual ecologies," which is about the translational texts and the shaping process of the translated texts, including the transplants, transformations, reconstructions, and innovations of the linguistic, cultural, and communicative ecologies of the source and target systems, the related Eco-translational strategies and methods, etc.
4. "the thread of 'translator community' ecologies," which is about translator actions. Considerations could be given in the textbooks to the relevant "participants," such as the translator, the writer, the reader, the patron, the critic, and the publisher. Of course, the translator is the representative and the dominant factor of the "translator community."

The "five characteristics" to be highlighted are as follows: (1) the authenticity of the materials; (2) the operability of translation; (3) the practicability of the content; (4) the suitability for different levels; and (5) the pertinence of the tasks.

The author believes that translation textbook research and compilation from the perspective of Eco-translatology will have strong feasibility because novel concepts and appropriate methods can be combined with the existing textbooks' references and utilization.

7.2.5 Studies on Translators

In recent years, the research on translator review from the perspective of Translation as Adaptation and Selection or Eco-translatology has been valued and has become a highlight of the applied studies of Eco-translatology. Liang Qichao is a milestone hero in the earlier period of modern Chinese translation history whose translation activities not only elaborate the classic proverb "History makes the hero" but also pioneer the use of translation to propel social reformation. He also led the literary reformation in three areas: translating political novels, introducing Byron and his poems into China, and promoting vernacular Chinese, which in some ways determined the development of modern Chinese translation history and proved another Chinese notion, "The hero makes the history." Bian Lihong of the School of Foreign Languages of Changsha University of Science and Technology explores "Selection and Adaptation: A Systematic Study of Liang Qichao's Translation Activities." In this paper, on the basis of the Translation as Adaptation and Selection approach and the core concepts of Eco-translatology, she carries out a comprehensive study of three aspects of Liang Qichao's translation activities and thoughts: Liang Qichao as the leader of the age, his translation strategy of "new wine in old bottles" and his drastic heroic translation style. She concludes that Liang Qichao's translation activities, in which enlightening historical meanings are embedded, proved to be his adaptation and selection of the translational eco-environment of his age.

Huang (2009) of Heilongjiang University verifies in his article "Adaptation and Selection: the Ideological Roots of Yan Fu's Translation" that Yan Fu's translation activity lasted 38 years, from 1878 to 1916, that his translation thought and strategies typically embody "Translation as Adaptation and Selection" (Hu, 2003, pp. 283–291), and that the ideological roots of his translation lie not only in external elements but also in internal elements, including the influence of Chinese classics and the integration of Chinese and foreign cultures. Focusing on the ideological roots of Yan Fu's "Translation as Adaptation and Selection" from the perspective of Eco-translatology, he concludes that Yan Fu's translation reflects the concept of adaptation and selection and experiences the process of adaptation and selection, which is proven to be derived from ancient China, to follow the example of ancestors and predecessors, to benefit from the great scholars and talents of his time and to differ from the translation practices of the West.

Zhu H. (2010) of the School of Foreign Languages of Beijing Forestry University discusses "An Ecocriticism of Lu Xun's Translation of Fairy Tales: A Case Study of Der Kleine Johannes." In this paper, she points out that Lu Xun spent more than half of his lifetime on translation, steadily applying cultural considerations to his selection of translation sources as well as translation styles. Contrary to the popular espousal of Western scientific progress and industrialization during the May 4th period, as the epitome of China's modernism, Lu Xun's translation was informed by an ecologically apocalyptic concern for the human future. For him, an important question was "Whither goest human civilization?" His ardent interest in the translation of fairy tales mirrors his dubious longing for a world as innocent as a "child's heart." From his translation of Vasili Eroshenko's fairy tales in the 1920s to his deathbed translation of Russian tales, Lu Xun went beyond his original "children-oriented" thought and began to see the contaminated and alienated adult world through children's eyes. For example, Der Kleine Johannes, one of the Lu Xun's favorite works, was a fairy tale by Dutch writer F. Van Eeden, who presents his melancholy doubts about the value of the science-driven human exploitation of nature. In Van Eeden's Der Kleine Johannes, Lu Xun found an expression of his own reflection on the interactions between humankind and nature, reason and emotion, ethics and scientificism, etc.

Fu Lei is one of the outstanding modern literary translators, foreign literature researchers, and literary and artistic critics in China. He devoted himself to French literary translation and introduction (mainly the masterpieces of French literary giants Balzac and Romain Rolland) from 1929 on. Over the succeeding decades, owing to his perseverant, conscientious, and meticulous efforts and attitudes, Fu Lei produced as many as 34 translated works, many of which have achieved perfection. Together with his handed-down *20 Lectures on World Masterpieces of Art* and *Family Letters*, Fu Lei has gained renown both in China and abroad. Fu Lei's translation philosophies and achievements have attracted much attention. Much research on him has been done in the domestic translation world, but most of these studies are from the linguistic, cultural, and literary perspectives. The author Hu (2009a) explores Fu Lei's resourceful translation philosophies in terms of Eco-translatology and points out that embodied in Fu Lei's translation activities are the following ideas: (1) "harmonious unity" of translational ecosystems; (2) the "translator-centeredness" of translation subjectivity; (3) the translator's action of "adaptation and selection"; (4) "selection through elimination" in translation; (5) the translator's pursuance of "doing things with translations"; and (6) "survival of the fittest" in translated texts.

Prof. Sun Yingchun of the School of Translation and Interpretation of Shandong University at Weihai has some knowledge of the life, translation achievements, translation philosophies, and translation techniques of translation master Zhang Guruo due to his previous monograph *Studies on Zhang Guruo's Translation Art;* thus, he probes Zhang Guruo's translation practice and notions by employing the core concepts of the Translation as Adaptation and Selection approach and Eco-translatology. He holds that the translator's "selection" is no other than his/her reaction or specific translation methods and techniques to adapt to the "real eco-environment"; that is, the selection is for adaptation, and adaptation is for survival. He draws the conclusion that the translated work's influence can reflect the following: (1) whether it is adaptive to the translational eco-environment is of vital importance, as it concerns the survival, effect, and impact of the translator and the translated version; (2) "selection through elimination" and "survival of the fittest" are rules that have existed since ancient times; and (3) the law of "survival of the fittest" requires that the translator should do things with translation, pursue spiritual similarity, and not abide strictly by the literal meaning but attach importance to the holistic effect rather than mechanical correspondence in every detail. His findings conform to the holistic effect and the overall balance of translational ecologies, which are central concepts in Eco-translatology.

Xu Chi is famous in China's literary world for his poetry, essays, and reportage. He received the three-word motto "Poems convey ambition" (诗言志) written by Mao Zedong in 1945, gained the honor of "Modern and Contemporary Chinese Goethe," and won the title of "Father of Reportage." This eminent writer has another little-known identity as a translator who produced and published not only poems, essays, reportage, novels, and commentaries but also translated many foreign works into Chinese, amounting to ten million words in total. However, as a translator, Xu Chi did not receive the attention he deserved in the translation world. Based on her study on Xu Chi's translation philosophies and translated works, Liu (2010) of Qufu Normal University states that from the perspective of Eco-translatology, certain foreign works selected Xu Chi as a translator who attaches great importance to the psychological eco-environment. In turn, Xu Chi, who attaches great importance to the psychological eco-environment, makes a selective adaptation to this specific translational eco-environment with these foreign works as its typically important components and makes multidimensional adaptive selections and transformations, including the linguistic, cultural, and communicative dimensions, which is the real reason for the success of his translated works. Xu Chi's translation activity follows the exploratory course of adapting the dynamic translational eco-environment and adaptive selection to achieve the greatest possible degree of holistic adaptation and selection in the target text.

At present, apart from Liang Qichao, Lu Xun, Fu Lei, Zhang Guruo, and Xu Chi, studies on translators from the perspective of Eco-translatology also include research on the following translators and the interpretation of their translated works: Yan Fu, Lin Shu, Lin Yutang, Gu Hongming, Zhang Yousong, Wang Zuoliang, Cao Minglun, Arthur Waley, Zhao Yuanren, James Legge, Pearl S. Buck, and Zhang Ailing.

7.2.6 *Studies on Interpreting*

Interpreting is an important part of translation studies, which is naturally part of the theoretical applications and applied studies of Eco-translatology. Huang (2007) probe the interpreting eco-environment and the interpreter's multidimensional adaptation and selection within it from the perspective of "adaptation and selection." The authors first expound the interpreting eco-environment and determine that the interpreting eco-environment contains primarily the source discourse; the source and target languages; the speaker; the audience; the client; the setting; and the communicative, cultural, and social factors, which are the elements that should be carefully considered by the interpreter prior to and during the interpretation in addition to changing and uncertain factors resulting from differences in the speakers, scenarios, and interpreting modes. For instance, factors, such as the speaker's speech style, voice and pronunciation, speech speed, and even eye contact and gestures as well as the level of the foreign language, the mood of the audience, the interpretation channels, and the communication effect can affect the quality of interpretation. If the speaker uses obscure words and vague pronunciations, and the interpreter has no idea of these issues beforehand, the interpreter is very likely to produce misinterpretation in the interpretation process, or the interpretation may have to stop. For another example, if the interpretation is carried out outside the interpretation site, any noise and disturbance on the site will inevitably affect the interpretation. The interpreter must consider all the above factors of the interpretation eco-environment; otherwise, any errors in the interpretation process will affect the interpreter's adaptive selection. According to Translation as Adaptation and Selection, if the interpreter cannot adapt to the interpretation eco-environment, he/she will be "dismissed and replaced," that is, be "eliminated" by the interpretation eco-environment. Thus, the authors illustrate the interpreter's adaptation and selection in the linguistic, cultural, and communicative dimensions of the interpretation eco-environment and state that the approach to Translation as Adaptation and Selection can explain the interpretation phenomena and that a study on the interpreter's multidimensional adaptation and selection in the interpretation eco-environment is a new perspective for interpretation studies. (See "*On the Interpreting Eco-environment and the Interpreter's Multidimensional Adaptation/Selection*" [J]. The Journal of Macao Polytechnic Institute, 2007(3); 93–104.)

Runjuan (2011) of Hebei Normal University studies "Interpretation Quality Assessment from the Perspective of Eco-translatology." In this article, the author intends to prove that interpretation quality can be assessed by judging whether the interpretation is harmonious with its ecosystems. The author is tentatively looking for a new perspective in determining what is reasonable in interpretation.

Zhang (2012) of the School of Foreign Languages of Southwest Petroleum University discusses "Variation in C-E Consecutive Interpretation from the Perspective of Eco-translatology." In this paper, she holds that consecutive interpretation is playing an increasingly critical role in international communication, such as in diplomatic meetings, bilateral conferences, escort interpreting, business negotiations, press conferences, and public speaking. China's increasing status in the international community has resulted in growing global attention to China, with more international and intercultural communication than ever before, which entails demands for interpretation from Chinese to English with content related to social, cultural, moral, and spiritual life. Consecutive interpretation, as one essential type of interpretation in international communication, is an impromptu activity that occurs under unpredictable and stressful circumstances. How to interpret Chinese into English efficiently and effectively is of great significance and poses challenges for all interpreters and scholars. As a cultural mediator, the interpreter must make choices of the proper language according to the pragmatic conventions and specific expressions of the target language. When interpreting, the interpreter tends to consciously base adaptations on the purpose of enabling communication. Variations are common in this situation for better acceptance from the audience or for aesthetic reproduction, and they enable the interpreter to make optimal selections of language and expression that will assist in making the communication smooth. Eco-translatology is an ecological approach to translation studies. The affinity and isomorphism between the translational and natural ecosystems enables the establishment of fresh descriptions and explanations of the nature, procedure, principles, and phenomena of translation from an ecological perspective. During the last decade, Eco-translatology research has been developing and has obtained remarkable achievements domestically and abroad. However, the research on C-E consecutive interpretation is far from sufficient. Prominent eco-features are obviously reflected in both natural ecosystems and translational ecosystems, including stressing holism and relevance, presenting eco-aesthetics, and seeking dynamics and balance. In the eco-environment of interpretation, the interpreter is the balance coordinator in both the aesthetic dimension and the cultural dimension; he/she aims to prove the practicality of Eco-translatology for consecutive interpretation to facilitate cultural communication and foster mutual understanding, in turn offering new guidance for interpretation practice.

Yang Chunli of Zhengzhou University intends to research how translators develop their translation abilities and enhance their competence for better living conditions from the perspective of anthropology. She believes that according to the theoretical framework of Translation as Adaptation and Selection, to reach a high level of translation and improve their translation competence, translators must dynamically adapt to the eco-environment of translation, which covers the natural environment, social surroundings, normalized environment, and translators' own physical and psychological conditions. Only in this way can they produce qualified translated versions that harmonize with certain communicative aims according to different people, times, and places. Recently, studies on interpretation have tended to increase every year.

7.2.7 *Studies on Internet Translation*

Internet translation refers to translation activity carried on through the Internet platform of operation and transmission. According to the CNNIC (China Internet Network Information Center) 30th statistical report about China Internet development, the number of Chinese netizens had reached 538 million, and Internet penetration had reached 39.9% by the end of June 2012. Internet information has a broad coverage and develops quickly, which makes it an effective platform for communication and exchanges between translators and readers; the Internet has become a crucial tool for translators to improve their translation abilities and the quality of translated texts.

The network era has features such as instant messaging, online multiplayer platforms, vast amounts of information, hypertext, and anonymous communication, all of which endow Internet translation with strong modern characteristics.[2]

Chen Yi of Changsha Social Work College discusses the traits of Internet translation from the perspective of Eco-translatology. In the paper "On the Translational

[2] See Chen (2011).

Ecology of the Internet Era," he points out that statistics show that 80% of the information on the Internet is in English. Although Chinese people increasingly understand English, they cannot achieve barrier-free reading and absorbance of English materials, as they can with Chinese materials. Therefore, online translators have come into being. Some people single-handedly translate their favorite works in their blogs, but many translation teams with dozens, hundreds, or even thousands of people are carrying out collective translations regardless of pay. On July 21, 2007, the Chinese Internet's most flamboyant folk translation action emerged. Within 9 days after the worldwide synchronized release in English of the novel *Harry Potter and the Deathly Hallows*(hereafter "Harry 7"), a network translation organization named "Hogwarts translation school" translated the more-than-700-page "Harry 7" into Chinese, while the official Chinese version from People's Literature Publishing House was not launched until 3 months later. A commentator joked that "the 'people's' translation era is coming into competition with People's Literature Publishing House."

Regardless of the copyright issues caused by this translation activity, the linguistic dimension is divorced from the traditional form of translation. Solo translation has been replaced by team translation. With the aid of the power of the web, translators in different spaces can come together to achieve a common goal. With multiplayer real-time online communication in the translation process, the terms, style, etc. of the translated version can be unified, and open-source software can ensure that the translated versions produced in this way can be corrected by all participants to minimize "bugs" in the translation. Translators' different cultural backgrounds and knowledge structures are no longer barriers to the unity of the language style of the translated texts and instead provide the encyclopedic background knowledge required for translation. This may be the most significant difference between Internet translation and traditional translation performed behind closed doors.

In Internet translation, the mass information of the network itself becomes a treasure house of information for the translator. The network itself becomes a comprehensive dictionary. The search engine is the most direct approach to consulting this dictionary. If a translator wants to look up the translation of a mechanical term, "无极调速" (wu ji tiao su), the "inducing method" can be used, namely, using a word to induce a web page in both Chinese and English. Generally, the "inducing term" should be the target word and should be directly related to the query words. The "inducing term" for the English translation of "无极调速" (wu ji tiao su) should be an English word. According to our experience, the English translation of this term should contain "速度," so if "无极调速 (wu ji tiao su), speed" is input, a variety of corresponding English versions, such as "stepless speed variation," "stepless speed regulation," etc., will emerge in the search results.

Moreover, in the mode of traditional translation, verifying the quality of translated texts is also a headache for the translator. Generally, it is challenging for the translator to objectively judge whether his/her expression in the translation is aligned with the target language expression conventions, especially when the target language is not the native language of the translator. However, on the Internet, this problem can be easily addressed. For instance, if we want to determine whether "convenient traffic" is the best English translation of "交通便利" (jiao tong bian li), we can directly input "convenient traffic" into Google to search and obtain 42,400 results. However, after the analysis of the websites, we find that the majority of these results come from web pages edited by Chinese people, which means they are typical Chinese English.

Therefore, in the Internet era, the latest data and the original data exist synchronically and are open to all translators for consultation. Thus, in Internet translation, different translators can synchronically share vast resources that cannot be matched in the traditional translation mode, providing the best judgment standard for cultural adaptive selection.

In Internet translation, the initiative of the target readers is more significant than ever before. First, the chance for readers to be exposed to the original has significantly increased, and, more importantly, readers can directly read the original with the help of translation software and an e-dictionary. For the majority of readers, this method has very strong feasibility. Readers and the translator can understand and expose themselves to the original synchronically. With the help of their understanding of the original, the readers can comprehend the translation more profoundly and provide a more targeted reaction to it.

Second, under the conditions of the Internet, easy access to the original will stimulate more people's desire to translate. The tremendous openness of the Internet can make various translated versions of the same original emerge at the same time for readers to read, compare, comment on, and judge them. Moreover, on the Internet, readers can take advantage of anonymity to genuinely and objectively assess the translated texts. Additionally, all of these operations need not be carried out after the completion of the translation, as in traditional translation. Therefore, translators in the Internet era must have a clear translation pursuit and translation goal and constantly strive for perfection in translation to win readers and enable their translations to stand out from numerous other versions.

In summary, on the Internet, a brand-new translation context has been constructed in which translators from different areas and with different ability levels can gain a "feeling of field translation" and write in the conventional "translation field." With the development trend of Internet technology, a new landscape for the development of translation will inevitably emerge. At the same time, this trend will also bring about new research topics for translation studies.

7.3 Applied Cases of Practical Studies (2)

7.3.1 Studies on Translation History and History of Translation Theory

Translation history study is an integral part of translation studies. However, studies on translation history from the perspective of Eco-translatology are currently rare. Recently, Bian (2012) of the School of Foreign Languages of Changsha Technology

Institute has conducted related research. She holds that as the eyewitness and memory of a certain nation's translation activities, translation history cannot only provide sound and profound evidence and a basis of thought for the studies of translators but also can function as a pioneer of the nation's cultural and literary evolution. Owing to the differences in geographical environment, cultural and political systems, ways of thinking, aesthetic tastes, and time of occurrence, translation history displays a clear difference in national features, cultural patterns, and spatial and temporal features. From the perspective of Eco-translatology, the underlying reason for such differences lies in the difference between the eco-translation environment of the translator's translation activity and that of the translator as an eco-body. Thus, it is necessary to take a holistic perspective in studies of translation history. Eco-translatology prioritizes the translator's adaptation and selection in a certain Eco-translation environment, the ecological paradigm in the process of translation activities, the formulation of translation strategies, and certain translation phenomena that refer to the integration in the Sequence Chain of language, culture, society, and nature. She takes the development of both Chinese and Western translation history as cases and aims to establish the Eco-translatological paradigm for the studies of translation history, including the translator's history, the text's history, the ecological views of the national translation as a body and the specific translation cultural view during a certain period, and generalizes the features of translation history from the perspective of Eco-translatology. Ultimately, she concludes that for studies of translation history, it is necessary to take a holistic approach and probe the specific development regulations with reference to the ecological research methodology to establish an ecological paradigm for translation history research.

The author thinks that related studies on the history of translation/translation theory could be carried out using the following clues within the theoretical framework of Eco-translatology:

1. the clue of the "Sequence Chain"

   The "Sequence Chain" is the theoretical path of human cognitive development. As a human activity, translation follows this path. Taking "the Sequence Chain" as a research clue and translation, language, culture, human society, the natural world, etc. as the perspectives or horizons to investigate and extend is a choice in "rewriting" the history of translation/translation theory.
2. the clue of the "translational eco-environment"

   Translational eco-environments are different at different times. A history of translation/translation theory can be said to be a history of the continuous development and change of the translational eco-environment. Taking the "translational eco-environment" as a research clue in the sequence of the macroenvironment, mesoenvironment, and microenvironment of translation ecology is also a choice of "rewriting" the history of translation/translation theory. The author believes that the "translational eco-environment" is the most appropriate clue in studies on the history of translation/translation theory under the theoretical framework of Eco-translatology.
3. the clue of "translator-centeredness" (i.e., the clue of "translation philosophy")

The translator plays a dominant role in translation activity; the translation philosophy of a period is, in fact, the collection of different translators' philosophies in that period. Taking "translator-centeredness" in translation activity as a research clue is equal to taking "the human element" as a clue, taking "schools of translation" as a clue, and taking translation philosophy as a clue. As far as studies on the history of translation/translation theory are concerned, especially studies on the history of translation theory, taking "the human element" as a study clue can be said to be a good choice.

4. the clue of "Doing Things with Translations"

   "Doing things with translations" can be the "motivation" and also the "effect" of translation activities. Taking this kind of translation activity with "cause and effect" as a study clue conforms to the reality of the translation process and to the law of the development of things. Therefore, taking "doing things with translations" as a clue for studies on the history of translation/translation theory may produce originality and topics for new monographs.

5. the clue of "synergic fusion"

   "Synergic fusion" is one of the methodologies of Eco-translatology that involves more academic foci and research perspectives. It is an enormous project to construct a synergic fusion of the history of translation/translation theory from different perspectives and in different categories. However, "the greater the obstacle, the more glory in overcoming it."

There is no doubt that different research clues, different observation perspectives, and different theoretical bases for exploration can provide more room and possibilities for discoveries, new explanations, and new ways of writing. We look forward to studies on the history of translation/translation theory from the perspective of Eco-translatology to not only communicate the history but also link it with the future. If such monographs as "*A New Interpretation of Chinese Translation History/History of Translation Theory from the Perspective of Eco-translatology*" or "*A New Interpretation of World Translation History/History of Translation Theory from the Perspective of Eco-translatology*" emerge in the near future, we will be delighted.

7.3.2 Studies on Translation Criticism

As an important part of the system of translation studies, translation criticism serves as the link between translation theory and translation practice. On the basis of central theoretical tenets and research foci such as the Eco-translation process, translator-centeredness, eco-ethics, the translation eco-environment, and the degree of holistic adaptation and selection, Yue (2012) of Henan University of Science and Technology explores different Eco-translation criticism approaches to systematize the paradigm of Eco-translation criticism, which includes process criticism, translator criticism, ethics criticism, environment criticism, text criticism, and reader criticism.

The author believes that translation criticism from the perspective of Eco-translatology could focus on specific theoretical perspectives, viewpoints and arguments.

For the moment, translation criticism could focus on the following "ten approaches" of Eco-translatology: (1) the approach to "balance and harmony" of translational ecology; (2) the "textual transplants" approach to translational texts; (3) the "translator responsibility" approach to the translation subject; (4) the "adaptation and selection" approach to translation activities; (5) the "multidimensional transformations" approach to translation methodology; (6) the "preservation of the strong and elimination of the weak" approach to the translation process; (7) the "multidimensional integration" approach to translation criteria; (8) the "survival of the fittest" approach to translated texts; (9) the "doing things with translations" approach to the translator's pursuit; and (10) the "Sequence Chain" approach to translatology development. Eco-translatology is an overarching discourse system of translation theory. Studies on translation criticism carried out from different perspectives, dimensions and profiles are sure to be successful.

7.3.3 Studies on Translation Methodology

Studies on translation methodology from the perspective of Eco-translatology can be roughly divided into three categories: first, traditional and conventional methodological research (such as studies of descriptive translation, the inductive method, the deductive method, the comparative method, or the corpus-based method); second, the methodology research of Eco-translatology itself (interdisciplinary method, analogy method, transplantation method, systematic and synoptic method, etc.); third, other methodology studies deduced by eco-reason.

At present, there have been some studies on translation methodology. As we know, the construction and expansion of any academic paradigm not only embodies the further exploration of a particular discipline but also shows the shift of its cognitive perspectives or the expansion of its cognitive horizons. As an emerging paradigm of Eco-translation, Eco-translatology's evolving cognitive perspectives and horizons indicate these tendencies. Wen (2010) of Southwest University comments "On the Norm of Eco-translatology from the Perspectives of Eco-linguistics and Cognitive Linguistics." In this paper, he holds that the emergence of Eco-translatological norms reflects the necessity of human cognition in translation studies in a specific cultural context. On the one hand, Eco-translatological norms are established in accordance with a doctrine that is the same or similar to that of eco-linguistics, thus shedding light on the construction of Eco-translatological norms. On the other hand, the latest developments in cognitive linguistics will also affect the cognitive dimension of Eco-translatology.

Wen (2010) proposes in a paper entitled "The Methodology on Eco-translatology" that the research methods of Eco-translatology should be discussed on three levels, philosophical methodology, scientific methodology, and specific methodology, and that in an era when science is becoming increasingly complex, philosophical studies concerned with the problems of Eco-translatology are of great significance in terms of epistemology and methodology.

Hua Xianfa and Hua Manyuan of Central China Normal University suggest in the paper "An Eco-translatological Perspective on the Chinese Translation Methodology" that in comparison to Western views on translation methods, Chinese views of the same topic are the product of a two-valued orientation. The values are represented both in theory and practice by disputes over such pairs of terms as "elegance" and "plainness," "similarity in form" and "similarity in spirit," and "literal translation" and "liberal translation" and dominated our understanding and application of Chinese translation methods for a long time. Born in traditional Chinese cultural soil and entailing the rich Chinese eco-philosophy, they cannot be considered synonymous with Western translation categories such as "translation as an interpreter" and "translation as an orator" and "foreignization" and "domestication." Instead, they are closely connected with such traditional Chinese philosophical and cultural concepts as "plainness" and "elegance," and "spirit" and "body." Therefore, tracing and analyzing their roots from an Eco-translatological perspective is of considerable significance, as such study can help us better understand not only the relationship between the social eco-context and the birth of Chinese translation methodology but also the prominent role of these factors in discussing the reconstruction of contemporary Chinese translation theory.

7.3.4 Studies on Translation Ethics

Eco-translatology is a holistic study of the ecosystem of translation that focuses on the relationships between the internal elements of the ecosystem of translation, including the study of the coordination between translation activity and language, culture, human society, and even the natural world; the study of the relationship between the translator and the translational eco-environment; the study of the relationship between the "participants," including the author, the reader, the patron, the publisher, and the critic, and the translational eco-environment; and the study of the relationship between the essence of translation, the process of translation, the principles of translation, the methods of translation, the criteria of translation, etc. Macro-, meso- and microlevel studies on Eco-translatology involve the relationships between the internal elements of the translational system. All of this will inevitably be restricted by certain social norms and follow all the presuppositions that are relevant to the value concept system. As a result, to reflect on studies on Eco-translatology from the perspective of ethics is an effective extension of Eco-translatology research that will promote its sound development.

At present, translation ethics studies are based mostly on the analysis of literary texts, mainly from linguistic, cultural, and ideological perspectives. Wang (2012) of the School of Foreign Languages of Guangxi Financial College provides "An Ecological Interpretation of Literary Translation Ethics." Specifically, he uses the ecological tenets of Eco-translatology to view existing translation ethics, integrates an ecological conception into translation and ethics, and proposes that ethics theory should be based on the new Eco-translatological tenet that the translator assumes responsibility for all others to underline his/her subjectivity. He believes that translation should be researched in the eco-environment to solve the limitations of the translator's selection of translation strategies; that is, others within and outside the text, including all related to the translator's behavior, should be considered in translation. Moreover, translation ethics can be connected to the natural ecological environment because translation can be regarded as reproduction. In practice, translation is related to both nature and society, while the relation between humankind and nature is embodied in relationships among humans. He also discusses whether translation ethics with ecological tenets can be treated as universal ethics and seeks to identify the ecological significance of translation ethics; that is, other views in translation ethics can be tolerated in or compatible within different sub-ecosystems.

The author deduces from the application of the analogy of ecological ethics aiming at the reality of translation along with the research orientation of Eco-translatology that the ethics and moral principles of Eco-translatology cover the following four basic principles: (1) the principle of "balance and harmony"; (2) the principle of "multiple eco-integration"; (3) the principle of "symbiosis and diversity"; and (4) the principle of "translator responsibility" (Hu 2011). These details can be found in Sect. 3.7.

Eco-translation ethics is an emerging research area of open-ended, systematic study. Certainly, studies in this area will continue. Moreover, any related studies should be welcomed and encouraged.

7.3.5 Studies on Schools of Translation

"Genres," also called "schools," refer mainly to research groups with common theoretical guidance, a unanimous ideological trend, similar purport and interests, and similar research style. The picture of translation theories in the West clearly shows that schools play a vital role in theoretical development and that innovative theories emerge from interaction between different schools. Schools indicate that scientific research has reached a certain depth, scale and level of maturity, as they are platforms for academic communication and contention as well as the driving force for promoting the in-depth development of academic research.[3] "Schools play a significant role in inheriting the cultural essence, bringing up academic masters, refining academic tradition, replenishing academic universality with diverse individuality, and promoting academic progress" (Zhang & Xin, 2005, p. 69).

[3] Prof. Qian Guanlian, a linguistic philosopher, pointed out that the greatest academic achievement of an individual is to create a school and that the most prosperous academic sign of a country is the emergence of brilliant and varied academic schools. (Quoted from Sang and Mu (2008, pp. 31–36).

Since Cicero and Horatius in ancient Rome, there have been many schools of translatology with numerous and complicated theories. Since the latter half of the twentieth century, an unprecedented number of studies on Western translation theories have appeared, and the research mode of plural and interdisciplinary reference, exchange, confrontation, and penetration has provided a new foundation and point of growth for the development of this discipline. Domestic and foreign recommendations, introductions, and evaluation of translation theories have been uninterrupted. Most of them divide studies on translation theory into the following main genres: the philological school, linguistic school, school of translation studies, and cultural school of deconstruction.

Some Chinese studies in this field focus on average introduction and pectination. Research on the causes and characteristics of these issues is insufficient, and the monographic study of schools of Chinese translation theory is still nonexistent.

Macao Polytechnic Institute's monographic study from the perspectives of the research levels, the research objects, the research methods, the understanding of the nature of translation nature, etc. classifies and examines the existing translation theories to determine the main causes of the formation of schools and their basic characteristics, mainly demonstrating and expounding the reality and embryonic development of "the school of Eco-translatology" in China. Based on this, the study proposes the main criteria for the judgment of the formation of translation theory schools with "the school of Eco-translatology" as an example and explores the basic mode of contemporary theoretical research in translatology. The study mentioned above provides a theoretical basis for "the school of Eco-translatology" in China earning a place in the international academic community of translation and comprehensively analyzes it.

The research project on academic schools of translation, as mentioned earlier, is based on the overall research approach and principles of "from international to domestic" and "from a comprehensive survey to a case study." Specifically, research methods, comparison methods, narrative research methods, logic speculation methods, and holistic coordination methods from the literature are mainly adopted to carry out the research and to make the research results more scientific and convincing.

This study elaborates on the research models of the contemporary theoretical study of translation, from depending mainly on individual effort to condensing the collective wisdom and from emphasizing the individual breakthrough to emphasizing the collective breakthrough,[4] to fulfill the historical mission of major innovations of translation studies. This study also demonstrates that the shaping of "the school of Eco-translatology" has an orientation significance for translation studies, especially the theoretical study of translation.

[4] It is similar to a short message sent to me by a friend: The twenty-first century is the era in which the individual hero defers to the collective team. To conquer the difficulties in teams or in pairs has become a tendency. The one who wins the masses wins the market. "1 + 1 = 2" is called mathematics; "1 + 1 = 11" is called economics. The acronym TEAM is the very spirit of the international Eco-translatological research "team": "together, everyone achieves more ".

Based on the case study of "the school of Eco-translatology," this research proposes primary standards for the shaping and judgment of the schools of translation. Research shows that (1) a school should have a leader; (2) a school should have a core team with solid achievements; (3) a school should have a group of like-minded followers and supporters that forms gradually at the initial stage; (4) a school should have representative monographs and a systematic theoretical system that gradually produce a significant impact on academic research and at the same time have their own research style; (5) a school should have certain academic organizations; (6) a school should have corresponding academic positions and communication platforms; (7) a school should have a safe and stable society[5]and the appropriate academic environment; and (8) a school should gradually obtain social recognition and approval.

The above project conducts a prospective investigation into the shaping of "the school of Eco-translatology," and Eco-translatology's academic significance and practical significance for the development of China's translation theory and the enrichment of worldwide translation studies cannot be underestimated.

Through Eco-translatological approaches, other relevant research projects on academic schools can be carried out, based not only on the "translational eco-environment" approach with philosophic thought as its study clue but also on the "translator-centered" approach with a representative person as its study clue, not only on the "translation community" approach with country/community as its study clue but also on the "eco-paradigm" approach with paradigm transformation as its study clue. Of course, the relevant research on schools of translation can also be carried out based on the "holistic relevance" approach with synergic fusion as its study clue.

7.3.6 *Other Related Studies*

In recent years, some articles have appeared that explore the relationships between Eco-translatology and the relevant disciplines in which the research objects, theoretical goals, and theoretical bases, etc. of Eco-translatology are clarified.

Fang (2010) of Shanghai University states in his paper "On Translational Eco-Environment—Construction of Eco-translatology from the Perspective of Its Key Terms" that the translation eco-environment can be broken down into the translation ecosphere and translation environment. This paper describes the formation of the translation ecosphere and translation environment and their harmonic existence. He introduces the globalism, relevance, equilibrium, gradation, and epochal characteristics of the translation eco-environment. The paper shows that translation activities require a natural, harmonic relationship of coexistence and co-development between

[5]For instance, if a society suffers from the long-term chaos caused by war, and the masses have nothing to live on and can hardly keep themselves alive, it is difficult for the whole society, science, education system, etc. to develop, let alone achieve academic prosperity and the shaping of genres.

the translation ecosphere and translation environment. Zhu L. (2010) of Hengshui University investigates multidimensional perspectives of the conditions, layers, goals, modalities, and results/values of interdisciplinary translation, from ecology to translatology, in light of certain rules of interdisciplinary transplantation to discover possible ways in which an ecological approach can facilitate translatological advancement. In addition, Zeng (2010) of Guangdong University of Finance & Economics analyzes the relationships between pragmatic adaptation and Eco-translatology. Wang (2010), a doctoral candidate at Northeast Normal University, comments on "Sustainable Development of Trans-disciplinary Research on Ecology and Linguistics: Translatology." Gousty (2011) of the Sino-French Institute for Technological Development points out in a speech entitled "On Translational Eco-environments of Translation for Science and Technology" at the second International Symposium on Eco-translatology that due to globalization and to the complexity and cost of developing advanced technologies, one of the specificities of science and technology (S&T) projects is that they are increasingly international. Because the S&T methodology consists mainly of identifying facts, implementing concrete experiences, and using mathematical modeling, it could appear that international S&T project teams do not have translation problems. After interviewing engineers and scientists from scientific project teams (aeronautics and space, telecommunications), however, it appears that even in strictly scientific exchanges, misunderstandings and mistakes can appear due to translations. To date, analyses of these difficulties and mistakes have attributed them to differences in the pedagogy of scientific education and in the presentation of concepts between different cultures and languages. The consequences of this phenomenon are important in terms of costs and human life. He proposes that a group focusing on S&T translations could study this problem from the point of view of Eco-translatology, taking into account the interdisciplinary orientation of this new field and its focus "on the relationship between the translator and the translational eco-environment." Li (2011) of Xiyasi International Studies College of Zhengzhou University finds through research that the comparison between the survival environment for the fighter and the translational eco-environment for the translator and between the fighter's adaptation and selection infighting and the translator's adaptation and selection in translation can help us further understand what translation is and improve the translation process.

7.4 Overviews and Comments on Applied Studies

7.4.1 Overviews and Comments on Theoretical Applications

If we browse from different perspectives all the articles related to Eco-translatology officially published in various journals in the past 10 years, we observe the following main features.

From the perspective of the article contents, extensive cases are covered, dealing with the applications of Eco-translatological ideas to the translation of literary works, translation of philosophical and sociological literature, translation of sci-tech materials/commercial materials/legal documents/public signs, and Internet translation/interpretation translation/dubbing translation/subtitling translation as well as studies of the translator, translation teaching, history of translation/translation theory, translation methodology, translation ethics, schools of thought in translation, translation criticism, translation terminology, and other related subjects.

For instance, articles have been published that deal with the application of Eco-translatological ideas in the translation of literary works and masterpieces, including Lin Yutang's *Six Chapters of a Floating Life*; Yan Fu's translation of *Evolution and Ethics* (*Tian Yan Lun*); Gu Hongming's translation of *The Analects of Confucius*; the English translation of Tao Yuanming's poetry; Buddhist scriptures translation; the translation of the *Bible*; *Suwen of Huangdi Neijing* (*The Yellow Emperor's Classic of Internal Medicine*); Lin Shu and Wei Yi's translation of *Uncle Tom's Cabin*; the English versions of *Mooring on the Qinhuai River*; the English translation of *The Analects of Confucius*; two translated versions of Du Fu's *Wang Yue*; Hu Shi's translation of *The Isles of Greece*; the translation of *Wolf Totem* (*Lang Tuteng*); Gary Snyder's English translation of Hanshan's poems; the English translation of *Fortress Besieged* (*Wei Cheng*); the translation of Lu Xun's *The New Year's Sacrifice*; the translation of *Dawn Blossoms Plucked at Dusk* (*Zhao Hua Xi Shi*); Zhang Yousong's English translation of *The Adventures of Tom Sawyer*; four English versions of *Chuang Tzu*; Wang Zuoliang's translation of Francis Bacon's *Of Studies*; two Chinese versions of *David Copperfield*; four English versions of *The Book of Songs*·(Biao You Mei); the dubbing translation of *Mulan*; three English versions of Zhu Ziqing's prose work *The Transient Days*; Hawkes's translation of *A Dream of Red Mansions* (*Hong Lou Meng*); Zhang Ailing's translator identity; Xu Chi's Eco-translation; the subtitling translation of *If You Are the One II*; Arthur Waley's translation of *The Book of Songs*; the English versions of the poem *Chun Xiao*; Liang Qichao's translation activities; Zhao Yuanren's translation of *Alice's Adventures in Wonderland;* James Legge's translation of *Tao Te Ching*; Pearl S. Buck's translation of *All Men Are Brothers*; Yang Xianyi's Chinese translation of *Pygmalion*; the analysis of the actual translation examples of *The Return of the Native*; two English versions of Zhu Ziqing's prose work *He Tang Yue Se*; the actual translation examples of *Desperate Housewives*; the exploration of the Chinese translation of *Gospel of Mark*; the translation of *Jitanjiali*; three translated versions of *Jane Eyre*; the English translation of *The Art of War*; and the like.

Regarding the article authors, most of the researchers are young and middle-aged scholars and postgraduates.

Regarding the theoretical focuses, these articles center on the Eco-translatological theories of the translational eco-environment, three-dimensional transformations, adaptation and selection, and translator-centeredness.

Regarding the theoretical basis, the Translation as Adaptation and Selection approach is most prominent, followed by Eco-translatology. The transition to Eco-translatology is gradual. Specifically, the titles and contents of articles published from 2001 to approximately 2007 are closely related to the Translation as Adaptation and Selection approach, while those published from 2008 to the present day are related to Eco-translatology, and many refer directly to Eco-translatology in the title, for example, "from the perspective of Eco-translatology."

7.4.2 Overviews and Comments on Critical Reviews

Critical reviews can be divided into three categories. The first is articles of general discussions and comments on the theories and development of "Translation as Adaptation and Selection" and "Eco-translatology," for example, "The Translation as Adaptation and Selection Approach: An examination and a case study" (Gao, 2009); "Reflections on Translation as Adaptation and Selection" (Zhang, 2010); "On the Interpretation and the Status Quo of Eco-translatology" (Lv, 2011); "An Overview of Eco-translatology" (Zeng & Zhao, 2012); "Theoretical Reflections on Eco-translatology" (Zhu, 2012); and "The Development and Prospects of Eco-translatology" (Yang, 2012).

With the continuous in-depth development of Eco-translatology, this kind of article has gradually increased in recent years. There is no lack of complementary and/or critical content in some articles of general discussions and comments.

The second category consists of articles reviewing the book *An Approach to Translation as Adaptation and Selection*, which are primarily complimentary, such as "A New Perspective of Translation Studies" (Yang, 2004); "A New Book of Translation Studies with an Exploring Spirit: A comment on and analysis of *Approach to Translation as Adaptation and Selection*" (Liu & Xu, 2004); "Reviewing Hu Gengshen's *Approach to Translation as Adaptation and Selection*: A unique theoretical build-up" (Li & Huang, 2005); "A Theory 'Turning to the Translator: A comment on *Approach to Translation as Adaptation and Selection*" (Hu, 2006); "Bookreview: *An Approach to Translation as Adaptation and Selection* written by Hu Gengshen" (Song, 2007); "Turn, Surpass, and Return: Theoretical Reflection on *The Theory of Translation as Adaptation and Selection*" (Meng, 2008)[6]; and "On Professor Hu Gengshen's Scholarly Contributions: A Western scholar's view" (Dollerup, 2007/2013). This kind of article appeared annually a few years ago, centering mainly on *An Approach to Translation as Adaptation and Selection*, but has gradually decreased in recent years.

The third category consists of consulting and challenging articles, for example, "A Critical View of the *Approach to Translation as Adaptation and Selection*" (Hu, 2010); "Is the Translator Really the 'Center' in Eco-translation Theory? A discussion with Prof. Hu Gengshen" (Leng, 2011); "Eco-translatology and Its Three Ecological Dimensions: A discussion with Prof. Hu Gengshen" (Zhang & Wenning,

[6] See Meng, F. *Turn, Surpass, and Return: Theoretical Reflection of the Theory of Translation as Adaptation and Selection*. http://www.language.international.com. Retrieved on August Meng, 2008.

2011); "A Probe into Some Key Concepts of Eco-translatology" (Wang, 2011); "A Research on *Approach to Translation as Adaptation and Selection*" (Zhan Xiaofeng, 2011); and "Research on *Approach to Translation as Adaptation and Selection* in China: retrospect and prospect" (Zhao, 2012).

This type of consulting and challenging article was published every year over the past few years. However, an analysis of the related documents shows that some of the present questions and objections[7] are directed mostly at the book *An Approach to Translation as Adaptation and Selection*, which was published 8 years ago. With regard to Eco-translatology, most of these articles discuss independent and individual papers published in the previous period. However, in the past decade, with in-depth research on the relevant issues, we have constantly furthered our understanding and studies of Eco-translatology, and the related presentations and statements have also gradually improved.

7.4.3 Overviews and Comments on Applied Studies in General

According to incomplete statistics, by the end of 2012, hundreds of articles of applied studies using the related theory of Eco-translatology had been published in different levels of journals. These studies explore concrete translation problems and interpret translation phenomena, and some propose opinions on the clarification of Eco-translatological research areas, theoretical goals, theoretical foundations, etc. These applied Eco-translatological studies, in a sense, have enriched the research content of Eco-translatology and expanded the research teams, thus promoting the continuous study and development of Eco-translatology.

However, on the whole, the limitations of some applied studies are also undeniable. Take the problem of "a mixed bag," for example. After reading some abstracts of the published articles and the full text of some papers, theses, and dissertations, the author feels that some of them are superficial and that the phenomenon of an "isolated" relationship between them and the theories of Eco-translatology exists. At the same time, some related studies and discussions lack the combination and interaction with translation instances and therefore seem unconvincing.

Take the problem "taking a part for the whole" as another example. Some of the published articles (including some articles with questions and worries about Eco-

[7] The reasons are presumably as follows: First, the theoretical system of Eco-translatology was still being constructed in the previous years; some philosophies needed further development and elaboration; and some expressions and statements needed further improvement. Second, there was no relatively comprehensive, systematic and holistic discourse of Eco-translatology some time ago, so questions and doubts inevitably emerged, which was a natural and normal response. Third, some people still do not fully understand the basic contents and ideas of Eco-translatology. Of course, it cannot be ruled out that some people are simply contemptuous of China's local theory and that some individuals even have "mentalities" of pride and prejudice, dismissiveness, etc. However, from the perspective of eco-reason, all these reactions are natural and normal, and the author has an "ecological" attitude towards them.

translatology) appear to be lacking in-depth knowledge of the related theories of Eco-translatology. In other words, only after gaining in-depth knowledge of the ideological system and the essence of Eco-translatology can people develop targeted problems and questions: practical applications, transformations and innovations, or questions and criticisms. Otherwise, the proposed problems and questions seem superficial and repeatedly discuss confusions and misgivings that were resolved many years ago. Some articles cannot avoid "interpreting out of context"; therefore, it is difficult to engage in equal and in-depth dialogues and exchanges at the academic level.

For a third example, consider the "single perspective" problem of some applied studies resulting from the above problems and causes. An excessive focus on the perspective of the minimal theories of Eco-translatology is a case in point. Meanwhile, the class of journals in which the research papers are published needs improve to avoid the phenomenon of repetition.

Additionally, based on the recent research and the practical applications, we have noted two extremes. One is the view that translation studies have nothing to do with eco-reason or the eco-approach. The other is too much "copying" or "mechanically applying" the terminology and content of Eco-translatology ontology in translation studies; "too much is as bad as too little."

However, the author believes that everything has a process of development. The problems and phenomena mentioned above are all facts, and some of them refer to the early stages of the development of Eco-translatology. Thus, the author is afraid that this is also an inevitable stage.

We believe that with the in-depth development of research and mutual communication, these problems will gradually be addressed and solved. We look forward to that time. In fact, to improve and consolidate the theoretical system of Eco-translatology, some scholars have become dissatisfied with the simple use for reference of its terms and concepts. They have started to gain an in-depth systemic understanding and reflection at the level of theoretical principles and laws and to scientifically prove the necessity and rationality of Eco-translatology and its academic relevance to other disciplines (Song & Liu, 2012, pp. 69–72).

7.5 Some Phenomenological Interpretations: An Eco-Translatological Perspective

What phenomena and problems can Eco-translatology reveal and explain in translation studies? To solve this problem, all the contents of this book can be considered more or less relevant. However, this section focuses on how to interpret several translation phenomena and issues that have long been heatedly discussed from the perspective of Eco-translatology: translatability, retranslation, translation strategy employment, translation styles, and "turns" of translation studies.

7.5.1 Translatability

In the book by Yanzi Chunqiu (晏子春秋), "Spring and Autumn of Master Yan," a collection of stories around Yan Ying (晏婴), a lower-nobility minister of the dukes of Qi who lived in the late sixth century, a passage reads, "Ju that grows in the area south of the Huai River becomes ju (mandarin orange), and becomes zhi (something that resembles ju but tastes totally different) in the area north of the Huai River. Only their leaves look similar; their flavors differ. Why? They grow in different environments (shuitu, meaning water and earth)" ("Nei-pian Za Xia" in Yanzi Chunqiu).

A crop will be transformed due only to the geographical differences resulting from separation by the Huaihe River. Similarly, translation faces two completely different languages at the lexical, syntactic, grammatical, and textual levels, spanning hundreds of thousands of years of history and countless mountains and rivers and serving as a communicator between different countries, different races, different cultures, and different faiths.

"The water and the soil are different" indicates that the eco-environmental conditions such as geography, climate, water quality, soil, and biological chain are the key and premise of species formation. Everything is the product of the environment. Naturally, different species will be created by different eco-environmental qualities. Generally, the same is true of the textual ecology and the textual transplantation of translation.

Expressed in the terminology of Eco-translatology, the source text is one textual ecosystem, while the target text is the other textual ecosystem. The textual ecosystem of the source language (hereafter referred to as the "source-text ecology") involves the linguistic ecology, the cultural ecology, and the communicative ecology in the source-language system. The textual ecosystem of the target language (hereafter referred to as the "target-text ecology") involves the linguistic ecology, the cultural ecology, and the communicative ecology in the target-language system. The "bilingual" ecologies (referring to the "source language" ecology and the "target language" ecology) are illustrated in Fig. 7.1.

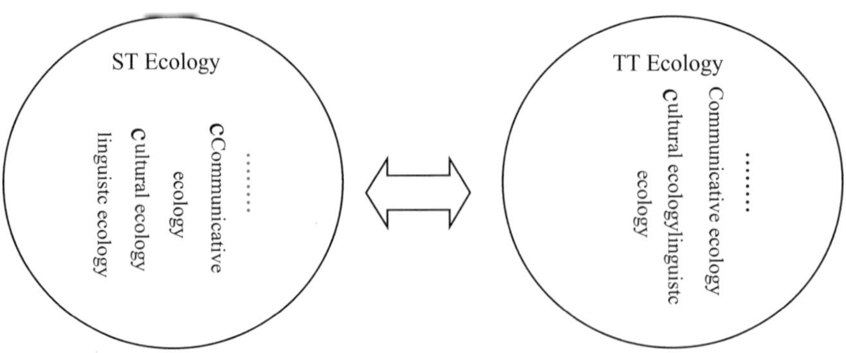

Fig. 7.1 ST ecology vs. TT ecology

From the perspective of Eco-translatology, translatability or untranslatability results not only from the differences between the source text and the target text but also from the degree of difference between the linguistic ecology, the cultural ecology, and the communicative ecology of the source-text and target-text ecologies. In other words, it is a problem of the degree of difference between the source-text ecology and the target-text ecology.

In general, the smaller the degree of difference is between the linguistic ecology, the cultural ecology, and the communicative ecology of the source-text and target-text ecologies, the larger the degree of translatability, that is, the smaller the degree of untranslatability. Otherwise, the larger the degree of difference between the linguistic ecology, the cultural ecology, and the communicative ecology of the source-text and target-text ecologies, the smaller the degree of translatability, that is, the larger the degree of untranslatability. The inverse relationship is expressed with the equation in Fig. 7.2.

In Fig. 7.2, the equation indicates that as a translational action, when the source text remains unchanged, the larger the degree of difference between the source-text and target-text ecologies is, the smaller the degree of translatability is; the smaller the degree of difference between the source-text and target-text ecologies is, the larger the degree of translatability is. When the degree of difference between the source-text and target-text ecologies is the smallest, or when the source-text and target-text ecologies are closest to each other or even basically the same, in other words, when the ratio of the two is close or equal to 1, the degree of translatability is the largest.

The above relation between the degree of difference between the source-text and target-text ecologies and the degree of translatability and the change of "the degree of difference" and the change of "translatability" is further illustrated in Fig. 7.3.

Figure 7.3 indicates that the smaller the degree of difference between the source-text and target-text ecologies is, the larger the degree of translatability, and the larger the degree of difference between the source-text and target-text ecologies is, the smaller the degree of translatability. At the same time, it shows that the degree of difference between the source-text and target-text ecologies and the degree of translatability from small to large is a continuum, a process of gradual increase or decrease.

Because translatability or untranslatability is closely related to the degree of difference between the language ecology, the cultural ecology, and the communicative ecology of the source- and target-language systems, to improve the degree of "translatability," the translator can perform his/her duties to maintain the source-text ecology and the target-text ecology, to use his/her mind to coordinate the source-text

$$\text{translatability} = \frac{\text{ST ecology}}{\text{the degree of difference between ST and TT ecologies}}$$

Fig. 7.2 The inverse relationship of the degree of difference between the ST and TT ecologies and translatability

ecology and the target-text ecology, to work on the balance between the source-text ecology and the target-text ecology, and to make great efforts to create (reconstruct) the eco-environment of the source text in the ecosystem of the target text.

Here, for the ecologies of the source and target texts, the first step is "maintenance." If it fails, then "coordination" is needed. The objective of coordination is "balance." When it is not easy to balance and cannot be done without textual transplantation or translation transformation, then creation and "reconstruction" are needed. That is, the eco-environment adaptive to the source-text ecology must be created or constructed in the target-text system. Through the translator's efforts or even through technical means, the world of the source-text system is made to "resurge," which crystallizes the source text, thus making the target text survive, ideally forever, in the new eco-environment of the target text.

As far as how to specifically "maintain," "coordinate," "balance," and "reconstruct" the ecologies of the source-text and target-text systems in the process of translation, it depends on the translator's ability to make a particular analysis and settlement according to particular cases. Therefore, the discussion space is broad. However, in principle, on the one hand, the key to "maintaining" and "balancing" the ecologies of the source and target languages is to maintain and balance the "genes" and the "blood" of the source text and the target text. Consequently, the "genes" and the "blood" of the source text can still flow in the target text and can be embodied in it. On the other hand, when the translator understands the source text, he/she usually takes the world he/she is familiar with as the starting point, and he/she often understands the source text on the basis of his/her own "prior knowledge," "prior views," and "prior grasp." As a result, to maintain, coordinate, balance, and reconstruct an eco-environment that is adaptive to the source-text ecology, the translator, in most cases, must "transform" or even "empty" as much as possible the prior or existing "ecology" in his or her mind. Only in this way can the new eco-environment that is adaptive to the source-text ecology be transplanted (see

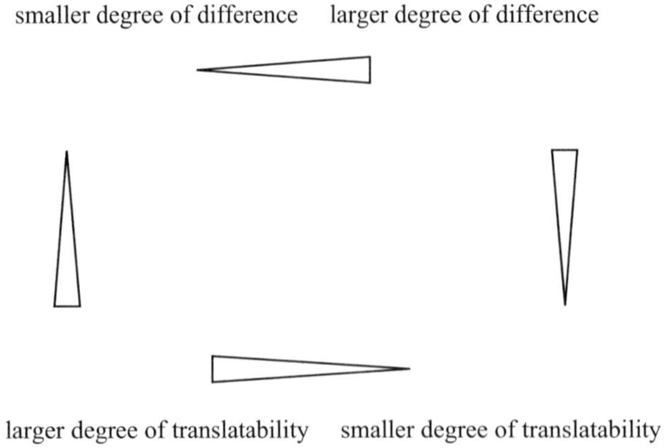

Fig. 7.3 The correlation of "the degree of difference" between the source-text and target-text ecologies and "the degree of translatability"

Sects. 6.2–6.7). In addition, as an Eco-translational strategy choice, the application of the "leaning on" translation strategy to the "untranslatable" text is effective (see Sect. 6.5).

In summary, most of the previous related research interpreted translatability or untranslatability from the perspectives of language, culture, and criteria for translation. This section interprets and deduces translatability or untranslatability based on the principle of "balance and harmony" in Eco-translation ethics and the dimension of the transformation and transplantation of the source-text ecology vs. the target-text ecology. Therefore, we conclude that untranslatability results from varying degrees of "unconformity," "imbalance," and "discrepancy" between the source-text ecology and the target-text ecology. The eco-reasons for "translatability or untranslatability" lie in the "degree of differences" between the textual ecologies.

7.5.2 Retranslation

The so-called retranslation refers to the phenomenon of a work being translated again or being repeatedly translated. Cases of retranslation abound. It has been a heatedly debated topic in the translation field. For example, the Chinese version of *The Plays of Shakespeare*, translated by Zhu Sheng-hao (1912–1944), a famous Chinese poet and translator, first appeared in 1947. In 1967, the retranslated version was published in Taiwan by Liang Shih-chiu (1903–1987), a renowned Chinese educator, writer, translator, literary theorist, and lexicographer. In 1999, it was translated again in poetic style by Fang Ping (1921–2008), a famous Chinese translator and authority on Shakespeare. As a second example, there have been as many as 17 English versions of *A Dream of Red Mansions* (Hong Lou Meng, sometimes translated as *The Dream of the Red Chamber*), one of the four great classical Chinese novels. In addition, there have been nearly 20 Chinese versions of *Walden*, a masterpiece by Henry David Thoreau (1817–1862), the nineteenth-century American transcendentalist writer.[8] There have been nearly two dozen Chinese ver-

[8] (1) Liu Fei's version published by Huashan Art and Literature Press in 1996; (2) Xu Chongxin and Lin Benchun's version published by Shanghai Sanlian Publishing House in 1996; (3) Wang Guanglin's version published by China Writers Publishing House in 1998; (4) Dai Huan's version published by The Contemporary World Press in 2006; (5) Mu Zi's version published by Wuhan Press in 2009; (6) Zhang Zhiyao's version published by Harbin Press in 2003; (7) Yang Jiasheng's version published by Tianjin Education Publishing House in 2004; (8) Su Fuzhong's version published by People's Publishing House in 2004; (9) Zhang Zhiyao's version published by Tianjin Education Publishing House in 2005; (10) Tian Ying and Zhu Chunfei's version published by Shanxi People's Publishing House in 2005; (11) the version entitled with 《湖滨散记》 (hu bin shan ji) Trans. edited by Zeng Guanghui published by Chinese Books Publishing House in 2005; (12) Zong Huazheng's version published by China Film Press in 2005; (13) Wang Guanglin's version published by Changjiang Literature and Art Press in 2005 and 2007; (14) Lin Zhihao's English and Chinese version published by Hainan Publishing House in 2007; (15) Li Mu's English and Chinese version published by Shanghai Sanlian Publishing House in 2008; (16) Wang Yiguo's version published by Beijing Yanshan Press in 2008; (17) Xu Chongxin and Lin Benchun's version

sions of *The Red and the Black*, a novel published in French in 1830 as Le Rouge et le noir by Stendhal (Marie-Henri Beyle) (1783–1842), the nineteenth-century outstanding French critical realist writer, that has caused endless debate among French experts.

From the point of view of the "three ecologies" and the relationships among the "three ecologies," retranslation is mainly caused by changes in textual ecologies, translation ecologies, and "translation community" ecologies, which can be specified as follows:

1. Caused by Changes in Textual Ecologies.

   Changes in textual ecologies include changes not only in the source-text ecology but also in the target-text ecology. The changes in the source-text ecology contain the changes in diverse versions of languages, different languages, and so on.[9]

   The changes in the target-text ecology include such "intra-semiotic translation" as changes in textual genre and form, textual languages, and textual expression. In addition, changes in the target-text ecology include cases such as outdated language, many problems or defects in the original translated versions, mistranslation, translation errors, omissions in the translation, and various other issues. Relatively speaking, retranslation resulting from changes in the target-text ecology is more common than retranslation for other reasons.
2. Caused by Changes in Translation Ecologies.

   Changes in translation ecologies cover variable factors and a wide range, such as changes in the social history environment and value orientation, ideology, new problem-solving ability provided by high and new technology, and degree of cultural acceptance (fusion and convergence). Although the translation products are separate from the age, they still have a certain charm. However, the seasons change; waters and mountains move. With distance in time and space, previously vivifying translation products may become increasingly difficult for present-day people to understand. Thereupon, this provides the possibility and necessity of retranslation.
3. Caused by Changes in "Translator Community" Ecologies.

   The "translator community" is an aggregation of "participants" involved in translation activities, with the translator as their representative. A great variety of changes may occur in it. For the translators themselves, the same translator experiences a continuous maturation process as he/she changes from youth to middle age and to old age, and he/she will be likely to translate and retranslate the same works differently in different maturation stages. For example, among the most

published by Yilin Press in 2009; (18) Wang Jiaxiang's version published by Beijing October Literature and Art Press in 2009.)

[9] For instance, "The discovery of ancient books is dynamic as it cannot rule out the possibility that unearthed or lost versions will be discovered or found as a result of archaeological excavation, which provides the possibility of re-examining the extant translations and continuing to produce new translations" (Cui, 2012, p. xvi).

successful translation works of Mr. Xu Chi (1914–1996), a famous Chinese translator, poet, essayist, and critic, are his Chinese versions of *Walden*, a masterpiece by Henry David Thoreau (1817–1862), the nineteenth century American transcendentalist writer. He first translated it in 1949 and retranslated it in 1982 and 1996. These translated versions still shine in the vast body of translated literature in China.[10]

Regarding readers, from the perspective of reception aesthetics, the reception awareness of readers of different ages is different. Moreover, with changes due to the development of history, it is impossible to produce a "final version" that once and for all meets the needs of readers. In addition, retranslation resulting from changes in publishers is a common occurrence. For example, the changes in agents, publishing methods, and sponsoring intensity can become motivations for retranslation.

4. Caused by Changes in Other Factors of the Translational Eco-environment

   Changes in academic ideas, translation theory, translation criticism orientations, etc. could provide possible conditions and reasons for retranslation.
5. Ultimately, Caused by Various Motivations of "Doing Things with Translations"

In the final analysis, a retranslation is still a translation. Moreover, translation has its own purposes, that is, a translation needs to "do things." "Doing things with translations," which is applicable in the explanation of the retranslation phenomenon, has been specially discussed in the previous chapters and sections (see Sects. 5.6.1–5.6.3). Therefore, the author will not give unnecessary details about it here.

In short, according to Eco-translatology, retranslation results from the translational eco-environmental changes of the original translated versions. Therefore, translators will select retranslation to adapt to the changed translational eco-environment. In the terminology of Eco-translatology, retranslation is inevitable to maintain the balance of translational ecologies. It is a "natural phenomenon" in translation activities that can be considered a normal state of translation activity.

7.5.3 *Translation Strategy Employment*

Translation strategies are indispensable aspects of both translation practice and translation teaching. When translation strategies are discussed, dichotomies are often employed to endlessly debate literal translation versus free translation, semantic translation versus communicative translation, formal translation versus functional translation, foreignizing translation versus domesticating translation, etc.

From the perspective of Eco-translatology, confronted with the restrictions of the source-text and target-text ecologies ("bilingual" ecologies), the translator's selection of translation strategies can be regarded as, under the premise of his/her adaptation to

[10] For example, Liu Aihua took these retranslations as an illustration to carry out a monographic study in her doctoral dissertation for Shan Dong University.

the translational eco-environment, the degree of choosing to obey or subvert the restrictions of the "bilingual" ecologies—from complete obedience or subversion to partial obedience or subversion—which can also be regarded as the translator's selection of different strategies to adapt to the translational eco-environment, or the selection of the degree of his/her adaptation to the translational eco-environment.

Therefore, according to the terminology of Eco-translatology, whether foreignizing translation or domesticating translation, whether literal translation or free translation, and whether semantic translation or communicative translation, all translation strategies and translation methods can be seen as the translator's selection of translation strategies or translation methods to adapt to the translational eco-environment. As mentioned in the previous sections, how to specifically "maintain," "coordinate," "balance," and "reconstruct" the ecologies of the source and the target languages in the process of translation depends on the translator's judgment and selection. For example, for the sake of maintaining and balancing the "genes," "blood," and "flavor" of the source and the target texts to make the genes and blood of the source-text flow and be embodied in the target text—in other words, to retain the original unique flavor of the source text in the target text—the translator can employ the "leaning on" strategy of ecological translation to manage and settle the texts. According to Eco-translatology, the "leaning on" strategy refers to the fact that the translator tries to adapt to and lean on the source-text ecology (the "original ecology" of the source text) to select a translation. Alternatively, the translator tries to adapt to and lean on the target-text ecology (the "original ecology" of the target text) to select a translation. As another example, for the sake of maintaining, coordinating, balancing, and reconstructing the eco-environment to adapt to the source-text ecology, the translator could first try to "transform" and even "empty" the prior existing "ecology" in his/her mind to transplant the new eco-environment to the target text, which is adaptive to the source-text ecology.

The TAS theory defines translation neither from the angle of the author/the source text nor from the angle of the target text/the reader but from the angle of the translator. That is, it defines translation as "the translator's selective activity to adapt to the translational eco-environment and transplant texts with the translator as its dominator, text as its basis and cross-cultural information transfer as its purpose." Accordingly, with regard to problems such as which is better, foreignizing translation or domesticating translation, literal translation or free translation, we can explain them as choosing and following what is right, i.e., as a translator's "selection" for his/her "survival." We can say that it is easy to explain the above problems from the perspective of Translation as Adaptation and Selection because translation is defined as "the translator's selective activity to adapt to the translational eco-environment and transplant texts with the translator as its dominator, text as its basis and cross-cultural information transfer as its purpose." The translational eco-environment, which includes society, culture, and all "participants," changes in an endless and dynamic way. To adapt to the dynamically and endlessly changing translation eco-environment, it is natural for the translator to make a selection that is adaptive to the translational eco-environment between foreignization and domestication, literal translation and free translation, etc., that is, to "let nature take its course."

7.5.4 *Translation of Styles*

In regard to the translation of styles, the basic principle is the same as that of the translation strategy employment described above. In the final analysis, the translation styles are still done through "bilingual" textual transplants, textual eco-equilibrium, and the translator's adaptation/selection. If the text of one language ecosystem can be successfully transplanted into another language ecosystem, the styles of the source language can naturally be maintained in the target language. Equally, if the textual ecologies, including the linguistic, cultural, and communicative ecologies, can maintain the balance between the source and target systems, the styles of the source language can naturally be maintained in the target language. If the translator is able to make multidimensional "selective adaptation" and "adaptive selection" of the translational eco-environment and the final translation text, the styles of the source language can naturally be best reflected in the target language.

In other words, the key to retaining the styles of the original in the translation lies in retaining the ecologies of the source text, and vice versa. Retaining the ecologies of the source text is equivalent to retaining the styles of the source text. Consequently, the degree of verisimilitude of the translation of styles is determined by the textual transplantation and the degree of retaining the "bilingual" ecologies in the transformation.

Translation practice shows that, concretely, the translation of styles consists of coordinating and balancing the relationships among the "three" ecologies, i.e., the textual ecology, translational ecology, and "translator community" ecology (see also Sect. 3.4.4).

7.5.5 *"Turns" of Translation Studies*

The "turns" in translation research fields have been a "hot topic" in China and abroad in recent years. Apart from the well-known linguistic turn and cultural turn of translation studies, in recent years, many kinds of "turns" have been discussed in foreign and domestic translation studies and works that directly use the term "turn"[11]: the translator's turn (Robinson, 1991), the sociological turn (Pym, 2001), the power turn (Tymoczko & Gentzler, 2002), the practical philosophy turn (Sun, 2003), the creative turn (Loffredo & Perteghella, 2006), the empirical turn and the ideology turn (Snell-Hornby, 2006), the overt and covert turns (Xu, 2007), the pragmatic turn (Zeng, 2007), the cognitive turn (Liu, 2008), and the reality turn (Huang & Du, 2008), among others, in which the author's "ecological turn" (2007)[12] is not

[11] For instance, Loffredo and Perteghella (2006), Pym (2001, pp. 129–138), Robinson (1991), Snell-Hornby (2006), Tymoczko and Gentzler (2002), Huang and Du (2008, pp. 18–21), Li J. (2009), Liu (2008, pp. 88–93), Luo (2008, pp. 61–63), Xu (2007, pp. 139–141), and Zeng (2007).

[12] See Hu Gengshen. The "Ecological Turn" of Translation Studies from the Perspective of "the

included. These turns demonstrate an inevitability of translation studies' seeking new study methods directions and exploiting new research fields and also symbolize the translation discipline's dynamic and prosperous development.

From the perspective of Eco-translation ethics, all the types of "turns" of translation studies are due to the exploration of translation theory for the eco-ethical principle of "symbiosis and coexistence." The eco-ethical principle of "symbiosis and coexistence" points out that according to ecological principles, symbiosis is a basic condition of biological survival, i.e., a state of interdependence and mutual development between organisms. Like biological diversity and symbiosis in natural ecology, the pluralism of translation theory research should become a kind of normal state of translatology development. Translation theory research is a kind of academic research, and academic research should emphasize the coexistence of "commonness and difference." China already has the ancient precepts that "all things of creation under heaven in their natural communion develop toward the common goal along different roads and converge hundreds of thoughts and considerations into one unified concept" by Confucius in the Survey of *The Zhou Book of Change*, one of the Chinese classics, and that "the superior man seeks wider common ground while reserving small differences" in Image of the Kui Hexagram of *The Zhou Book of Change.* In this sense, the fact that translation theory research emphasizes pluralism accords not only with the reality of translation theory research but also with the tradition of Chinese academic ethics. The "plural or multiple" ethics of translation theory research reflects respect for the "construction right" of translation theory researchers. In this sense, the "turn" is a kind of "right" for theory constructors or builders (see Sect. 3.7.3).

From the perspective of biological "diversity," all types of "turns" of translation studies are an inevitable result of the "diversity" of Eco-translation research. Diversity reflects a tremendous difference between the personalities of various things. We know that "ecology" is the study of relationships between organisms and the environment. Translation studies from the perspective of ecology include all aspects of translational ecosystems and natural ecosystems. To begin a research project from different aspects and different priorities, the focus of different studies will probably be transferred. It is very natural for "diversified" and different "turns" of translation studies to arise. In this sense, the "turn" can be said to be a "condition" of theory construction (see also Sects. 3.6.6 and 4.4.1).

From the cross-disciplinary perspective of Eco-translation, all types of "turns" of translation studies are the natural reflection of the cross-disciplinary characteristics of translatology. Because cross-disciplinary characteristics are a major feature of translation studies, the cross-disciplinary emphasis can be viewed as one direction for the focus shift of translation studies. In this sense, this feature of the development of translation studies can also provide the "turn" with the "possibility" of theoretical construction (see also Sect. 4.2.3).

Sequence Chain" [A]. The keynote speech at the second Cross-Straits Symposium on Translation and Intercultural Communication, 2007-11-22/24, Macao, published in Hu (2009b), pp. 25–35).

From the perspective of Eco-translation research, all types of "turns" of translation studies are the process of translation research attempting to draw "closer to the truth," which means the transformation of perspective, the expansion of vision, the change in thinking.... The "turns" here are also translation theorists' generalization of translation from different angles, including the expansion of the research vision and the transformation of the research perspective; the sublation of traditional translation concepts and the shift of translation focus; the introduction of research modes of thought and the renewal of research methods; etc. In this sense, the "turn" has its own "motivations" and methods.

From the perspective of Eco-translatology, the "turns" of translation studies have rules to follow and coordinates to consult; that is, they expand and broaden in the turn of the "Sequence Chain." In light of the fact that the "Sequence Chain" of "translation ↔ language ↔ culture ↔ society/humankind ↔ nature" reflects the elementary path of the epistemic vision of the expansion and rational progress of humankind, which conforms to the basic law of the evolution of human epistemic ability and to the research approach of "sticking to the basis just as the river is the source of the sea," this "Sequence Chain" has the obvious functions of "inference" and "prediction," which can infer the future from the past and predict the unknown from the known world. In this sense, the "turn" can be said to have a theoretical basis (see also Sects. 2.2.1, 3.7.3, and 8.3.2).

Finally, there are indeed many "turns" of translation studies. However, from the perspective of Eco-translatology, it remains to be seen whether they can be accepted and recognized in the translation world, and particularly whether they can be practiced and follow eco-reason and the law of nature to "preserve the superior and eliminate the inferior" and ensure the "*survival of the fittest*" and the "long survival of the strongest."

7.6 Summary

In recent years, with the continuous in-depth development of Eco-translatology, various studies of its theoretical applications have been increasing every year. Reviews and complimentary and/or critical comments have appeared in many journals and periodicals. Many researchers in applied studies are not satisfied with simply applying one principle of Eco-translatology to one or two illustrations. They have started to explore the relationship between Eco-translatology and related disciplines and perform in-depth research on its concepts, generating many constructive opinions and views for the resolution of such problems as Eco-translatological studies and the further development of the discipline. However, this research has also led to questions and objections.

The author believes that it is always beneficial and harmless to review and tease apart problems, objections, and questions. On the one hand, some questions and doubts can be clarified. On the other hand, some views and expressions can be reaffirmed. Third, some necessary revisions can be made to perfect the theoretical system of Eco-translatology. The author believes that all these results accord with the fundamental law of understanding and exploring unknown things.

References

Ames, Roger T., & Henry Rosemont, Jr. (trans.) (1998). *The Analects of Confucius: A Philosophical Translation*. New York: The Random House Publishing Group.

Bian, L. (2012, November 24). *Studies on translation history from the perspective of eco-translatology*. Paper Presented on the Third International Eco-Translatology Symposium, Chongqing.

Cao, Z. Y. (2007). A study on translation of tourist English from perspective of translation as adaptation and selection. *Journal of Hubei University, 6.*

Chen, Y. (2011). On translational ecology of internet era. *Journal of China Newspaper Industry, 6.*

Cui, Y. H. (2012). *On English translation of Lao-zi (founder of Taoism).* Unpublished Ph.D. Dissertation, Shandong University.

Dollerup, C. (2007/2013). On Professor Hu Gengshen's scholarly contributions: A western scholar's view. *Journal of Eco-Translatology* (1–2).

Fang, M. (2010, November 9–10). *Construction on eco-translatology from the perspective of core terminologies*. Paper Presented on the First International Eco-Translatology Symposium, Macau.

Gao, H. X. (2009). *The approach to translation as adaptation and selection- an examination and a case study*. Jinan: Shandong University.

Gousty, Y. (2011, November 11–13). *On translational eco-environments of translation for science and technology*. Presented at the Second International Symposium on Eco-Translatology. Shanghai.

Hu, G. S. (2003). Translation as adaptation and selection. *Perspectives: Studies in Translatology, 4*, 283–291.

Hu, G. S. (2008). Eco-translatology: A primer. *Chinese Translators Journal, 6*, 11–15.

Hu, G. S. (2009a). Fu Lei's philosophy on translatology. *Journal of Foreign Languages, 2*, 47–53.

Hu, G. S. (2009b). *Translation and intercultural communication: Integration and innovation* (pp. 25–35). Shanghai: Shanghai Foreign Language Education Press.

Hu, G. S. (2011). Eco-features of eco-translatology and the implications to translation studies. *China Foreign Languages, 6*, 96–99/109.

Hu, G. Z. (2006). A theory of "translator's turn": A review of "translation as adaptation and selection". *Chinese Book Review Monthly, 2.*

Hu, L. (2010). *A critical view of the approach to translation as adaptation and selection*. Chongqing: Sichuan International Studies University.

Huang, C. (2010, November 9–10). *Legal texts translation from the perspective of eco-translatology*. Paper Presented on the First International Eco-Translatology Symposium, Macau.

Huang, D. X., & Du, X. J. (2008). Translation reality turn in translation studies. *Shanghai Journal of Translators, 3*, 18–21.

Huang, M. H. (2007). Interpreters and translators multidimensional environment "adaptation/selection". *Journal of Macao Polytechnic Institute, 3*, 106–115.

Huang, Z. L. (2009). Research on Yan Fu's translation. *Reading Journal, 2*, 116–120.

Jiang, X. H. (2009). Translator's selective adaptation and adaptive selection. *Shanghai Journal of Translators, 4*, 11–15.

Jiao, W. H. (2008). Comparative analysis of "Hao Liao Ge" and its English version. *Journal of Macao Polytechnic Institute, 2*, 124–131.

Jiao, W. H. (2010). An analysis of "evolution and ethics" translated by Yan Fu from "the sequence chain". *Shanghai Journal of Translators, 4.*

Ku, H.-m. (1898). *Discourses and sayings of confucius: A new special translation, illustrated with quotations from Goethe and other writers*. Shanghai: Kelly and Walsh, Ltd..

Legge, James. (trans.) (1861). *Confucian analects, the great learning, and the doctrine of the mean. The Chinese Classics. I.* London: Trübner.
Leng, Y. H. (2011). Is the translator the centre in the eco-Translation theory? *Shanghai Journal of Translators, 3*, 71–73.
Li, G. S. (2011). Analysis on Chinese- English translation textbooks used in Chinese Universities. *Technology Information, 28*, 29–30.
Li, G. S. (2012). comparison studies between Kung Fu theory and the idea of translation from the perspective of Eco-translatology. *Journal of Eco-Translatology, 1*, 47–53.
Li, J. (2009). *The pragmatic turn of translation studies*. Shanghai: Shanghai Translation Publishing House.
Li, Y., & Huang, Z. L. (2005). A unique theoretical build-up: On reading HU Gengshen's monograph titled: An approach to translation as adaptation and selection. *Foreign Language Education, 6*, 95–96.
Li, Y. Z. (2009). *English translation of TCM disease names in Huangdi Neijing from the viewpoint of adaptation and selection*. Nanjing University of Chinese Medicine.
Lin, Yu-tang. (Editor, Translator & Annotator) (1938). *The wisdom of confucius*. New York: Random House/The Modern Library.
Liu, A. (2010). A critical study on translation from ecological perspective. *Journal of Xiudy on Translation from Ecologi, 3*, 75–78.
Liu, J. P. (2008). Re-conceptualize the cognitive map and initiate "cognitive turn" in translation studies. *Journal of Hubei University for Nationalities (Philosophy and Social Sciences), 4*, 88–93.
Liu, Y. F. (2009). On translation of metaphorical idioms in news report from the perspective of translation as adaptation and selection. *Shanghai Journal of Translators, 9*, 16–19.
Liu, Y. H., & Xu, J. (2004). A new translation book with spirits of exploration---book review of *translation as adaptation and selection. Chinese Translators Journal, 6*, 40–43.
Loffredo, E., & Perteghella, M. (2006). *Translation and creativity: Perspectives on creative writing and translation studies*. London: Continuum.
Luo, H. (2008). Translation ethics turn and reflection of translation studies. *Journal of Xiangtan Normal University (Social Science Edition), 5*, 61–63.
Lv, Z. F. (2011). On eco-translatology and its status quo. *Journal of Shangqiu Vocational and Technical College, 10*(3), 91–92.
Mao, L. (2010, November 9–10). *The reflections of eco-translatology in MTI education*. Presented at the First International Eco-Translatology Symposium, Macao.
Meng, F. J. (2008). *Turn, surpass, and return: Theoretical reflection of the theory of translation as adaptation and selection*. Retrieved August, 2008, from http://www.language.international.com
Pound, Ezra. (trans.) (1951). *Confucius: Analects*. London: Peter Owen Limited.
Pym, A. (2001). The return to ethics. *The Translator, 7*(2), 129–138.
Robinson, D. (1991). *The translator's turn*. Baltimore: Johns Hopkins University Press.
Runjuan, S. (2011, November, 11/14). *Interpreting quality evaluation from eco-translatology perspectives*. Paper Presented at the Second International Eco-Translatology Symposium, Shanghai.
Sang, Z., & Mu, L. (2008). Today's Chinese translation field: Overviews on China's translation studies in 2007. *Journal of Tianjin Foreign Studies University, 4*, 31–36.
Snell-Hornby, M. (2006). *The turns of translation studies: New paradigms or shifting viewpoints*. Amsterdam: John Benjamins Publishing Company.
Song, Z. P. (2007). *Choice and adaptation: A study of translation process from the pragmatic perspective*. Northeast Normal University.
Song, Z. P. (2010, November 9–10). *Translation pedagogy ecosystem: An Eco-translatological perspective*. Paper Presented at the First International Eco-Translatology Symposium, Macau.
Song, Z. P., & Liu, A. (2012). A summary of the second international eco-translatology symposium. *Chinese Translators Journal, 2*, 69–72.
Sun, N. N. (2003). The impact of practical philosophy turn in translation studies. *Journal of Hehai University (Social Sciences), 3*, 76–78.

Sun, W. (2010, November 9–10). *A study on E-C translation of Lunyu from the eco-translatology perspective*. Paper Presented at the First International Eco-Translatology Symposium, Macau.

Sun, Y. C. (2004). *Zhang Guruo's translation art*. Beijing: China Translation & Publishing Corporation.

Tymoczko, M., & Gentzler, E. (2002). *Translation and power*. Amherst: University of Massachusetts Press.

Waley, Arthur. (trans.) (1938). *The Analects of Confucius*. New York: Vintage Books.

Wang, D. (1996). The value of theory lies in the call of the reality. *China Book Review, 4*, 56.

Wang, H. (2011). On core concepts of eco-translatology. *Shanghai Journal of Translators, 4*, 10–11.

Wang, H. (2012, November 23-25). An ecological interpretation of literary translation ethics. Paper Presented at the 3rd International Eco-Translatology Symposium, Chongqing.

Wang, Y. (2010, November 9–10). *On sustainable development of ecology and translatology*. Paper Presented at the First International Eco-Translatology Symposium, Macau.

Wen, X. (2010, November 9–10). *Eco-translatology paradigm from perspectives of cognitive linguistics eco-linguistics*. Paper Presented at the First International Eco-Translatology Symposium, Macao.

Wenxiong, Z. (2007). *Pragmatic approaches to translation*. Wuhan: Wuhan University Publishing House.

Wu, Y. Q., & Li, J. P. (2006). Role of translator in the translation of *cat in the rain*. *Journal of Anhui Normal University, 6*.

Xu, J. (2007). On the overt and covert turns of contemporary translation studies. *Journal of Yunmeng, 1*, 139–141.

Yang, G. H. (2012). The development and prospects of eco-translation. *Journal of Hubei University of Education, 5*.

Yang, Z. J. (2004). A new perspective of translation studies. Preface. In G. S. Hu (Ed.), *An approach to translation as adaptation and selection* (p. 1). Wuhan: Hubei Education Press.

Yingzhen, G. (2010, November 9–10). *Tourism translation in Henan Province from the perspective of eco-translatology*. Paper Presented on the First International Eco-Translatology Symposium, Macau.

Youlan, T. (2010, November 11/13). *On translation textbooks studies: Perspectives from eco-translatology*. Paper Presented at the Second International Eco-Translatology Symposium, Shanghai.

Yue, Z. S. (2012). On hybrid interpreted by eco-translatology. *Journal of Eco-Translatology, 2*, 60–66.

Zeng, W. X. (2007). *Pragmatic approaches to translation*. Wuhan: Wuhan University Publishing House.

Zeng, Y. (2010, November 9–10). *Linguistic adaptation theory and eco-translation*. Paper Presented at the First International Eco-Translatology Symposium, Macao.

Zeng, Y. T., & Zhao, Q. L. (2012). Comment on eco-translation studies. *Journal of Eco-Translatology, 1*.

Zhai, H. M., & Zhang, D. (2005). Translator-centered theory and choice of Lin Yutang's translation of *Fu Sheng Liu Ji*. *Journal of Anhui Normal University (Humanities & Social Sciences), 1*, 115–119.

Zhang, B., & Xin, H. J. (2005). Establish our own translation schools in China. *Chinese Foreign Language, 5*, 69–73/79.

Zhang, L. F., & Wenning, J. (2011). Eco-translatology and its three ecological dimensions --a discussion with Prof. Hu Gengshen. *Journal of University of Shanghai For Science and Technology (Social Science), 4*, 261–266.

Zhang, Q. X. (2010). Thoughts on the theory of translation as selection and adaptation. *Journal Gansu United University, 6*, 22–26.

Zhang, S. J. (2012, November 23-25). *Variation in C-E consecutive interpretation from the perspective of Eco-translatology*. Paper Presented at the 3rd International Eco-Translatology Symposium, Chongqing.

Zhao, H. (2012). On domestic translation as adaptation and selection: Retrospect and prospect. *Science and Technology Information Journal, 27.*

Zhongsheng, Y. (2012). On hybrid interpreted by eco-translatology. *Journal of Eco-Translatology, 2.*

Zhu, H. (2010, November 9–10). *The ecological thinking in Lu Xun's fairy translation –with the "Little John" as an example*. Paper Presented at the First International Eco-Translatology Symposium, Macau.

Zhu, L. (2010, November 9–10). *Rethinking the theoretical interdisciplinary construction issues in translation from "eco-turn" perspective (1)*. Paper Presented at the First International Eco-Translatology Symposium, Macao.

Zhu, X. (2012, November 23–25). *Reflections on eco-translation theories*. Paper Presented at the Third International Eco-Translatology Symposium, Chongqing.

Chapter 8
Implications for Translation Studies and the "Road Map" for Further Development

8.1 Introduction

Over more than ten years, Eco-translatology has both germinated and steadily developed. At the start of the next decade, scholars will keep a watchful eye on the following points: What is the road map for Eco-translatology? How can Eco-translatological research be deepened and broadened in the future? What can be done to promote the sustainable prosperity of Eco-translatology? How can more scholars, particularly energetic young scholars, be attracted to participate in the study of Eco-translatology and move it steadily forward?

To answer the questions above, this chapter further elaborates on the developmental pattern and describes the academic pursuit of that pattern. We intend to reveal the implications and significance of the "road map" of Eco-translatology with predictions and expectations on a global level.

8.2 Developmental Framework of Eco-Translatology Research

As research from the ecological perspective, Eco-translatology probed investigations into translation studies, translation theories, and translation performance in the first decade of the twenty-first century. A developmental pattern of the macroscopic translatological structure (particularly emphasizing translation studies), the mesoscopic theoretical system (emphasizing translation theories), and the microscopic textual operation (emphasizing source/target texts) has been gradually formulated.

(Hugs) G. S. Hu, *Eco-Translatology*, New Frontiers in Translation Studies,
https://doi.org/10.1007/978-981-15-2260-4_8

8.2.1 Studies at the Macro-, Meso-, and Microlevels

1. Research on Translation Studies

At the macrolevel, Eco-translatology emphasizes an overall view of ecological rationality and is devoted to the ecological system as an associated, balanced, and harmonious whole. The interrelationships and interplay of the elements in the entire translational ecosystem will eventually lead to the formation of an interactive, balanced, stable, dynamic, and harmonious whole.

Observed from an Eco-translatological perspective, the whole translational ecosystem is composed of "4+1" subsystems of different ranges: the translation management ecosystem, translation market ecosystem, translation educational ecosystem, and the translational ecosystem by itself, all of which are integrated into the eco-environment (see details in Sect. 4.3).

During the process of maintaining and developing the body of the overall translational ecosystem, the translational ecosystem by itself is the core; the translator educational ecosystem is the foundation; the translation market ecosystem is a platform; and the translation management ecosystem is a guarantee, while every individual ecosystem, without exception, relies on the eco-environment. If we analyze them on the basis of the nature of translational and cultural values, in addition to the external translation environment, the rest of the linear order ranges from low to high as follows: translation management ecosystem—translation market ecosystem—translator educational ecosystem—the translational ecosystem by itself. These "4+1" ecological subsystems interplay with and influence each other, integrating and constituting an organic whole of translational ecology.

On the other hand, "the Sequence Chain" (translation → language → culture → humans/society → nature) (Hu, 2003, pp. 298–299; Hu, 2004a, 2004b, 2004c, pp. 60–61) reveals not only a mutual connection between translation and nature but interactive and complementary features between nature and human beings. It also reflects a necessary path to expand human beings' cognition and make rational progress. Once relevant and correlated studies of "the Sequence Chain" are conducted, it will be possible to create an overview and integrate the ontology of translation theories and even of translation studies as a whole with the help of Eco-translatology.

Therefore, at the macrolevel, Eco-translatology is a paradigm that offers a comprehensive and integrated view. Given that translational ecology is a complex system, Eco-translatology, to ensure and promote the healthy and stable development of the entire ecosystem of translation, pays close attention to coordinating and incorporating each sub-ecosystem with a hope of attaining the maximum value of resources from the translational ecology by promoting their complementary advantages and enabling them to be fully effective.

In other words, the macro-level study of Eco-translatology focuses on a holistic vision. Vision comes from a height, and height contributes to the layers and forms the system of the final whole. The macro-level rational features of Eco-translatology originate in the "upper," "middle," and "lower" processes in the operation of

Eco-translatology theory and lead to macro-, meso-, and micro-level research. Therefore, the meso- and microlevels are positioned as the banner and marker as well as the important basis for Eco-translatology.

2. Research on Translation Theories

At the mesolevel, Eco-translatology focuses on the cognition of the source text, which refers to the systematic theory of translation. Eco-translatology regards translation as a "translation community" consisting of contexts, texts, and translators. Correspondingly, translational theory tends to regard translation as eco-balance, as adaptation and selection, and as textual transplantation. Eco-translatology defines translation as "selective activities of the translator when adapting to the translational eco-environment and transplanting texts, with the translator acting in the leading role, the text as its basis, and cross-cultural information transmission as its ultimate purpose," based on the theory of Translation as Adaptation and Selection.

The process of translation can be interpreted as a translator-centered circle for the translator's adaptation of the natural ecosystem and selection of the final text. At the same time, the principles of "survival of the fittest" and "preservation of the strong" express the intrinsic and timeless motive of translators and translations. These concepts show that meso-level research is based on and improves the theory of Translation as Adaptation and Selection (Hu, 2001, 2004a, 2004b, 2004c).

Eco-translatology is a self-contained translation theory. It is under the overall guidance of the ecological metaphor in which human survival is universally accepted as "survival of the fittest," "preservation of the strong and elimination of the weak." It also relies on fundamental principles such as "the unity of man and nature," "people orientation," "imitation of nature," and "moderation." In addition, Eco-translatology constructs a whole translational ecosystem that reveals the rationality of translational ecology and of ecological translation ethics. On this basis, it answers the questions of what translation is, how we translate, who translates, and why we translate. Meanwhile, Eco-translatology discusses the entire translational ecosystem and such matters as principles, criteria, strategies, and ways of translation from a rational ecological perspective.

In short, from a functional point of view, research at the mesolevel not only focuses on the understanding of translation itself, which is a rational description of translation, but links macro- and micro-level studies. Research at the mesolevel, which is based on the theory of Translation as Adaptation and Selection, is the backbone of the overall study of Eco-translatology.

3. Research on Translation Texts

At the microlevel, Eco-translatology focuses on the impact of basic concepts on the translation and its text. The selected translation strategies can also reflect the translator's recognition and understanding of the nature and criteria of translation. Therefore, the study of text conversions at the microlevel and the analysis of ecological translation operations will also contribute to the cognition and understanding of Eco-translatology. Not only the management of the text, including translation principles, methods, and criteria, but also specific operations and reflections of translation are at the center of Eco-translatology. Therefore, from the perspective of the applicability and feasibility of translation theories, micro-level research is the foundation for the survival and development of Eco-translatology.

In summary, this section briefly lists and explains the three levels of Eco-translatology. If we say that the macrolevel focuses on communication, society, and systems, and the mesolevel lies in cultural, interdisciplinary, and some theoretical issues, then the microlevel encompasses language conversion as well as strategies and practices for ecological translation. In a word, the three levels interplay with and associate with each other harmoniously. The "three-level" study can be summarized as follows (Fig. 8.1):

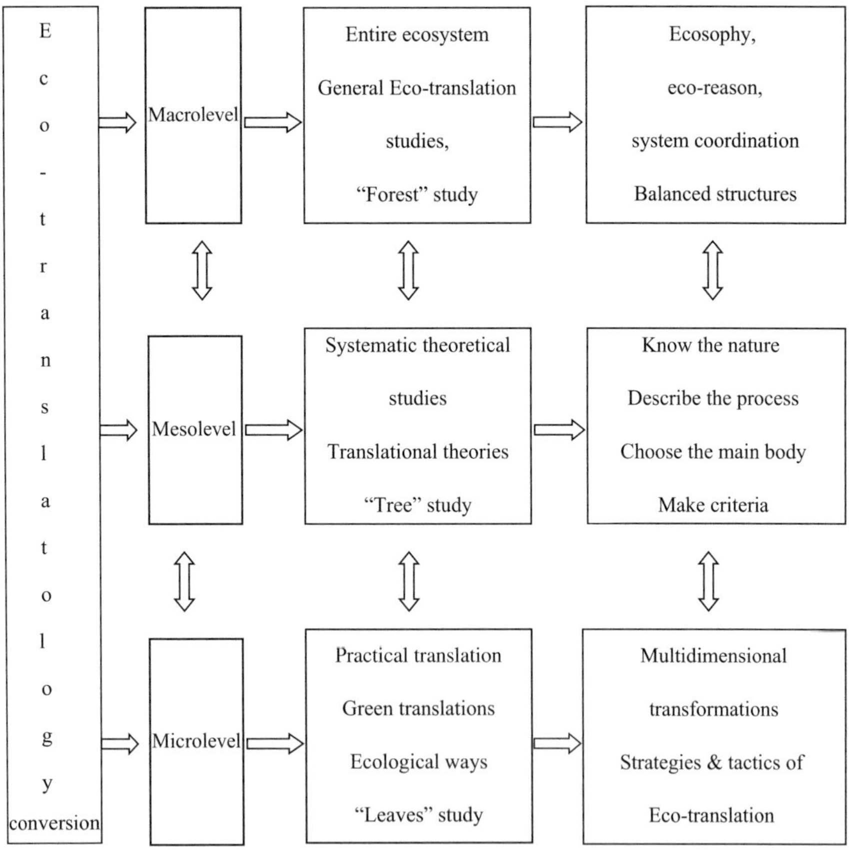

Fig. 8.1 The three levels of studies of Eco-translatology

The research on the "three levels" shows that a developmental pattern of the macroscopic translatological structure (emphasizing translation studies), mesoscopic theoretical system (emphasizing translation theories), and microscopic textual operation (emphasizing source/target texts) has been gradually formulated. The lack of macro-level research means an absence of the overall framework; the absence of meso-level research refers to the lack of the systematic theory; missing microscopic operations will separate Eco-translatology from translation practice.

The "three levels" outlined above not only show the internal logical relations but also contribute to the further understanding of the overall development of Eco-translatology and lay a foundation for a systematic Eco-translatological discourse.

8.2.2 *Academic Pursuit of Integrating "Present with Past," "Eastern with Western," and "Social with Natural Sciences"*

The "three pursuits," namely, the "fusion of Eastern and Western cultures," the "junction of ancient and contemporary civilizations," and the "channeling of arts and science" are the academic pursuits targeted by Eco-translatology in the process of constructing the entire theoretical system of discourse, which is the goal of every academic field. In fact, Eco-translatology, based on the theory of Translation as Adaptation and Selection, has been seeking and insisting on integrating "present with past," "Eastern with Western," and "social with natural sciences," which makes it a new way to study translation after a "cultural shift."

1. Academic Pursuit of Integrating "Eastern with Western"

Integrating "Eastern with Western" is the product of combining Chinese and Western cultures. Closer relations and mutual penetration owing to increased globalization and internationalization contribute to the necessity of "becoming part of each other." Under such circumstances, in terms of academic research, it is hard to distinguish research in China from that in Western countries; the perspectives and results of research from both China and Western countries are promoted and respected in many fields, such as research on translation studies. Professor Gu Zhengkun of Peking University (Gu, 2001, p. 8) regards "an ability to blend Chinese and Western perspectives while taking a national stance" as one of the three features of Chinese theoretical works on translation (the other two features are "theoretical depth" and "the organization of ideas into a systematic whole"); he also "deems it an urgent affair to open up an approach that allows the strengths of Chinese and Western translation theories to complement each other." Professor Meng Fanjun of

Southwest University (Meng, 2002, p. 17) states that "the trend of development for China's modern Translation Studies must be compatible with both Chinese and Western cultures ... both be compatible and parts of *Yin* and *Yang*." During the process of research on Eco-translatology, integrating "Eastern with Western" is a consistent academic pursuit that can be embodied as follows.

On the one hand, Eco-translatology, first proposed by Chinese scholars, initially made use of Chinese discourse in narration. Therefore, so to speak, Eco-translatology reveals a great "Chinese complex." It is evident that although abstraction and universality do not differentiate theories' nationality and specific cultural direction, theorists are associated with their individual nationality and cultural environment. Similar to many humanities and social sciences theories, such as "postcolonialism" and "feminism," which reflect their cultural backgrounds and researchers' subjectivity to varying degrees, Eco-translatology, first proposed by Chinese scholars, also interprets the "Chinese discourse" and way of thinking.

On the other hand, although Eco-translatology shows a potent "Chinese complex," it does not ignore or reject Western academic thought and translation theories. It not only absorbs the spirit and methods of modern Western translation theories but also acts as a pedestal for two other important philosophical theories, modern ecology and eco-holism, that originated in Western scientific thought. Therefore, it is natural that the academic pursuit of integrating "Eastern with Western" can be found in the theoretical foundation and research methods of Eco-translatology.

Moreover, the birth and growth of Eco-translatology have benefited from various opportunities, conditions, and contributions. It germinated in Hong Kong, China, where East and West converge, and it expanded in Macao, where Chinese and foreigners rub shoulders. These realities are not a coincidence, and the emergence and development of Eco-translatology undoubtedly gained colors and support from the geographical environment, cultural environment, and other aspects of its strong philosophical origin.

As seen above, the proposal and construction of Eco-translatology in a certain sense embody the features of integrating "Eastern with Western." Thus, we also hope that in the field of translation studies, Eco-translatology can increase an equal exchange and act as a new topic for "Chinese discourse" and "Western discourse."

2. Academic Pursuit of Integrating "Present with Past"

The emergence and development of Eco-translatology did not occur in isolation. Eco-translatology comes from the combination of Chinese traditional philosophical wisdom and the modern philosophy of translation. The combination is revealed as follows.

First, the research on Eco-translatology reflects "modernity" because "this multidisciplinary approach reflects the trend toward theoretical diversification in translation studies and may mark a new point of departure for the discipline as a whole"

(Hu, 2008, p. 1). Meanwhile, Eco-translatology is traditional because "it is naturally related to the traditional Chinese philosophy, which adopts a holistic approach and is ecologically oriented. The Chinese cultural elements characterized by such theories as 'the unity of Nature and Humanity,' 'the doctrine of the Golden Mean,' and 'the principle of human-centeredness' reflect 'the relations, manifestations, connotations, and relationships between the Chinese traditional culture and the ecological approach" (Hu, 2008).

Second, the thought of "adaptation" in traditional Chinese translation and the concepts of "adaptation" and "selection" come from the same theory. Studies have shown that translators have had ideas about adaptation since ancient times. For example, in 382 AD, the famous translator Dao'an (314-385) pointed out his adaptation in his masterpiece translation of Buddhist scriptures (Chen, 2000, p. 18; Ma, 2001, p. 37), which showed that the term "*Shengren* (wise man)" came from an ancient custom that was different from its current meaning. Therefore, it had to be changed to adapt to contemporary situations. This strategy not only affected the translation of Buddhist scriptures from the Han and Jin to the Sui and Tang Dynasties but also promoted the spread of "adaptation" in the period of the Ming and Qing Dynasties (Liu, 2006, p. 229).[1] In modern times, the Chinese scholar Ma Jianzhong (1845–1900) also noted that "translations should adapt to the source text" (Ma, 1894, p. 2). The concept of adaptation and selection is best embodied in translations from the Chinese translator Yan Fu. He adapted his translation to the local ecological translation environment and then made use of alternative related strategies during the translation process (Huang, 2009). These examples show that research on the basic theory of Translation as Adaptation and Selection normally follows the concepts of "adaptation" and "selection" in ancient translation theories.

Third, Eco-translatology is consistent with ecological ideas from Western countries. Eco-translatology based on the theory of Translation as Adaptation and Selection is an integrated research from the perspective of ecology. In other words, what Eco-translatology emphasizes is the integrality of the ecosystem of translation, which has existed since ancient times. Ancient Greek thoughts of "unity" and "the existence of a continuous whole" can be described as the initiative of the earliest ecological holism (Wang N, 2005, p. 88). Modern scholars' discussion of "holism" and "relevance" has never stopped. For example, the famous French scientist P. C. Laplace noted that everything that is seemingly unrelated is actually associated by the universal law of nature (Laplace, 1835: also c.f.: the Chinese version by Li et al., 1978, p. 305). The American ecologist Barry Commoner also pro-

[1] See more from Liu (2006, University of Madrid, Spain), who explores the application and impact of "adaptation strategies" in the sixteenth and seventeenth centuries in translation. It is mentioned that Defense of the True Religion in the Chinese Language was first published in Chinese in 1593.

posed four laws of ecology in *The Closing Circle* in 1971.[2] As seen above, the basic concepts of adaptation, selection, life, ecology, ecosystems, and ideas are all consistent and inheritable. They reflect the academic pursuit of integrating "present with past."

3. Academic Pursuit of Integrating "Social[3] with Natural Sciences"

Translatology belongs to the humanities. Eco-translatology, which studies translation from the perspective of ecology, despite its features of "interdisciplinarity" or "multidisciplinarity," is also undoubtedly a part of the humanities. Thus, it is not necessary to further explain the social science of Eco-translatology.

Now, we will discuss the obvious natural science of Eco-translatology. First, the groundbreaking basis of Eco-translatology is the theory of Translation as Adaptation and Selection, which borrows and adapts Darwin's theory of adaptation and selection. Darwin believed that adaptive evolution between creatures and the environment contributed to biological evolution. Therefore, he is known as the "founder of ecology." In addition, his masterpiece *The Origin of Species* is considered "the pioneer of ecology" (Wang & Zhou, 2004, p. 77). The theory of Translation as Adaptation and Selection, which borrows the basic principle of "survival of the fittest" and related discussions from Darwin's *The Origin of Species*, discusses and explains the nature, process, criteria, principles, and methods in translation by the narrative method of Eco-translatology.

Second, ecology is one of the theoretical foundations of Eco-translatology. Ecology is considered a sub-branch of biology that studies the interrelation between organisms and between organisms and their environments based on holism. Along with the importance of all the related elements and knowledge as well as the ecological principles of integrity, connection, and balance, we can study the logical relation between organic bodies and their surrounding environments.

In fact, almost every era is dominated by a specific theory of natural sciences that governs the social way of thinking as well. In modern times, "ecological science" is such a subject with ultimate meaning for human existence and development. In this sense, Eco-translatology has a solid philosophical foundation and broad prospects for development.

[2] The American ecologist Barry Commoner proposed the following four laws:

1. Everything is connected to everything else. There is one ecosphere for all living organisms, and what affects one affects all.
2. Everything must go somewhere. There is no "waste" in nature, and there is no "away" to which things can be thrown.
3. Nature knows best. Humankind has fashioned technology to improve upon nature, but such change in a natural system is, says Commoner, "likely to be detrimental to that system."
4. There is no such thing as a free lunch. The exploitation of nature will inevitably involve the conversion of resources from useful to useless forms.

[3] Here, the word "social" refers to the knowledge, ways of thinking, or research methods of the liberal arts, and the word "natural" refers to the knowledge, ways of thinking, or scientific research methods of the natural sciences. The term "integrating social with natural sciences" means the communication, connection, and mixture of the social and natural sciences.

Third, the concepts and methods of the natural sciences require a certain way of thinking, research methods, and language science. For instance, we have always compared the relationship between adaptation and selection in translation as follows.

This "dual" identity of the translator can be interpreted through mathematical language: in the first stage of the translator's adaptation, the translational eco-environment (HJ) is the independent variable, and the translator (YZ) is the dependent variable. YZ is a function of HJ; that is, YZ = (HJ). In the second stage, the translational eco-environment (HJ) is still the independent variable, while the dependent variable is now the translation.... The mathematical expression is as follows:

$$\mathrm{HJ} \xrightarrow{f} \mathrm{YZ}(\text{'}\ \mathrm{HJ}) \xrightarrow{g} \mathrm{YW}$$

Hu (2004a, 2004b, 2004c, pp. 73–74)

For another example, over a long period, people discussed how the outside world constrains translators or how translators withstand outside pressure. However, few studies discussed how translators resist outside pressure or react to external force. According to the laws of motion proposed by I. Newton (1642–1727), action and reaction are equal and opposite (LECSD, 1994, p. 1131) (Hu, 2004a, 2004b, 2004c, p. 125).

The cultural background, social experience, knowledge structure, mode of thought, and values of any founder will inextricably connect with the theory he/she creates. Eco-translatology is no exception. Though I have been engaged mainly in teaching and research in language, culture, translation, international exchanges and other fields and originally studied "liberal arts," I have also worked at the Chinese Academy of Science and Technology and other research institutions for over 20 years. In recent years, I have experienced influence from the atmosphere of scientific research. As a result, I gradually accumulated my practical experience of translation, increasingly gained scientific knowledge and imperceptibly cultivated a scientific way of thought and research. My mixed knowledge of and background in the social and natural sciences therefore have something to do with the feature of integrating the social and natural sciences in Eco-translatology.

In fact, integrating "social with natural sciences" is not uncommon in many fields. The twentieth century's outstanding scientist, Albert Einstein (1879–1955), pointed out that problems cannot be solved by thinking within the framework in which they were created (Boone, 2011, p. 1). Wang Hongyin also draws our attention to the following:

> Even in literary translational criticism, comprehensive criticism integrating various methods of social and natural sciences is also very common. In specific translational reviews, neither method can be effectively adopted. The reality is the combination of social and natural sciences instead. (Wang H. Y., 2005, p. 125)

Therefore, it has become an important feature of contemporary science in "post-modern" academic studies that the natural and social sciences are both involved in communication. As the scientist Budiar said, "scientific development[4] will emerge where there is no discipline limit." There has long been a call for interdisciplinary research in the field of translation. In this regard, the integration of social with natural sciences may provide references for interdisciplinary research in translation studies.

In summary, scholars in Eco-translatology, for more than ten years, have attempted to achieve the academic pursuit of integrating "present with past," "Eastern with Western," and "social with natural sciences." These attempts play an important role in bridging the gap between the Chinese and Western translation fields and breaking the boundaries of research in the social and natural sciences. Therefore, Eco-translatology may be regarded as one of the cases that successfully insists on the academic pursuit of integrating "present with past," "Eastern with Western," and "social with natural sciences." As XU Jun, the executive vice president of the Translators Association of China, said in a speech at the closing ceremony of "The Second International Symposium of Eco-translatology,"

> Eco-translatology, which is based on the basic principles of modern ecology and benefits from traditional Chinese ecological wisdom, is a crossover study of translation and ecology. This sentence involves the modern times and Western countries; the middle part is related to ancient times and China, while the last words involve the social and natural sciences. Therefore, it is encouraging that Eco-translatology, the product of integrating present with past, Eastern with Western and social with natural sciences, has developed in China. (Xu, 2011)

It will also be our unremitting academic pursuit.

8.2.3 *Three Paradigmatic Features of Universality, Openness, and Practicality*

As an integrated study of translation from an ecological perspective, Eco-translatology has demonstrated its three basic paradigmatic features of universality, openness, and practicality. Eco-translatology is in pursuit of a combination of theories and practice in academia. Therefore, Eco-translatology displays a distinct practicality. It also shows that its essence lies in practicality in terms of the spread of its predecessor—the Translation as Adaptation and Selection (TAS) approach—in China. Since the first decade of this century, the concepts that Eco-translatology has advocated, such as translational views from ecological perspectives, the definition of translational eco-environments, the concept of Translation as Adaptation and Selection, the translation methods of selective adaptation and adaptive selection, and translation criticism standards of the degree of holistic adaptation and selection,

[4]This developmental trend is the same in the translation field (see Sect. 7.4.4).

are gaining increasing recognition and acceptance, attracting the attention of translation scholars and being applied, gradually expanding to different translation fields and research topics, achieving an increasing number of results in applied research, and experiencing success in theoretical application. Symposiums have been held in a continuous sequence, and many researchers responded after Eco-translatology was initially proposed (Prof. Fang Mengzhi; see Preface 2). The attraction and vitality of Eco-translatology, therefore, are reflected not only in its advanced ecological concepts but also in its practical application. In addition, it is not merely a macrotheory but has gradually permeated almost every field of translation studies, from theory to practice. Currently, Eco-translatology has become a research direction and topic for an increasing number of translation scholars (see Sects. 2.4.5 and 8.4.2). The development of Eco-translatology should be oriented towards strengthening its practicality, continually intensifying its practical applications, and promoting the incessant development and perfection of ecological translation theory in practical application.

We have built an open and integrated theory of Eco-translatology. "Open" refers to the ability of two sides being able to exchange materials, energy, and information to continually develop, which is a survival law as well as an ecological principle. When this principle is applied to Eco-translatology, it implies that Eco-translatology has an unfixed but open paradigm through which it constantly improves itself in the process of the absorption and tolerance of extraneous cultures and other translation theories.

Western translation theories have many wonderful and reasonable aspects that are new theoretical creations and the results of a particular era. However, due to the limitations of time, space, geography, culture, and vision, these theories are also incomplete or cannot be applied in all instances (Hu, 2002, pp. 80–104). Therefore, when constructing and interpreting the Eco-translatology framework, we closely followed the basic principles of assimilating or discarding, synthesizing or surpassing, inheriting or developing as guidelines for constructing the Eco-translatological discourse system. Through these guidelines, we endeavored to assimilate the rational core of different translation schools while discarding the partially successful elements, to override the restrictions of different schools by synthesizing their standard views, to transmit the traditional wisdom of Eastern and Western translation studies, and to regard translation studies as the starting point and the end result of Eco-translatological development, which means theoretically synthesizing translation schools and thus constructing a pluralistic and unified theoretical paradigm. Since ecology is a metascience, and the ecological orientation is interdisciplinary, Eco-translatology, a sort of methodology, dominates general social thinking modes. It is a discipline that not only exerts an extensive influence on society but also has ultimate meaning for human survival and development. Translation studies from an ecological perspective have the significance of philosophy and methodology in that they are under the guidance of ecological rationality or apply ecological philosophy (such as the holistic equilibrium principle, the law of symbiosis, or the law of diverse unity) to translation studies. However, as a general theory of translation studies from an ecological perspective, Eco-translatology is an integrated study that touches upon "translation studies," "translation theories," and "translation performances." Therefore, its universality will become increasingly conspicuous.

8.2.4 *Three Developmental Stages of the Theory, the Paradigm, and the School of Thought*

For the past 10 years, Eco-translatology has experienced roughly three different stages.

The first stage can be termed the period of initial exploration from 2001 to 2004, which was marked by the paper "An Initial Exploration into an Approach to Translation as Adaptation and Selection" presented at the 3rd FIT Asian Translators' Forum in 2001 and by a monograph titled *An Approach to Translation as Adaptation and Selection* published in 2004. This stage is a meso- and micro-level study of constructing Eco-translatology.

The second stage can be defined as a period of deep integration from 2005 to 2010, which is indicated by the paper "Understanding Eco-translatology" presented at the international conference *Translating Global Cultures: Towards Interdisciplinary (Re)Construction* in 2006, and the paper "Eco-translatology: A Primer" in 2008. Eco-translatology was viewed as an approach to translation studies in the first period and as a branch of general translation studies in the second period.

The third stage, beginning in 2010, can be labeled the period of systematic expansion, which has witnessed the founding of the "International Association for Eco-translatology Research," the release of the *Journal of Eco-translatology* (JET) and the holding of the "International Symposiums on Eco-translatology" in 2011 and the publication of the monograph titled *Eco-translatology: Construction and Interpretation* in 2013. This stage centers around the research paradigm of Eco-translatology, the ecological paradigm of translation studies or translation studies of the ecological paradigm, hovering between an approach to and a branch of translation studies.

The evolution of definitions is the best representation of theoretical development.[5] However, no matter how it is defined and evolves, Eco-translatology, a new and systematic discourse hierarchy of translation theories, will undergo further thorough and extensive studies and will continue to proceed in the proper order.

[5] Several years after the three developmental stages of "theory," "paradigm," and "the school of thought," it is still a "Chinese dream" of Eco-translatology explorers—whether Eco-translatology could prospectively build a consensus within a larger scope, and even develop into a real "ecological turn," and whether this trend, which will develop into the so-called phase of a leading diffusive trend, will be led by Eastern scholars. With respect to this point, Eco-translatology emphasizes "natural evolution," and the ecological research paradigm that follows is the principle of "letting nature take its course."

8.3 Trends and Enlightenment of Eco-Translatology Studies

The definition of Eco-translatology emphasizes the organic entirety of translational environments, source and target texts, and translator actions; advocates the interaction, balance, and harmony of the translational ecosystem; and pursues ecological aesthetics and diverse unity. Therefore, the development of Eco-translatology will contribute to promoting and guiding the evolution of many aspects of translatological studies. In other words, Eco-translatology follows a path of ecological rationality that embodies some fundamental features, such as emphasizing holism and relevance, seeking dynamics and balance, reflecting ecological aesthetics, identifying the "translator-community," adhering to Eco-translation ethics, and highlighting the unity of diversity. Accordingly, the development and evolution of the following aspects will likely become a trend.

8.3.1 *Enriching the Philosophical Basis*

Philosophy is a "theory on world outlook, values, and methodology" that generalizes and summarizes natural and social knowledge. Based on the scientific knowledge of disciplines, it is characterized by abstraction, reflection, and universality. The history of translation theory studies has indicated that translation theories are always built on philosophical thoughts whose influence and penetration underlie the development of translation theories. In addition, translation philosophy is meant to "study the general natures and state of translation from philosophical perspectives" (Wang, 2001, p. 2). Moreover, translation philosophy is "a high-level theory of translatology which studies the essential nature of translational activities by means of philosophical thoughts" (Yang, 2009, p. 8). In view of the analysis mentioned above, interpreting the evolution and development of translation theories from the perspective of philosophical reasons should be a meaningful "penetration point."

Philosophy has benefited greatly from "the linguistic turn" of the twentieth century. The "turn" inspired us with a meaningful mode of thought to gain an overall understanding of Western philosophy and to realize a new level of dialogue between and fusion of eastern and western philosophy (Chen, 2003, p. 3; Piao, 2006). There exists a close relationship between research and development in translation theories and the philosophical "linguistic turn." Since the middle of the last century, the international translation community has proposed translation theory studies and systems based on philology, structuralism, deconstruction, cultural anthropology, and eco-holism. In other words, the philosophical thoughts and theoretical bases of translation theoretical studies are influenced by epistemology and methodology in every stage of translation studies and systems. The philosophical foundation and translation theory orientation of contemporary translation theories are shown in the following table (Table 8.1):

Table 8.1 Comparison between the philosophical foundation and translation theory orientation of contemporary translation studies

Theoretical nature	Philosophical reasons	Theoretical orientations
Linguistics	Linguistics, semantics	Semantic equivalence
Literary theories	Literature, aesthetics, philology	Aesthetic value
Cultural studies	Literature, cultural studies, Politics	Recipient culture
Communication studies	Communication, information theory	Communicative effect
Deconstruction	Deconstruction, postmodernism	Source text yielding to target text
Polysystem	Functionalism, system theory, formalism	Literary system
Purpose of action	Action teleology, functionalism, communication theory	Purpose of action
Eco-translatology	Eco-holism, eastern eco-wisdom	Holistic translational ecology

As shown in the above table, the philosophical foundations of translation theories have been unceasingly enriched by a range of concepts from linguistics to theories of literature and art, cultural studies and information theory, cultural anthropology, eco-holism, etc. In addition, the conceptual scope of philosophical reasons has gradually expanded from local application to global application.

The abovementioned translation theoretical research based on different philosophical foundations, if expressed in the cognitive terms of philosophy, can be said to have evolved from "intuitionism" to "structuralism" and from "structuralism" to "pluralism." In accordance with the direction revealed by the cognitive extension of "the Sequence Chain" (see Sect. 2.3.1), however, the trend of development in the future will continuously evolve from "pluralism" to "holism." More importantly, the above judgment is based on such development: since the twentieth century, the subject–object dichotomy has been diverted to intersubjectivity, and anthropocentrism has been transformed to holistic ecology in the thoughts and philosophy influenced by Western ecological philosophy and literary ecological criticism (Zeng, 2007, p. 143). This diversion of philosophical thought will undoubtedly affect the choice and orientation of the philosophical foundation of translation theories. Therefore, from the perspective of philosophical reasons, the philosophical foundation of translation theories is in transition from local application to global application and is evolving from a single orientation to an overall orientation. In conclusion, the emergence and development of Eco-translatology have demonstrated these thoughts and pioneered ways to carry out the transformations.

8.3.2 *Broadening Academic Vision*

The so-called academic vision refers to the horizon of the field. Moreover, there is a causal relationship between academic vision and philosophical reasons. In terms of translation theory research, the evolution of philosophical reasons will inevitably bring about changes in academic visions and trends. Changes in the trend of translation studies, especially the expansion of academic vision, are mainly embodied in two directions: "reviewing the past" and "understanding the present." "Reviewing the past" means observing the present in the light of the history of translation theories, and "understanding the present" refers to considering the future by following "the Sequence Chain" (see Sect. 2.3.1). The translation community has reached a consensus on "reviewing the past" (see relevant comments by Venuti, 2000; Snell-Hornby, 2006; Liao, 2001a, 2001b; Liu, 2005; Xie, 2007, etc.), that is, from the "translation self-rotation" to "the linguistic turn" and from "the linguistic turn" to "the cultural turn." "Understanding the present," which means considering the future in accordance with "the Sequence Chain," might be of great concern to the translation community. The history of translation theories has proved the perspective conversion of "translation—language—culture," while the logic sequence "translation—language—culture—society/humankind—nature (natural ecology)" may predict the development trend. In accordance with "the Sequence Chain," the expanding horizon of the "ecological turn" in translation studies is a possible direction (Meng, 2009, pp. 48–54). Nevertheless, the practice of translation studies has also indicated that "every turn has offered us a new possibility to make a comprehensive understanding of translation" (Xu, 2007, pp. 232–233).

Overall, different components of "the Sequence Chain" intersect and overlap, connect and interact, and bear a relationship of progressive extension. On this basis, we need to emphasize that every new expansion of the academic vision and the original vision should maintain an inclusive and transcendent relationship rather than isolation or forsakenness, much less subversion or replacement (Xie, 1984, p. 63). In brief, this approach is of paramount importance for the "turn" of translation studies.

Ecology, the metascience based on holism, is inclusive of and takes the lead in other disciplines. Meanwhile, ecological orientation is an interdisciplinary orientation with a research method that emphasizes the integrity of interaction (Krohne, 2001, p. 11). Consequently, from the perspective of academic vision, translation theory maintains consistency on the whole with the path of the expansion of human cognitive vision. Moreover, the vision of Eco-translatology has surpassed the single dimension and instrumental rationality and has extended from a single discipline to interdisciplinary integrity.

8.3.3 Constructing the Ecological Paradigm

A research paradigm is closely related to the theoretical system, but the scope of the former concept is wider than that of the latter (see Sect. 3.2). Translation studies will benefit greatly from a more integrated discourse so that all researchers believe that this discourse, even if it is not at the center of their research, has a close bearing on their research (Lefevere, 1993, pp. 229–230). In recent years, the domestic and overseas translation communities have endeavored to pursue "a more integrated discourse." Quite a few scholars have touched upon some features of translation studies, such as "interdisciplinarity," "pluralism," "synthesis," and "integrity," in publications such as *Translation Studies: An Integrated Approach* (Snell-Hornby, 1988), *Redefining Translation: The Variational Approach* (Lance & Jacky, 1991), *Translation Studies: An Interdiscipline* (Pöchhacker & Kaindl, 1994), *Interdisciplinarity in Translation Studies* (Wilss, 1999, pp. 131–144), and *Multidimensional Translation: A Game Plan for Audiovisual Translation in the Age of GILT* (O'Hagan, 2005). These studies provide translation theory researchers with an excellent theoretical reference system utilizing the integration of different modes and different theories.

In China, academia has long advocated the principle of "letting a hundred flowers blossom and a hundred schools of thought contend," emphasizing the combination of Chinese and Western elements, connection between the past and the present, and the pursuit of diverse unity and integration. Only by drawing upon the strength of others can we can develop new ideas and innovatively create a new school of thought (Jiang, 1993, p. 47). "Diversity" reflects individuality and different theories, and "unity" embodies commonality and the holistic connection of theories.

Given the above understanding and theories, the global outlook, values, and methodology that are formed under the guidance of the traditional cultural thought of "diverse unity" will inevitably affect Chinese translational theories. Additionally, following the process of mutual reference, grafting, adaptation, penetration, confrontation, replacement, and evolution, different translation theoretical concepts will inevitably develop in the direction of "distinctive convergence" through comparison and integration at all times and in all countries. Therefore, they will eventually move from "being largely identical with only minor differences" to "integrated unity," ultimately achieving the state of "supreme harmony" (Meng, 2002, p. 17). The underlying unity gives rise to the development of potential pluralism and integrates pluralism and diversity into actual harmony and perfect unity (Cheng, 2005, p. 243). In fact, scholars in the Chinese translation community have made painstaking efforts to realize the integration of translation theories in their deliberations and practice. Eco-translatology, evolving from the continuation and development of the "approach to Translation as Adaptation and Selection," is a research paradigm that makes an integrated study of translation from ecological perspectives. It is a kind of multiple integration. Some scholars have proposed an idea of related study of Eco-translatology from the perspective of different disciplines, such as translatology, linguistics, cultural studies, anthropology, and ecology. From the perspective of different disciplines, they have also pointed out "interdisciplinary integration" that is eventually incorporated into their mutually reliant ecosystem and thus constitutes the organic whole of the translational ecosystem (Hu, 2006; Hu, 2009, pp. 3–8). Furthermore, Zeng Lisha has carried out a systematic and integrated study on systems of translation theories (Zeng, 2008, pp. 1–5). Based on the current situation of applied translation theories at home and abroad, he holds, it is necessary to conduct a systematic and integrated study from different angles, the key of which lies in building an applied logic category from the macro- to the meso- to the microlevel. In addition, Wang Kefei proposes that translation is the cross-language and cross-cultural integration of internal and external factors, macro-, meso-, and micro-level thoughts, source texts, translators, and target texts (Wang, 2009, p. 390). The development of translation studies domestically and abroad indicates that translation theories are developing from vertical self-containment to horizontal expansion and integration following the evolution from "self-rotation" in linguistics to interdisciplinary "revolution." Moreover, studies are presenting "differentiation" and "synthesis" as coexistent. However, from the perspective of research paradigm transformation, translation studies are at present in a "pluralism" stage following the process of "intuitionism" and "structuralism" and will finally evolve into "diverse unity" through the spread and practice of "holism." Eco-translatology, as an eco-paradigm of translation research, is taking the lead in turning this development trend into practice.

8.3.4 Stimulating Cross-Disciplinary Research

Albert Einstein, the most distinguished scientist of the twentieth century, pointed out that "problems cannot be solved by thinking within the framework in which they were created" (Boone, 2011, p. 1). This viewpoint is applicable to both natural science research and social science research. The unanimous communication trend in social science and natural science has become one of the distinguishing features of postmodern academic research because the humanistic spirit and scientific spirit are unified, indivisible, and interactive in the dialectical category. The methodologies of natural science, such as Newton's HDV (hypothesis, deduction, verification), Whewel's ACV (analysis, colligation, verification), and Einstein's $E = mc^2$, have become the leading scholarly paradigms in modern civilization (Blake, 1966, p. 217). It is therefore quite popular to apply the thoughts, methods, and achievements of natural science research to social science research. The revolution of science impels social science and natural science to draw closer to each other by revealing the universality of things and breaching the artificial barriers between various disciplines.

This development momentum also appears in translation studies; for instance, the metatheory, which belongs to mathematics, is borrowed by Fan (2003), pp. 5–9 to study translation. Li (2009, pp. 149–162) proposes the tension of translation, with the concept of tension stemming from physics. Song (2009, pp. 111–120) explains nonlinearity in the selection of translation processes. Meng (2009, pp. 48–54) points out that the division of the Western cultural era, the flourishing of cultural trends, and even the shift of translation studies are all associated with the corresponding development of physics. First, the atomic theory of the "undivided material entity" during the research period of conventional physics not only underwent the cultural trends of substantialism but also connected with "translation theory for the sake of translation." Second, subatomic theory in the modern period, together with unified field theory, conformed to the cultural trends of structuralism in conjunction with translation studies after the linguistic turn. Third, the nonunified field theory of contemporary physics corresponds to the cultural trends of structuralism in conjunction with translation studies after the linguistic turn. Therefore, the shift in translation studies is a necessary reflection of scientific cognitive tendencies and undulations in cultural trends.

In the field of literary translation criticism, it is common for criticism to combine various research methods of both social and natural sciences. In translation practice, each method cannot function well in isolation. There must be a certain form or extent of combination (Wang H., 2005, p. 125). I once used the Delphic method to set up a mathematics-based CREDIT model to appraise interpretive effects (Hu, 1991); I also proposed the concept of Translation as Adaptation and Selection by borrowing this idea from the theory of biological evolution (Hu, 2001, 2004a, 2004b, 2004c). In addition, I initially explored Eco-translatology from the modern ecological perspective of holism (Hu, 2006, 2008, 2013). These developments, as mentioned earlier, indicate that translation studies are moving towards the combination of social science and natural science. Translation activity is an interdisciplinary integration of various social and natural factors. Thus, it is of great importance to break the limitations among different disciplines to enable translation studies to achieve their original goal. As G. Sarton puts it, the universal significance of natural science comes into being only by coordinating it with social science, the scientific spirit, and the humanistic spirit (Deng, 2000), which is similar to how the French sociologist Pierre Bourdieu put it: In science, progress emerges from breakthroughs in the boundaries of science (Bourdieu, 1998, p. 197).

8.3.5 Nourishing New Translation Schools of Thought

Since the period of ancient Rome, the translation research initiated by Cicero and Horatius has experienced long-term development with various theories and research approaches. With the development of Western translation research, translation studies have been considered an independent discipline instead of a branch of linguistics. Meanwhile, the essence and complexity of translation practice give translation studies features such as pluralism and interdisciplinarity. Exchanges among different theories lay the foundation for the establishment and development of the translation discipline and provide new vision and new approaches for the further development of translation theory research. The practice indicates that the development of translation studies and the translation discipline is based on a combination of inheritance and innovation that is determined by the characteristics of translation studies. This kind of development propels translation theory research into a new stage.

Meanwhile, research on Eco-translatology has expanded rapidly. Scholars in the field of translation have pointed out that the formation and development of the ecological paradigm in translation studies, as an enrichment, innovation, and leap forward, have given domestic and overseas researchers new insights into theoretical research. The explorative and pioneering spirit that it demonstrates played an important role in orienting and stimulating the further scientific development of translatology. Eco-translatology began to take form, with an impetus motivated by the translatology of the international school, the R&D conducted by the translatology of the Chinese school, and the proactiveness of Eco-translatology school. Over the years, after Eco-translatology underwent three stages of free development, adaptation to the market, and targeted promotion, a group of people gradually came together, like-minded participants aspiring to conduct research in Eco-translatology, from the academic leaders initiating study in this field to the backbone force of young and vigorous talents who have made remarkable achievements. Moreover, under the leadership of the International Association of Eco-translatology Research, they have carried out various kinds of cooperation and communication, including holding a series of international academic seminars for several years, publishing periodicals, treatises, and theses, and opening an official website. By doing so, an academic community composed of people with the same beliefs and values, shared modes of thought, and established research techniques is gradually taking shape. The gradual progress in the school of Eco-translatology is not only a fact but also a process. Today, this vibrant school is gradually reaching maturity.

8.3.6 Paying Back to Related Disciplines

Modern languages and translation studies are well established largely because of the development patterns of various disciplines, benefiting from the research approach and the development of research modes, concepts, approaches, and ideas. With the deepening of the research on translatology, translation studies, especially the research paradigm of Eco-translatology, can in turn give back to other disciplines with their findings. In addition, Eco-translatology also takes advantage of donor disciplines to obtain theoretical support, such as pragmatics, cross-cultural communication, dissemination, cultural studies, and linguistics, as the theoretical models and research techniques of these disciplines can contribute to sparking new thoughts.[6] Wang Ning comments:

> For researchers specializing in translation studies and theoretical construction, our mission is not only to contribute our due share to the construction of our own discipline but also to the humanities at large with theory-minded innovation. Now that we have borrowed a lot of

[6]The comprehensive demonstration and integrated study of the ecosystem may have an inevitable bearing on other disciplines because they are important aspects of research in Eco-translatology. Such comprehensiveness and integrity are not confined to the internal theoretical system of Eco-translatology but take a macroperspective to apply the achievements of related disciplines to Eco-translatology, which is beneficial to its theoretical development. In fact, many disciplines such as pragmatics, cognitive sciences, psychology, sociology, anthropology, and culturology are inextricably linked with translation studies, so it is natural for translation studies scholars to integrate the study of those disciplines into Eco-translatology. Such integration is consistent with the concepts of systematicization, comprehensiveness, and integrity that are stressed in ecology.

> ready-made theories from other disciplines, why not apply our own theory or fresh paradigm to such disciplines in translation studies? (Wang, 2009, p. 241)

Wang continues:

> In the past, we simply followed cutting-edge advances in research in other related disciplines in an effort to introduce new and applicable theories and research techniques. Today, when we carry out dialogue and communication on an equal footing against the backdrop of globalization, why not consider constructing our own theoretical frameworks so that we can "export" them to other disciplines? (Ibid., pp. 251–252)

Per XU Jun:

> We have been learning, borrowing from, and employing the theory, methods, and research achievements from other disciplines. For most of us, we are followers, translators, and demonstrators as well as practitioner of Western translation theories. However, when there rich and well-established discourse systems emerge in our translation circles through concerted efforts or more scientific and useful discourse systems in Chinese translation circles, does it mean that it is possible for us to provide studies on Chinese and Western translation theory with equal dialogue and in turn give back and feed other disciplines? For me, I do believe that through our concerted efforts, the answer to this question is positive. (XU, 2001)

In terms of research in Eco-translatology, we are delighted to find that according to the existing published papers, especially those written for international symposiums on Eco-translatology, scholars are now studying and discussing questions in their own specialties by learning and employing the terms, achievements and ideas of Eco-translatology. These scholars specialize mainly in linguistics, culturology, literature, cross-cultural communication, dissemination, diplomacy, terminology, editing science, medical science, ecology, and computer science. In the future, there will be much more such "giving back" and "feeding." Therefore, the question of how to make translatology "give back to" and "feed" related disciplines at length is facilitating our thinking, motivating our actions, and looking forward to our development of satisfactory answers.

8.4 Further Development of Eco-Translatology

8.4.1 From a "Theory Consumer" to a "Theory Producer"

Eco-translatology in China is giving the country a stronger voice in the development of translation theories. In a sense, China has moved from a "theory consumer" to a "theory producer." The move is of considerable significance because the comprehensive review and integration of translation studies from an ecological perspective will change the narrative angle and expression mode of translation studies. The internal logic and ultimate concern of "interdisciplinary" integration, in other words, the ecologic point of view, is expected to broaden the theoretical and academic view of future translation studies and enrich the theoretical discourse and expression modes of translation studies; furthermore, such integration may bring translatology to a new stage in which, instead of reviewing translation studies in a single discipline, researchers will move towards a more advanced and efficient interpretive system.

Translation is the fruit of human civilization. Translation in China has a history of over 2000 years, from the translation of Buddhist scriptures that started in the Eastern Han Dynasty and thrived in the Tang Dynasty to the translation of Western social science and technological works in the late Ming and early Qing Dynasties, the translation of socialist works after the May 4th movement, and the translation of various disciplines and categories since the implementation of the reform and opening-up policy. Translation is a bridge connecting Chinese civilization with that of the world, improving the culture and social progress of humankind. The rejuvenation of China needs and promotes the flourishing of original translation theories. In this sense, Eco-translatology, as a new and vibrant "growth and expansion point" of translation studies in the twenty-first century, may gradually develop into a major field of Chinese translation studies from the globalization perspective. Owing to the development of Eco-translatology, China has evolved from a "follower" to a "continuer" and, eventually, a "leader" of translation studies.

The concepts of "ecology," "environment," "life," and "diversity" in Eco-translatology provide topics and conditions for equal exchange between the Western and Eastern academic worlds. As long as we have a topic of shared understanding with Western scholars, we can sit at the same table for equal exchanges, discussions, and even debates.

Most Eastern scholars are "followers," "proof seekers," or "practitioners" of Western translation theories. "Translation studies in Asia is developed under the manipulation of Western rules" (Kong & Yang, 2000, p. 5). "In most cases, Asian scholars are the 'slow followers' of Western scholars" (Meng, 2007). Eco-translatology, which evolved from Eastern philosophies and ecological thoughts, is expected to change the abovementioned status. In other words, its international development will end the "ecological imbalance" of translation studies. Moreover, it will help to build a platform in which Western and Eastern scholars can exchange thoughts on the same "topic" in an equal manner. A positive proof is the growing number of Western scholars presenting academic articles of Eco-translatology and discussing related theory and practice in the series of international symposiums initiated by Chinese translation scholars since 2010 (see the following section).

8.4.2 Progress of Internationalization

In recent years, renowned scholars, such as Cay Dollerup, Head of the Translation Centre at the University of Copenhagen in Denmark; Edwin Gentzler, Director of the Translation Center at the University of Massachusetts-Amherst, in the USA; and Kirsten Malmkjaer, Head of the Middlesex University Translation Institute, have been invited to China to discuss Eco-translatological research and development with Chinese scholars. In April 2010, scholars on Eco-translatology established the

International Association of Eco-translatology Research, which hosted the first International Symposium on Eco-translatology in Macao, China, in November 2010. More than fifty experts and scholars from around the world attended the symposium.

Following the first symposium, the second International Symposium on Eco-translatology was held at Shanghai Maritime University on November 11–14, 2011, and the third was held at Southwest University in Chongqing on November 23–25, 2013. Over one hundred fifty experts and scholars from all over the world attended these two symposiums. Marion Boers, President of the International Federation of Translators (IFT) and consultant for the International Association of Eco-translatology Research, gave full academic affirmation to the establishment and development of Eco-translatology. "As an emerging translation research field, Eco-translatology that enjoys vigorous development and prosperity draws the attention of academia in the world," she said. She believes that Eco-translatology is similar to a slender sapling that will germinate, bloom, and ultimately grow into a towering tree. Huang Youyi, Vice President of IFT, Vice Chairman of the Translators Association of China, and Deputy Chief of China International Publishing Group (CIPG), affirmed the theoretical innovation of Eco-translatology and complimented the founder of Eco-translatology for his innovative courage and spirit of development. He thought the establishment of the International Association of Eco-translatology Research and the opening of the International Symposium on Eco-translatology were the best reward for Chinese scholars' years of theoretical studies (Huang, 2010). Prof. Cay Dollerup, the former chief editor of *Perspective: Studies in Translatology* and Director of the Translation Centre at the University of Copenhagen, indicated that Eco-translatology as a translation theory based on the traditional Chinese doctrine of "harmony between Man and nature" was the strongest expression of translation theorists in the world. Dr. Radegundis Strolze from the University of Technology in Darmstadt, Germany, on behalf of the European Society for Translation, extended warm congratulations on the convening of the International Symposium on Eco-translatology. She said:

> Eco-translatology originated from the East, which is an effective supplement and enrichment for traditional Western translation theories. China today plays an increasingly important role not only in global politics and economy, but also in culture. I believe that Eco-translatology can benefit Western translation theories and contribute to expand the horizons of Western scholars. I hope that it can build a bridge between traditional Western translation theories and new Eastern ones.

What is more encouraging is the fact that in the successive international symposiums on Eco-translatology, scholars from outside China have shown their interest in and expectations of engaging in Eco-translatological studies and have suggested many original ideas. Professor Dollerup of Denmark submitted a paper titled "Eco-translatology in Translation Theory Contexts."[7] He first analyzes the social, cultural,

[7] Source: "New Probe into Theory of Eco-Translatology." *Chinese Translators Journal*, 2011(1): 34–36.

and historical reasons for the constant emergence of translation theories in Europe, on the basis of which he believes that Eco-translatology, growing outside the European context, is the first translation theory with real "originality." Professor Joanna Radwańska-Williams, the American-Polish linguistic expert, has discussed in a paper[8] the relation between eco-humanism and translation studies and suggested a problem in the relation between Kuhn's "paradigm" and Eco-translatology as an Eco-translatological research paradigm. Professor Yvon Gousty of the Université d'Aix-Marseille in France is the senior consultant of scientific and technological cooperation with China. Considering his working experience around the world, he points out in his speech "On the Translational Eco-environments of Translation for Science and Technology" that science and technology programs are becoming increasingly internationalized due to the globalization, complexity, and cost of high-technology development. Misunderstanding and mistranslation may emerge in science and technology translation, which requires the highest accuracy, and can even lead to severe consequences. Therefore, he proposes establishing a translation team for science and technology to draw on the interdisciplinary characteristics of Eco-translation from the Eco-translatological perspective to focus on the concept of the closed connection between translators and translational eco-environments, which may help improve the translational situation of international technology projects. The translation theorist Douglas Robinson, now teaching in Hong Kong, delivered a speech titled "Eco-translatology, Mengzi, and the Politics of *li*(礼)" to inspect Western individualism and Chinese opportunism within the frame of eco-holism. Dr. Radegundis Stolze of Germany, in her speech titled "Eco-translatology and the Translator's Growth,"[9] demonstrates a Western scholar's understanding of Eastern Eco-translatology that may enlighten Western translation theory. She focuses on the internal pattern of the translator's growth—translation practice is a human activity that embodies the Eco-translation condition, and the translator, who is at the center of the activity, continually improves his/her knowledge and linguistic skill. We must regard translators as living individuals, a perspective that is often absent from translation theories. Professor Zoya Proshina of Moscow State University in Russia, the President of the International World Englishes League, discusses the relation between World Englishes and linguistic translation. Roberto Antonio Valdeón of Orville University, Spain, the chief editor of the international translation journal *Perspective: Translation Studies,* discusses the sources of stability and instability in news reports through the perspective of the recycling of the translational ecosystem. In addition, Professor Joanna Radwańska-Williams, chief editor of the international academic journal *Intercultural Communication Studies,* discusses the ecological relationship between language, cross-cultural communication, and translation in view of the global ecosphere. Dr. Douglas McNaught of the University of London elaborates on the relationship between the application of a new eco-critical approach and environmentalism in

[8] Source: "summary, review, and exploration." *Chinese Translators Journal*, 2012(2): 69–72.

[9] Source: "summary, review, and exploration." *Chinese Translators Journal*, 2012(2): 69–72.

translation studies by using Eco-translatological theory. Professor Flora Roider of the University of Innsbruck, Austria, describes the phenomenon of language death from the species extinction perspective. Dr. Carmen Garcés of the University of Alcalá discusses eco-literature, eco-culture, and Eco-translation in the context of Eco-translatology in a paper titled "Green Translation of Landscape and Life in Different Cultures—An Eco-translatological Perspective." She also explicates that the study is conducted "with a view of following the new approach of Eco-translatology" and "with an attempt to make my contribution to the development of this new area." Additionally, Professor Kristiina Taivalkoshi of the University of Helsinki, Finland, includes Eco-translatology in the thematic content of her "modern translation theory" lecture series and gives courses in it at the university. A Center of Eco-translatology Study at the University of Copenhagen in Denmark is also proposed. Professor Tefeng Lee of the University of London instructs PhD and master's students who major in Eco-translatology. Many foreign experts and scholars have paid attention to the results of Eco-translatology and applied it to their translation research.

We can see from all these instances that although Eco-translatology originated in East Asia, in China, international attention has increasingly been drawn to it, and some international translation scholars have begun to pursue and participate in the project. In addition, well-known international translation scholars such as Professor Edwin Gentzler (USA),[10] Professor Kirsten Malmkjear (UK), and Professor Mary Snell-Hornby (Austria) have undertaken to support and participate in Eco-translatological research and accepted roles as consultants for the International Association for Eco-translatology Research. To some extent, all of these scholars have been drawn to the vigor, vitality, and charm of Eco-translatology and predicted its international trend in the future.

8.4.3 Further Research

With the progressive development of Eco-translatology, a wider range, depth, and precision are needed for the study. The research topics of work in the near future should begin with the following three aspects: first, the expansion of the external "application"; second, the in-depth exploration of internal "accuracy"; and third, the advance of "internationalization."

[10] The "call for papers" and "conference announcements" of previous sessions of the International Symposium on Eco-Translatology hosted by the International Association for Eco-Translatology Research were all posted by Professor Gentzler on his personal global network. Each time, he provides the research topics with high evaluations: "very good," "congratulations to the success" (reference to personal correspondence with the author, 2009–2013).

1. The Extension of the Superficial "Application"

The extension of the superficial "application" means that in addition to literary translation, applied translation, translation teaching, interpretation and study of the translator, Eco-translatology can be expanded in the following aspects:

(a) Translation History/History of Translation Theory

The study of translation history from the translatological perspective is not typical at present, although it is an integral part of translatological study. I believe that the study of the history of translation/translation theory can be undertaken in the theoretical frame of Eco-translatology from different dimensions, such as taking the "translational eco-environment" as a clue, because the translational environment varies at different times. A history of translation/translation theory can be seen as the continuing historical development of the translational eco-environment. Alternatively, taking the "translational eco-environment" as a clue to explore the macroenvironment, mesoenvironment, and microenvironment as a sequence of translational ecology is not an incorrect choice for "rewriting" the history of translation/translation theory. I believe that the study of the history of translation/translation theory in the theoretical frame of Eco-translation with the translational eco-environment as a clue is the best choice.

Additionally, the track of "translator-centeredness" (translation thought) can be followed because translation activity is translator-oriented. The translation thought of a particular period is the collection of different translators' thought at that time. Considering "translator-centeredness" as a clue means taking "people" or "schools" or "translation thought" as a clue, which is also a correct choice in terms of the history of translation/translation theory, especially the latter.

In addition, "doing things with translations" can also be used as a clue because "doing things with translations," or "acting according to translation," can be both the "cause" and the "effect" of translation activity. Taking translation activity with "cause and effect" as a clue conforms to translation practice and the law of the development of things. Therefore, to study the history of translation/translation theory by following such a clue may produce new ideas or new publications.

In addition, "multidimensionality" is one of the study approaches to Eco-translatology. Since Eco-translatology includes many academic foci and research perspectives, it is obviously a massive project to study the history of translation/translation theory in different categories and perspectives. As an old saying goes, "The more difficult it is, the greater success it will achieve."

There is no doubt that different clues, perspectives, and motivations can provide a possibility of discoveries, explanations, and "writing styles." We look forward to the study of the history of translation/translation theory not only communicating with history but also connecting to the future. If monographs such as *New Understanding of Chinese Translation History/History of Translation Theory: Towards Perspectives of Eco-translatology, and New Understanding of the World's Translation History/History of Translation Theory: Towards Perspectives of Eco-translatology* appear in the near future, we should praise them.

(b) Studies on Translation Criticism

As an important part of the translation system, translation criticism is the bond between translation theory and translation practice. I believe that studies on translation criticism from the Eco-translatological perspective can work in these specific theoretical fields: perspective, viewpoint, and argument.

Judged on the basis of the current situation, translation criticism can focus on the "ten arguments": (1) the "balance and harmony" of translational ecology; (2) the "textual transplantation" of the translational text; (3) the "translator's responsibility" for the translation subject; (4) the "adaptive selection" of translation practice; (5) the "multidimensional transformations" of translation methods; (6) "preserving the strong and eliminating the weak" in the translation process; (7) the "multidimensional integration" of translation standards; (8) the "survival of the fittest" in the existence of translation works; (9) "doing things with translations" in the translator's pursuit; and (10) the "Sequence Chain" of translation development. Eco-translatology thus covers a large scale of the discourse system of translation theory. It will absolutely make a great difference to work on translation criticism from different perspectives/dimensions/profiles.

(c) Research on Translation Methodology[11]

The establishment and development of any academic paradigm not only reflect the in-depth exploration of specific subject areas but also show the changes in the cognitive perspective or the extension of the cognitive vision. Eco-translatology is a new paradigm, and its cognitive perspective and cognitive vision show the same trend.

The methodology in Eco-translatology can be generally divided into three categories: first, traditional and conventional methodological research, such as descriptive translation study, induction, deduction, comparison, and corpus; second, the study of Eco-translatology itself, such as cross-disciplinary analysis, analogy, transplantation, and holistic integration; and third, other studies derived from eco-reason. Currently, studies related to methodology have been performed, but more improvements are needed.

(d) Studies of Translation Ethics

Eco-translatology is the comprehensive study of translational ecosystems and stresses the relation of internal factors, including the coordinated study of translation activities and language, culture, human society, and even the natural world; the study of the mutual relationship between translators and the translational eco-

[11] Professor Wen Xu of Southwest University presents his own opinions about the research paradigm of Eco-translatology in view of eco-linguistics and cognitive linguistics (Xu, 2010). In 2011, he proposes in his paper titled "Methodology of Eco-translatology" that methods of Eco-translatology should be expounded from the perspectives of philosophical methodology, scientific methodology, and concrete methodology. Hua Xianfa and Hua Manyuan of the College of Foreign Language at Central China Normal University also discuss Chinese translation methodology from the perspective of Eco-translatology.

environment; the study of the mutual relationship between those such as authors, readers, sponsors, and publishers and the translational eco-environment; and the study of mutual relations among the translation essence, translation process, translation principles, translation methods, and translation standards. The above Eco-translatological study at the macro-, meso-, and microlevels all involve the relation of many internal factors in the translation system. These will inevitably be restricted by certain social norms and abide by predictions of relevant value concept systems. Therefore, considering Eco-translatology from an ethical perspective is the effective expansion of Eco-translatology and will accelerate its sound progress.

After having made analogies between eco-ethics and translation practice, based on the orientation of Eco-translatology, I deductively develop four basic principles of the ethics and morals of Eco-translation: (1) the principle of "balance and harmony"; (2) the principle of "multiple eco-integration"; (3) the principle of "symbiosis and diversity"; and (4) principle of "translator responsibility" (Hu, 2011).

Eco-translation ethics is a new research field and an extensive systematic study. It is certain that further and more in-depth study of this facet will continue, and any relevant study should be welcomed and encouraged.

(e) Schools of Translation

The "school" refers to research groups that have common theoretical guidelines, consistent modes of thought, similar research objectives, and approximate research styles. The translatological development in the West indicates that the emergence of innovative translation theory is closely associated with the preparation, generation, and development of different schools. A school signifies that the research has reached a certain degree, scale, and level of maturity; offers a platform for academic exchanges and academic debates; and facilitates the further development of academic study. "Schools play a critical leading role in inheriting and developing cultural quintessence, creating academic masters, condensing academic tradition, filling the academic universality with diversity, and promoting academic progress" (Zhang & Xin, 2005, p. 69).

At present, a relevant study on the "School of Eco-translatology" shows that a school needs (1) leading figures; (2) outstanding kernel teams; (3) like-minded followers and supporters who form gradually in the early stage; (4) representative monographs and theoretical systems that have a significant influence on the academy and the school's own styles; (5) certain academic organizations; (6) corresponding academic fields and communication platforms; (7) a safe and stable society[12] and appropriate academic environment; and (8) recognition by academia and society, etc.

I hold that in the context of Eco-translatological ethics, the research topics of related translation schools can be based on the idea of the "translational eco-environment" with the help of philosophical thought; on "translator-centeredness,"

[12] For instance, if a society suffers from a long-term war, it will be difficult to develop the society, science, and education, much less academic prosperity and the generation of schools, among people who are living in misery and even struggling for survival.

with translators as the representatives; on the "translator-community" of a nation/society; on an "eco-paradigm" of a paradigm shift; or on the "holism and relevance" of multiple dimensions.

2. Internal Intensive and Meticulous Research

Internal intensive and meticulous research refers mainly to the possession of a series of directional features such as intercommunity, interactivity, and progressiveness based on the Sequence Chain,[13] which seeks a more precise and well-founded translational ecosystem within itself. The general idea is to study and describe from perspectives that are closely related to translation activities, such as language, culture, and human communication, and finally to return to translatology studies under the guidance and reference of eco-reason. Specific cross-disciplinary research on the translational ecosystem by itself may include the following aspects:

(a) Research from the perspective of Linguistics.

In the translationalecosystem by itself, because "translation is the conversion of language" (based on the Sequence Chain), study from the perspective of linguistics should include the functional/cognitive/pragmatic analysis of the ecological approach[14] in discourse (source language and target language); the relationship between language and translational ecology; the diversity of translation and language; the translation of ecological terminologies; and other relevant studies related to the translational ecosystem from the perspective of linguistics.

(b) Studies from the perspective of Cultural Studies

In the translational ecosystem by itself, because "translation is the conversion of language, and language belongs to part of culture" (based on the Sequence Chain), studies from the perspective of cultural studies should include studies on cross-cultural differences/conflicts/similarities/restrictions of the ecological approach; on the relationship between the life of the translated text and cross-cultural exchanges and dissemination; on the cultural context of the translational ecosystem; on cul-

---

[13] The Sequence Chain refers to the basic principles that are applicable to the changes in human recognition and the basic path (namely, Translation ↔ Language ↔ Culture ↔ Society/Humanity ↔ Natural World) that embodies vision expanding and the rational progress of human cognition (Hu, 2003, p. 298; Hu, 2004a, 2004b; 2004c, p. 61).

[14] It can be predicted and assumed that translation studies is bound to undergo a final shift from the natural world to ecology, indirectly and circularly. Can we conceive that the shift of translation or the association of ecology with translatology is the leading discipline of interdisciplinary and integrative studies on translation? On this basis, we try to realize a harmonious eco-environment in integrative studies of the translation system and then construct translatology. With studies from the perspective of translatology as the core, we aim to expand from the core outwards to interdisciplinary and integrative studies, from inside to outside, from part to whole, from single discipline to interdiscipline and multidiscipline, and to integrate and unify the systematicness as well as control the inner coordination and stability of the structure of the translational ecosystem to realize a two-way inner relationship from translation to the natural world, thus forming an ecological translation outlook (Zhang & Jin, 2011, p. 265).

tural diversity and the translational eco-environment; and other relevant studies related to the translational ecosystem from the perspective of cultural studies.

(c) Studies from the perspective of Anthropology

In the translational ecosystem by itself, because "translation is the conversion of language, and language belongs to part of culture; culture is the essence of human activities, and human beings belong to part of the natural world" (based on the Sequence Chain), studies from the perspective of anthropology should include research on changes in translation and human cognition; on human memory and translation (interpretation in particular); on the translator's needs/emotions/desires/capability; on the translator's survival environment and capability improvement; on translation and human communication; on translation missions and human civilization; on the relationship between Eco-translation and eco-civilization; on translation and globalization; and other relevant studies related to the translational ecosystem from the perspective of anthropology.

(d) Studies from the perspective of Ecology

In the translation ecosystem by itself, because "translation is the conversion of language, and language belongs to part of culture; culture is the essence of human activities, and human beings belong to part of the natural ecology/world" (based on the Sequence Chain) (considering that translatology is a science that studies the interrelationship between living bodies and their surrounding environment), studies from the perspective of ecology should include research on the interrelationship between translators and the translational eco-environment; on authors/readers/patrons/publishers/critics, etc., namely, the interrelationship between the translator-community and the translational eco-environment; on the inner structures of the translational ecosystem and their interrelationship; on the interrelationship among the translation essence/process/principles/methods/standards; on the status and functions of translation in the overall natural ecosystem; on the interrelationship and dependence of translation with and on other disciplines; and other relevant studies related to the translational ecosystem from the perspective of ecology.

(e) Studies from the perspective of Translatology

In the translational ecosystem by itself, because "translation is the conversion of language, and language belongs to part of culture; culture is the essence of human activities, and human beings belong to part of the natural ecology/world," or the other way around, because it circles from the natural ecological world back to translation (based on the reversed Sequence Chain), studies from the perspective of translatology should include research on the integrity of the translational ecosystem, on coordination among the translation/language/culture/humankind/natural world, and on the optimization of the translational ecosystem; comparative studies of the translatology approach and other approaches; and other relevant studies related to the translational ecosystem from the perspective of translatology.

In the process of the abovementioned studies, studies from the perspective of anthropology, ecology, and translatology have seldom been touched upon or have even been ignored in the past; therefore, specific exploration and discussion from these perspectives are the key to further study of Eco-translatology. Obviously, through the abovementioned future development and studies that take the Sequence Chain as the clue, disciplinary perspectives as crucial points and comprehensive integration as the goal, the connotations and meaning are likely to be presented more explicitly.

3. Driving of External Globalization

The so-called driving of external globalization refers mainly to laying a solid foundation for the global development of Eco-translatology. Chinese scholars have long had no discourse rights; one of the reasons is that they lack a discourse system that is independent, innovative, and dynamic. In other words, what they lack in the international circle of translation studies is the high ground for translation theories. China boasts a translation team of a million people and therefore is undoubtedly a translation giant. Chinese scholars are interested in pragmatic and specific translation projects and tasks. As a translation giant, China tops the world's translation efforts in terms of quantity (Wang, 2011, p. 14). However, in terms of the current international pattern, we are competing not only in terms of how many people are engaged in translation work or how many pragmatic translation volumes have been produced at the microlevel but also in terms of whether we have contributed widely recognized and systematical translation theories or whether we hold the high ground for theories. Although it is a much smaller territory, Israel has caught the attention of the world because of its invention of the polysystem theory, which has led scholars of other countries to follow suit. Small as the country is, its academic status in international translation circles is no lower than that of any large country.

At present, Eco-translatology, which originated in China, has made preliminary achievements and trended towards increasing dissemination after years of exploration and study. Eco-translatology not only has a comparatively complete theoretical discourse system of its own but also has gathered several talented research cohorts. Each year, domestically and abroad, dozens of degree dissertations and journal papers are published with Eco-translatology as the theme. Published by Foreign Language Teaching and Research Press, the *Journal of Eco-translatology* has begun publication, and an official website of Eco-translatology is now open to the public. Moreover, the International Association for Eco-translatology Research has been approved and established under the leadership of Chinese scholars, and International Symposiums on Eco-translatology led by the Chinese translation academia have been held in consecutive years, attracting an increasing number of scholars from China and abroad to attend and participate in relevant research. All these events show that Eco-translatology, invented by Chinese scholars, possesses the essential preconditions for sustainable development. If further studies are to be carried out on this basis, they will surely promote the further globalization of China's translation theories and be conducive to China making a substantial leap from a translation giant to a translation power.

During the process of globalization, the Chinese-created Eco-translatology has encountered numerous difficulties and challenges, but it has also achieved a hard-won opportunity. Based on the existing studies and the development of Eco-translatology, academic leaders and experts in China and abroad have also proposed pertinent evaluations and high expectations.

Eco-translatology is a research paradigm of Eco-translation that probes translational ecology, textual ecology, and translator-community ecology as well as their interaction and interrelationship. With eco-holism as the conception, Eastern eco-wisdom as the ultimate goal and adaptation/selection theory as the cornerstone, Eco-translatology is an integrative research with a comprehensive view of translation activities from an ecological perspective. In deepening and extending studies of Eco-translatology and as further research and development in Eco-translatology take place, if strategic research on the paradigms and development of translation studies is carried out in due time, or in the case of research and development of translation studies, if strategic research on the globalization of local translation theories is carried out, then the following specific topics can further explored:

1. development process and status quo of international translation theories;
2. local translation theories in China;
3. eco-paradigm of translatology;
4. comparative study on the Eco-translation paradigm and other research paradigms;
5. sustainable development of Eco-translatology;
6. dissemination, transformation, and effect evaluation of achievements of Eco-translatology;
7. international cooperation and communication of Eco-translatology;
8. foundation and development of Eco-translatology as a school of thought in East Asia;
9. academic status and influence of Eco-translatology and its development in the international translation community; and
10. globalization of local translation theories.

Predictably, carrying out and completing the abovementioned specific tasks is sure to be conducive to the further development of the globalization of Eco-translatology as a discourse system. Meanwhile, the prospective influence and academic contributions of the further development of Eco-translatology will improve, both in China and in the rest of the world.

8.5 Summary

Research and development in Eco-translatology have provided enlightenment in all aspects of contemporary translation studies. Specifically, Eco-translatology has promoted change and evolution in studies on translation theories to a certain degree: (1) From the perspective of philosophical reasoning, the philosophical basis of translation theories has been transferred from partial to universal application and from a unitary to an integral approach. (2) From the perspective of the research horizon, the horizon of translation theories has extended from a single discipline to interdisciplinary integration. (3) From the perspective of key points, translation studies have undergone an overall transformation from a long-term text focus to a translator focus and then to a translational ecological environment focus and from source-text centeredness and target-text centeredness to the balanced development of translator-centeredness and symbiosis. (4) From the perspective of the research paradigm, translation theories have shifted from rotation in the linguistics circle to revolution around interdisciplinary others, and after experiencing the movement from intuitionism to structuralism, translation studies are now at the developmental stage of pluralism and are moving towards the final holistic mode. (5) From the perspective of cross-disciplinary studies, translation theories are crossing the border between humanities and the social sciences and the natural sciences, moving towards the communication of the humanities and social sciences with the natural sciences as well as the integration of science and the arts. In conclusion, the research and development in Eco-translatology have to a certain degree promoted research on contemporary translation theories and gradually changed the research direction to reasonable universality, scientific integration, panoramic horizons, and diversified paradigms.

Eco-translatology has an invigorating vitality and promising developmental prospects; however, there is no reason for blind optimism. On the one hand, Eco-translatology focuses on a translation paradigm from an ecological perspective; therefore, it is a comprehensive view and description of translation, and it seeks to explore, discuss, and study translation activities; enrich translation theories; and extend translation studies rather than replace general translation studies. On the other hand, being still in a preliminary stage as a young branch of translation studies, Eco-translatology still needs constant enrichment and exploration as well as further developmental rules, which require undaunted efforts by all explorers. Meanwhile, like any other academic field, Eco-translatology has limitations and incompleteness. Despite the fact that Eco-translatology has compensated for certain deficiencies of existing translation research; offered a comprehensive and holistic view of translation studies, translation theories, and translation itself from new perspectives; and formed the three-level developmental pattern, the establishment does not mean perfection. Because all truths are not without limitations and relativity, all scientific fields are always in progress and all concepts need to be updated, the course on which people pursue truth has no bounds. Therefore, based on the law of development, Eco-translatology will be gradually exposed to its own weaknesses during the course of development. Only by facing this reality, constantly overcoming its limitations and consciously and rationally carrying out studies with ease and calm can we create a healthy environment for the unceasing development of Eco-translatology and carry it forward far and wide.

References

Blake, R. M. (1966). *Theories of scientific method.* Seattle: Gordon & Breach. Washington Paperbacks, UWP2.

Boone, T. (2011). What are they saying about exercise physiology? *Journal of Professional Exercise Physiology, 9*, 1–5.

Bourdieu, P. (1998). *Practice and reflection* (Li Meng & Li Kang, Trans.). Beijing: Central Compilation & Translation Press.

Chen, F. K. (2000). *A history of chinese translation theories. Shanghai*: Shanghai Foreign Languages Educatin Press.

Chen, J. Y. (2003). *Philosophy*. Beijing: The Peking University Publishing Press.

Cheng, Z. Y. (2005). *Standing up in C-E translation Chinese philosophy and new positioning of Chinese Culture*. Beijing: China Renmin University Press.

Deng, J. Q. A study on the development of social science. *Guangming Daily*, 2000/06/20.

Fan, S. Y. (2003). About theorization and thesis writing in translation studies: Some metatheoretical considerations. *Chinese Translators Journal, 2*, 5–9.

Gu, Z. K. (2001). On contemporary theoretical approaches to the establishment of translatology in China-Preface to *literary translatology. Chinese Translators Journal, 1*, 8–12.

Hu, G. S. (1991). A discipline needed to work hard at it. *Outbound Communication and Exchange, 8*, 28–32.

Hu, G. S.. *Study on transation as adaptation and selection.* Paper Presented on the Third Asian Translators Symposium of International Federation of Translators. 2001-12-6/8. Hong Kong.

Hu, G. S. (2002). Prosperity, confusion and exploration in translation theory. *Translation Quarterly, 25*, 80–104.

Hu, G. S. (2003). Translation as adaptation and selection. *Perspectives: Studies in Translatology, 11*(4), 283–291.

Hu, G. S. (2004a). An empirical study on evaluating the "Approach to Translation as Adaptation and Selection" through questionnaire surveys. *Foreign Languages and Their Teaching, 4*, 40–44.

Hu, G. S. (2004b). From "Translator's Subjectivity" to "Translator-centeredness.". *Shanghai Journal of Translators, 3*, 9–15.

Hu, G. S. (2004c). *Translation as adaptation and selection.* Wuhan: Hubei Education Press.

Hu, G. S. (2006). Understanding translation as adaptation and selection through case studies. *Foreign Language Education*, (4), 50–54.

Hu, G. S. (2008). Eco-translatology: A primer. *Chinese Translators Journal, 6*, 11–15.

Hu, G. S. (2009). Translation studies of eco-translatology: Inter-disciplinary integration. *Shanghai Journal of Translators, 2*, 3–8.

Hu, G. S. (2011). Eco-features of eco-translatology and the implications to translation studies. *Chinese Foreign Language, 6*, 96–99/109.

Hu, G. S. (2013). *Eco-translatology: Construction and interpretation.* Beijing: The Commercial Press.

Huang, Y. Y. (2010). Adventure and perseverance The opening speech on the first international eco-translatology symposium. *Journal of Eco-Translatology, 1*, 13–15.

Huang, Z. L. (2009). Research on Yan Fu's translation. *Reading Journal, 2*, 116–120.

Jiang, K. Y. (1993). *Latest studies on aesthetics*. Beijing: People's Literature Press.

Kong, H. Y., & Yang, C. S. (2000). *Translation in Asia: Past and present.* Beijing: Peking University Press.

Krohne, T. D. (2001). *General ecology*. Brooks: Thomson Learning.

Lance, H., & Jacky, M. (1991). *Redefining translation: The variational approach.* London: Routledge.

LECSD. (1994). *Longman English-Chinese science & technology dictionary* (p. 1131). Beijing: Tsinghua University Press.

Lefevere, A. (1993). Discourses on translation: Recent, less recent and to come. *Target, 5*(2), 229–241.
Li, H., et al. (1978). (tr.). *Cosmological system theory*. Shanghai: Shanghai Translations Press.
Li, Y. X. (2009). On tension of context in translation. In G. Hu (Ed.), *Translation and intercultural: Integration and innovation* (pp. 149–162). Shanghai: Shanghai Foreign Language Education Press.
Liao, Q. Y. (2001a). *Contemporary translation studies in UK*. Wuhan: Hubei Education Press.
Liao, Q. Y. (2001b). Research paradigms and translation studies in China. *Chinese Translators Journal, 5*, 14–18.
Liu, L. M. (2006). The application and influence of "Adaptation Strategy" in translation during 16th and 17thcentury. *Foreign Languages and Translation Studies Journal: Chinese Translation of Globalization (Taibei), 1*, 229.
Liu, M. Q. (2005). *Translation thinking: In China and in the West*. Beijing: China Translation & Publishing Corporation.
Ma, J. Z. (1894). A proposal for establishing a national translation academy. In the Editorial Board of China Association of Translators (Ed.), *Translation bulletin*.
Ma, J. Z. (1984). On establishing an institute of translation. In Chinese Translators Journal (Ed.), *Collections of translation research*. Beijing: Foreign Language Teaching and Research Press.
Ma, Z. Y. (2001). *A brief history of translation in China*. Beijing: China National Translation and Publishing Corporation.
Meng, F. J. (2002). The influence of the evolving Chinese cultural pattern on translation studies in China. *Chinese Translators Journal, 2*, 13–17.
Meng, F. J. (2007, April 9–12). Cultural reflections upon the present translation studies in Asia. Paper presented at the FIT 5th Asian Translators Forum, Bogor, Indonesia.
Meng, F. J. (2009). The new turn of translation studies after post-modernism. In G. Hu (Ed.), *Translation and interculture: Integration and innovation* (pp. 48–54). Shanghai: Shanghai Foreign Language Education Press.
O'Hagan, M. (2005). Multidimensional translation: A game plan for audiovisual translation in the age of GILT. Proceedings of the Marie Curie Euroconferences MuTra 'Challenges of Multidimensional Translation', Saarbrücken
Piao, J. B. (2006). Linguistic turn and its significance in philosophy history. *Jilin University Journal (Social Science Edition), 1*, 83–86.
Pöchhacker, F., & Kaindl, K. (Eds.). (1994). *Translation studies: An interdiscipline*. Amsterdam: John Benjamins.
Snell-Hornby, M. (1988). *Translation studies: An integrated approach*. Amsterdam: John Benjamins Publishing Company.
Snell-Hornby, M. (2006). *The turns of translation studies: New paradigms or shifting viewpoints*. Amsterdam: John Benjamins Publishing Company.
Song, Z. P. (2009). A nonlinear approach to translation selection. In G. S. Hu (Ed.), *Translation and cross-cultural communication: Integration and innovation* (pp. 111–120). Shanghai: Shanghai Foreign Language Education Press.
Venuti, L. (2000). *The translation studies reader*. London: Routledge.
Wang, H. Y. (2005). Research and criticism-Methodology in literary translation criticism. In L. Xuanmin (Ed.), *Culture criticism and translation studies* (pp. 117–128). Beijing: Foreign Language Press.
Wang, K. F. (2009). Integration of translation. In G. S. Hu (Ed.), *Translation and cross-cultural communication: Integration and innovation*. Shanghai: Shanghai Foreign Language Education Press.
Wang, N. (2005). Ecological ideas of confucianism and western eco-holism. In D. Y. Yue et al. (Eds.), *Cross-cultural dialogue* (Vol. 15, pp. 88–96). Shanghai: Shanghai Cultural Press.
Wang, N. (2011). Eco-literature and eco-translatology: Deconstruction and reconstruction. *Chinese Translators Journal, 2*, 10–16.

Wang, R. P. (2001). Literary translation appeal to philosophical thinking- Preface. In X. L. Cai (Ed.), *Arts philosophy of literary translation*. Kaifeng: Henan University Press.
Wang, R. S., & Zhou, H. (2004). *Human and ecology*. Kunming: Yunnan Peoples Press.
Wilss, W. (1999). Interdisciplinarity in translation studies. *Target, 11*(1), 131–144.
Xie, T. Z. (2007). *Introduction of medio-translatology*. Beijing: Peking University Press.
Xu, J. (2001, November 11–13). Closing speech: Adhere to the construction and perseverance. Delivered at the 2nd International Symposium on Eco-Translatology, Shanghai.
Xu, J. Insistence and construction - A speech at the closing ceremony of the "Second International Eco-Translatology Symposium", 2011-11-11/13. Shanghai.
Xu, J. (2007). *Light life and weighty translation*. Beijing: Literature and Art Press.
Xu, W. (2010). *Eco-translatology paradigm from perspectives of cognitive linguistics eco-linguistics*. Paper presented at the first international eco-translatology symposium, 2010-11-9/10. Macao.
Yang, Z. J. (2009). Understanding several translation problems. In G. Hu (Ed.), *Translation and cross-cultural communication: Integration and innovation* (pp. 3–13). Shanghai: Shanghai Foreign Language Education Press.
Zeng, F. (2007). The influence of Western literature in the new period on Chinese literature and art. *Literary Review*, (3), 140–146.
Zeng, L. (2008). Categorization extension of applied translation theory viewed from theoretical construction of tourism translation. *Shanghai Journal of Translators, 3*, 1–5.
Zhang, B., & Xin, H. J. (2005). Establish our own translation schools in China. *Chinese Foreign Language, 5*, 69–73/79.
Zhang, L. F., & Jin, W. (2011). Eco-translatology and its three ecological dimensions: A discussion with Prof. Hu Gengshen. *Journal of University of Shanghai for Science and Technology (Social Science), 4*, 261–266.

Appendices

I. The Author's Publications & Keynote Presentations on Eco-translatology

- From the Darwinian Principle of Adaptation and Selection to Translation Studies [P]. The first presenter of the *Translation Seminar*[1] organized by the Centre for Translation of the Hong Kang Baptist University, October 22, 2001.
- Initial Exploration on a Translation Theory of Adaptation and Selection [P]. Paper presented at the FIT: *3rd Asian Translator's Forum*, Hong Kong, 6–8 December 2001.
- Translated Texts: Products of the Translator's Adaptation and Selection [P]. *International Conference on Discourse & Translation,* Zhongshan University, Guangzhou, 2002.
- Adaptation and Selection: a Translator-centered Approach to Translation [P]. Paper presented at the *1st Tsinghua-Lingnan International Symposium on Translation Studies,* Beijing 14–16 June, 2002.
- Translation Theory: from the Development and Dissatisfactions to an Exploration [J]. *Translation Quarterly* (Hong Kong). 2002 (4): 1–26.
- Translation as Adaptation and Selection [J]. *Perspectives: Studies in Translatology*, 2003 (4): 283–291.
- A New Look at Translation: Adaptation and Selection [J]. *River Cam Breeze* (Cambridge, UK), 2004 (1): 21–23.
- Translator-centredness [J]. *Perspectives: Studies in Translatology*, 2004 (2): 106–117.
- From Translator's Subjectivity to Translator's Centredness [J]. *Chinese Translators Journal.* 2004 (3): 7–12.

[1]This monthly organized *Translation Seminar* has lasted over 18 years since 2001; many well-known scholars from different parts of the world have become speakers at over 200 regular translation seminars. For details, please refer to the website: http://www.trans.hkbu.edu.hk.

(Hugs) G. S. Hu, *Eco-Translatology*, New Frontiers in Translation Studies,
https://doi.org/10.1007/978-981-15-2260-4

- An Empirical Study on Approach to Translation as Adaptation and Selection [J]. *Foreign Languages and their Teaching*. 2004 (3): 40–44.
- The Theoretical Prop to Translation as Adaptation and Selection [J]. *Shanghai Journal of Translators*. 2004 (4): 1–8.
- Understanding Eco-Translatology [P]. Paper presented at the *International Conference on Translating Global Cultures–Interdisciplinary (Re)Construction*, Tsinghua University, Beijing, August 11–14, 2006.
- Interpreter's Adaptation and Selection in Consecutive Interpreting [J]. *Perspectives: Studies in Translatology*. 2006 (1): 3–12.
- Interpreter's Adaptation and Selection in the 'Triple Contextual Categories' (TCC) of International Communication [J]. *Journal of Macao Polytechnic Institute*. 2008 (3): 78–89
- Adaptation and Selection: A New Look on Translation Process [J]. *Journal of Sichuan Foreign Language University*, 2008(4): 91–95.
- Eco-Translatology: A Primer [J]. *Chinese Translators Journal*, 2008 (6): 11–15.
- Viewing the Translation Theory From Terms- Adaptation and Selection [J]. *Shanghai Journal of Translators*, 2008 (2): 1–6.
- Eco-Translatology: an Integrative Approach to Translation Studies [J]. *Shanghai Journal of Translators*, 2009 (2): 3–8.
- On Fu Lei's Translation Philosophies An Eco-translatological Perspective [J]. *Journal of Foreign Languages*, 2009(2): 47–53.
- An "Ecological Turn" Signified by the "Sequence Chain" [A]. In HU, G. S. (ed.). *Translation and Intercultural Communication: Integration and Innovation* [C]. Shanghai: Shanghai Foreign Language Education Press, 2009: 26–39.
- Translational Ecological System from the Perspective of Eco-Translatology [A]. In REN, D. S. (ed.). *Construction of Conceptual Framework of Translation Studies* [C]. Shanghai: Shanghai Foreign Language Education Press, 2010: 296–315.
- The Change and Interpretation of Translation Theory Study [A]. In HU, G. S. (ed.). *Translation and Intercultural Communication: Change and Interpretation* [C]. Shanghai: Shanghai Foreign Language Education Press, 2010: 10–30.
- Relevance, Affinity and Isomorphism between Translational Ecology and Natural Ecology [J]. *Shanghai Journal of Translators*, 2010 (4): 1–5.
- Eco-Translatology: Backgrounds and Bases for its Development [J]. *Foreign Languages Research*, 2010(4): 62–67.
- The Ecological Track, A Worthwhile Trek—Eco-Translatology Explained [P]. Paper presented at *the 6th Congress of the European Society of Translation (EST)*, University of Leuven, Belgium, September 23–25, 2010.
- Macro Vision, Meco Theory and Micro Operation: The *3-level* Research of [P]. Keynote presentation at *the 1st International Symposium on Eco-Translatology*, Macao Polytechnic Institute, November 8–10, 2010.
- A Macro Study of Eco-Translatology [J]. *Journal of Eco-Translatology* [J]. 2011 (1): 24–33.

- The Chinese Culture "Genes" in Eco-Translatology [J]. *Journal of Eco-Translatology*, 2011(2): 3–11.
- Eco-Translatology: Research Foci and Theoretical Tenets [J]. *Chinese Translators Journal.* 2011 (2): 1–5.
- 10 Years Research and Development of Eco-Translatology [J]. *Shanghai Journal of Translators*,2011(4): 1–6.
- Response to Translator-centeredness [J]. *Shanghai Journal of Translators*, 2011 (4): 7–9.
- Eco-features of Eco-Translatology and the Implications to Translation Studies [J]. *Foreign Languages in China*, 2011 (6): 96–101.
- Eco-Translatology: A Distinctive and Nascent Approach to Translation [P]. Keynote presentation at *the 2nd International Symposium on Eco-Translatology*, Shanghai Maritime University, Shanghai, November 11–14, 2011.
- Birth of Eco-Translatology and Its Global Spreading [N]. *Chinese Social Sciences Weekly* (Academic Frontier/Dynamics), 2011-11-17.
- Eco-Translatology: A Distinctive and Nascent Approach to Translation—A Comparative Study on Approaches to Translation Studies and a Brief Response to some Related Questions/Doubts [J]. *Journal of Eco-Translatology*, 2012 (1): 3–12.
- The Name and Nature of Eco-Translatology [J]. *Journal of Eco-Translatology*, 2012(2): 1–7.
- From Translator-centredness to Translator Responsibility—Thoughts on Ethical Principles of Eco-translation [P]. Keynote presentation at *the 3rd International Symposium on Eco-Translatology*, South-west University, Chongqing, November 23–25, 2012.
- Eco-Translatology: Internationalization and Its Trends [J]. *Shanghai Journal of Translators*, 2013 (4): 1–4/20.
- Eco-Translatology: The On-going Process of Problem-solving [J]. *Journal of Eco-Translatology*, 2013 (1–2): 18–31.
- Eco-Translatology: Construction &Interpretation *[M]*. Beijing: The Commercial Press, 2013. XXXVI + 512.
- Integration and Surpass – Paradigmatic Features of Eco-Translatology [P]. Keynote presentation at *the 4st International Symposium on Eco-Translatology*, Wuhan University, Wuhan, September 23–25, 2013.
- From 'Translator Centredness' to 'Translator Responsibility' [J]. *Chinese Translators Journal*, 2014 (1): 29–35.
- My pursuit in Academic Studies [J]. *Contemporary Foreign Languages Studies* [J]. 2014 (8): 1–7.
- The System Model of Eco-Translatology and the Representation of its Central Tenets [P]. Keynote presentation at *the 5st International Symposium on Eco-Translatology*, Chang Jung Christian University, Tainan City, Taiwan, June 26–28, 2015.
- *Eco-Translatology: Voices from Western Scholars* [C]. Taibei: Taiwan Shu Lin Publishing House, 2015. IV + 169.

- A New Orientation, A New Contribution—Eco-translation Studies in the New Era of Eco-civilization [P]. Keynote presentation at the *International Summit Forum on Translation/Interpreting Studies: New Orientations*, Guangzhou Foreign Studies University, Guangzhou City, Oct. 30 to Nov. 1, 2015.
- Eco-Translatology: A New Paradigm of Eco-translation—A Comparative Study on Approaches to Translation Studies [J]. *Translation and Interpreting Review*, 2016 vol. 26, 115–132.
- Integrating East and West, and Survival of the Fittest: the Emergence and Internationalization of Eco-Translatology [J]. *Foreign Languages in China*, 2016(5):92–98.
- Different Dimensions of Applied Research Inspired by Eco-Translatology [J]. *Shanghai Journal of Translators*, 2017 (5): 1–6.
- Globalizing Chinese Cultural Classics in the Light of Eco-Translatology [P]. Keynote presentation at the *12th International Forum on Translation Theory & Practice*, Changjiang University, Jingzhou City, October 15–16, 2017.
- Practising Neo-Ecologism: What, How, and Why? [P]. Keynote presentation at the *13th International Forum Translation Theory and Practice*, Wuhan University of Science and Technology, Wuhan City, October 19–20, 2018.
- Developmental Studies of Eco-Translatology in the New Era [P]. Keynote presentation at *The 6th International Symposium on Eco-Translatology & The 2nd Chinese Ph.D. Forum on Eco-translation Studies*, Zhengzhou University, Zhengzhou City, October 27–28, 2018.
- On Research Topics of Eco-Translatology and Eco-civilization Construction [J]. *Journal of PLA University of Foreign Languages,* 2019 (2): 125–131.
- On the Theoretical Construction of Eco-Paradigm in Modern Translation Studies [J]. *Chinese Translators Journal*, 2019 (4): 33–42.

II. The Author's Opening Addresses at the International Symposiums on Eco-translatology

1. Be Part of It; Be Proud of It!—Hugs Opening Speech at the 1st International Symposium on Eco-translatology (Macau, November 9, 2010)

MIP President Lei Heong Iok, Chairman of the Supervisory Panel of IAETR,
Ms. Marion Boers, President of FIT,
Professor Huang Youyi, First Vice-President of FIT, Vice-President of TAC,
Professor Cay Dollerup, Chairman of the Advisory Panel of IAETR,
Distinguished guests and scholars,
Ladies and gentlemen,
Good morning!

The year 2010 looms as a landmark in the history of Eco-translatology research. April of 2010 witnessed the founding of the International Association of Eco-translatology Research (IAETR); now, in November 2010, we are met here for the First International Symposium on Eco-translatology. By the end of this year or early next year, we will create the *Journal of Eco-translatology*.

Today, we are here on the spot of the First International Symposium on Eco-translatology with a view not only to advocating the communication and exchange among the scholars from various countries or regions all over the world as well as from China but to promoting the research and development of Eco-translatology. This symposium is composed of the academic discussions, achievement exhibition, and expert seminars by means of which we are intending to achieve four goals: to organize comrades, to centralize thoughts, to energize spirits, and to formalize the future.

To organize comrades is to communicate with and unite those who are going in for the studies on Eco-translatology (including the theory of Translation as Adaptation and Selection). At present, our academic comrades include the authors who have published research papers on Eco-translatological research, the students who have completed their MA or Ph.D. theses in the theoretical framework of Eco-translatology, and the distinguished authorities in the international field of translation studies who have for a long time offered support and concern in our Eco-translatology research. Obviously, our academic team is to be enlarged in size because sufficient highly talented participants can serve as the primary precondition for the success of our course, as in other walks of life. It is exciting that most of the participants in the symposium are in their young or middle ages; some are even MA or Ph.D. graduates. This is the very vitality upon which Eco-translatology relies in development, which can mirror the splendid tomorrow for Eco-translatology.

To centralize thoughts means embracing various thoughts and reflections on relevant themes and subjects. The discussion topics in this symposium are covered with the theoretical framework, application and empirical study of Eco-translatology, with the relationship between Eco-translatology and other approaches of translation studies, with the application and development of the theory of Translation as Adaptation and Selection, with the MA or Ph.D. theses written from the perspective of Eco-translatology, and with other aspects of Eco-translatology research, such as the puzzles, reflections and foresight of Eco-translatology research, etc. Only by this means can we, as it were, have a deepened understanding and comprehension of Eco-translatology before we are strengthened with further motives to engage in Eco-translatology research.

To energize spirits is to encourage our Eco-translatology comrades for the better. As we know, Eco-translatology did not exist in the world until it came into being in China; meanwhile, it offers the latest comprehensively systematic perspective in translation studies, with a lot of new terms and concepts, as a self-sufficient norm. Although Eco-translatology research has been developed for a decade, it is still in a preliminary stage as a young branch of translation studies. Its theoretical system needs to be optimized, its present applied study deepened, its team members enlarged, its theoretical ideas and achievements exchanged, and its comprehension and recognition shared. All this should depend upon all of you, explorers, researchers, and supporters, who have been not only participating in its research with lofty spirits but also witnessing its progress with screwed-up courage. Be part of it; be proud of it! We are cheering ourselves on and contributing to the future research and advancement of Eco-translatology.

To formalize the future is to map a panorama for the strategic development of Eco-translatology. The second 10 years are coming for Eco-translatology development, and we are working out the developing goal of Eco-translatology in the second decade to quicken our pace in the development of Eco-translatology so that a comprehensive development will be well obtained in various aspects, such as the theoretical perfection, academic cultivation, staff organization, and exterior communication as well as basic construction, etc. With the general strategy of "Keep to China and march towards the world for sustainable development", we, on the one hand, have had such works published in recent years as *An Introduction to Eco-translatology, Theoretical Application and Commentary Analysis of Eco-translatology, Theoretical Thoughts and Conceptions of Eco-translatology, Academic Perspectives of Eco-translatology from Various Schools of Translation Studies,* and *Eco-translatology, Another Turn of Translation Studies*. On the other hand, we will establish and improve the working system of the International Association of Eco-translatology Research, open the website of IAETR (http://www.Eco-trans.international), run the *Journal of Eco-translatology*, and hold annual symposiums for Eco-translatological studies at the international or national level. More importantly, during the second decade, we will take the opportunity of Macao Polytechnic Institute's upgrade to the academic level of a university and try hard through joint programs to recruit 30 Ph.D. graduates and 15 post-docs as a reservoir of talent for the sustainable development of Eco-translatology. Only by this means can we gradually make China, and Macao in particular, the research center, information bank, and training base of Eco-translatology and make "the school of Eco-translatology" march forward steadily before it enjoys a secure position in international translation studies. We can go as far as our thoughts can fly. By then, Eco-translatology will surely take on a more spectacular look.

All in all, the four doctrines mentioned above, i.e., to organize comrades, to centralize thoughts, to energize spirits, and to formalize the future, may, so to speak, summarize the major goals of today's symposium.

Ladies and gentlemen, the first decade of the twenty-first century is receding, and during the period of 10 years, Eco-translatology research has been moving on and on through fantastic innovations as well as cool contemplations, through puzzles as well as solutions, through down-to-earth explorations as well as emotional explanations, and through fierce challenges as well as cheerful successes. As the initiator and promoter of Eco-translatology, I am overwhelmed with seas of emotions at the very moment. Anyhow, what I can say is but as follows: I want to say "thank you" sincerely not only to those who care about, direct, support, and participate in my cause but also to those who, in particular, offer persuasions, questions, or criticisms to my work and also to those who are my leaders, teachers, classmates, and friends as well as comrades. Ladies and gentlemen, I'd like to end my speech with a line from one of my favorite songs—*Since the way to my goal is long, let me march ahead on and on.* And what is more, I want to go all out for it with all of you, my noble academic comrades.

Eco-translatology research has a long way to go before fulfilling our great mission. Therefore, let's march ahead on and on, hand in hand, and shoulder to shoulder!

2. Next Decade: A New Start from Scratch—Hugs Opening Speech at the 2nd International Symposium on Eco-translatology (Shanghai, November 12, 2011)

Ladies and gentlemen,

Comrades and friends,

Good morning!

November is a beautiful and pleasant season in Shanghai. In this beautiful autumn season and in this pleasant coastal city, we are here to celebrate the 10th anniversary of Eco-translatology studies. At this moment, please allow me, on behalf of all Eco-translatological workers, to express our sincere gratitude to all scholars and specialists, teachers and students, translators and interpreters, advisors and supervisors who have made their contributions to the development of Eco translatology in the process of its formation, communication, application and justification.

The theme of this 2nd International Symposium on Eco-translatology is 10 Years of Eco-translatologyanslatological workers, t

A Review of Its 10-Year Development

It can be seen more and more clearly that for the past 10 years, Eco-translatology has experienced roughly three different stages.

The first stage can be said to be a period of initial exploration from 2001 to 2004, which is marked by the paper "An Initial Exploration into an Approach to Translation as Adaptation and Selection" presented at the 3rd FIT Asian Translators Forum in 2001 and by the monograph titled *An Approach to Translation as Adaptation and Selection* published in 2004.

The second stage can be viewed as a period of deep integration from 2005 to 2009, which is indicated by the paper "Understanding Eco-translatology" presented at the International Conference on Translating Global Culture: Towards Interdisciplinary (Re)Construction in 2006 and the paper "Eco-translatology: A Primer" published in *Journal of Chinese Translators* in 2008.

The third stage can be labeled a beginning period of systematic expansion since 2010, which is witnessed by the founding of the International Association for Eco-translatology Research, the holding of the 1st International Symposium on Eco-translatology in 2010, and the opening of the *Journal of Eco-translatology* (JET) in 2011.

What we are going to do here is to understand the present by reviewing the past. In this symposium, the 10 years of development will be reviewed, the 10 years of achievements analyzed, the 10 years of difficulties and challenges revisited, and the 10 years of research directions checked so as to deepen and broaden the Eco-translatological studies in the future.

An Important Milestone for Eco-translatological Development

Eco-translatology is developing steadily and progressively. At present, facts and figures show that much progress has been so far achieved on Eco-translatology. They are: (1) Over 100 papers on Eco-translatology are published both at home and abroad; (2) More than 60 MA theses and Ph.D. dissertations, as well as research papers from different universities, colleges and other institutions, have been completed and published adopting Eco-translatology as their theoretical framework; (3) A series of paper presentations and discussions are conducted at the "Cross Taiwan-Straits Symposiums on Translation and Intercultural Communication" by scholars from the Chinese Mainland, Taiwan, Hong Kong, and Macao; (4) A national research project, *Eco-translatology: Translation Studies from an Ecological Perspective*, and a university project, *A Study on Schools of Translation Scholarship: with Reference to the School of Eco-translatology*, etc. are under way; (5) Special columns are established in core journals in China, and special issues on Eco-translatology research will be included in refereed international journals; (6) Keynote speeches on Eco-translatology are presented at different national top language conferences and translation forums; (7) Over 10 book reviews on the monograph *Translation as Adaptation and Selection* are published domestically and internationally; (8) The "International Association for Eco-translatology Research (IAETR)" is officially instituted; (9) The "First International Symposium on Eco-translatology" was held in Macao in November 2010; and (10) The *Journal of Eco-translatology* (JET) is founded in both Chinese and English editions. In addition, such monographs as *Introduction to Eco-translatology*, *Eco-translatology: Applied Studies and Their Reviews* will be turned out yearly. The above figures and facts show that 10 years of studies on Eco-translatology can be viewed as an important milestone for Eco-translatological development.

A New Start for the Next Decade

A new decade will start. Let's start from scratch again. Looking back, we are proud of what has been done; looking forward, we feel the responsibility and pressure. We will work still harder.

We are very glad to see that the quantity and quality of submitted papers this time have surpassed those of the previous sessions; more and more young scholars,

Ph.D. and MA graduates have participated in this symposium; the studies of the ecological approach have become more and more profound and extensive. This is the embodiment of vigor, liveliness and sustainable development of Eco-translatology. These highly motivated studies on Eco-translatology will certainly promise its future.

Ladies and gentlemen, the past 10 years have witnessed both the initial exploration and steady development of Eco-translatology. As a Western saying goes, "History makes a man wise." We are sure that when we look back to see what has been achieved in the past, our comrades-in-arms will also be prepared to welcome the future in different ways.

There is also an ancient Chinese saying, "Every ten to zero." To my understanding, it means a new round of starting from scratch. This is the case for research and development in Eco-translatology.

Let's share pride in your eyes. We can fly to a new height.

3. Challenges • Opportunities • Endeavors—Hugs Opening address to the 3rd International Symposium on Eco-translatology (Chongqing, Nov. 24, 2012)

All distinguished guests,

Members of Eco-trans. international,

Ladies and gentlemen,

Good morning!

I have been looking forward to meeting you at this 3rd International Symposium on Eco-translatology to be held in Southwest University, Chongqing, China. Eventually today, we are here on this dynamic, beautiful and ecological campus. I also have been greatly benefited and encouraged by all the previous speeches.

Since the main theme of this symposium is Eco-translatology: Challenges and Opportunities, please allow me to express my viewpoints directly to the point.

As is known to all, difficulties and challenges cannot be overestimated in blazing a new trail, and the pioneers are always with a sense of crisis. This is also true for the R&D of Eco-translatology.

Difficulties and Challenges

There are difficulties and challenges at present. They include:

(1) A new thing at its beginning is usually not perfect. So is the theoretical construction of Eco-translatology. The entire system of theoretical discourse is still under construction.
(2) The studies on theoretical application are somewhat superficial.
(3) Some backbone researchers have not been fully engaged in Eco-translatological studies.

(4) The International Association for Eco-translatology Research needs to be administered more effectively, there is much room for improvement on the *Journal of Eco-translatology*, and the official website should have been working more effectively.
(5) A kind of study named "Translation Ecology", to a certain degree, causes some confusion and produces some negative effects on Eco-translatology.
(6) The progress of translation studies in China is, especially its theoretical research achievements, often overlooked by some Westerners.
(7) Some Chinese scholars also traditionally look down upon domestic productions.
(8) And the internationalization drive is still far from enough, etc.

Usually, the greater the challenges there are, the greater the achievements will be. In addition, opportunities and challenges always arise at the same time. Facts prove that the R&D of Eco-translatology are concurrently faced with favorable conditions and opportunities.

Conditions and Opportunities

There are conditions and opportunities at present. They include:

(1) The global megatrend of ecological vision.
(2) The full swing of strategic endeavors of letting Chinese culture go global.
(3) The frequent calls for giant strides from a "huge country of translation" to a "strong country of translation".
(4) The great progress made in Eco-translatology for the past decade, etc.

With a situation characterized by the coexistence of both challenges and opportunities, Eco-translatological researchers need to strive earnestly for its down-to-earth-way development:

Efforts and Endeavors

Facing the challenges and opportunities, and looking back at what has been done in the past over a decade, we have to be sober-minded, draw a blueprint of our future and work hard at it.

First, we will streamline the regulations for the association, the journal, the website, and the annual symposium.

Second, we will continue to carry out at least 20 research projects on Eco-translatology.

Third, we will plan to publish annually a related monograph in the field.

Fourth, we will speed up the process of internationalization of the new paradigm by publishing papers in international journals and holding our symposiums in different countries and regions.

Fifth, we will gather and train people of talent to carry forward the cause of R&D of Eco-translatology.

Dear fellow workers in the field of Eco-translatology:

We will surely prefer the sense of participation and the sense of pride—"Be part of it, and be proud of it!"

We will certainly cherish the sense of responsibility and the sense of mission—"I grow up together with the new area!"

Let us work together and make our due contribution to the advancement of Eco-translatology, in the spirit of "TEAM"—

***T**ogether, **E**veryone **A**chieves **M**ore!*

4. Doing Right Things Is More Important Than Doing Things Right—Hugs Opening Speech at the 4th International Symposium on Eco-translatology (Wuhan, September 12, 2013)

Distinguished guests,

All participants,

Ladies and gentlemen,

In such a fruitful and pleasant golden autumn, we are gathering here in this Wuhan metropolis, beside of the Yangzi River and at the foot of the fragrant Guizi Hill, for the concurrent holding of the 5th Cross-Taiwan Straits Symposium on Translation and Intercultural Communication and the 4th International Symposium on Eco-translatology. On behalf of the two organizing committees of the two symposiums, I would like to extend our warm welcome and greetings to all of you.

1. A Few Words About the Cross-Taiwan Straits Symposium

To facilitate exchanges and cooperation among the scholars of translation studies and intercultural communication studies from the Chinese Mainland, Taiwan, Hong Kong and Macao, and the greater China region, we started to organize a "Cross-Taiwan Straits Symposium on Translation and Intercultural Communication" in October 2004. It is a biannual meeting and held by turn in the Chinese Mainland, Taiwan, Hong Kong and Macao.

The ***1st*** symposium was held in Beijing in 2005, with Tsinghua University as the host. The ***2nd*** one in Macao in 2007, with Macao Polytechnic Institute as the host. The ***3rd*** one in Taiwan in 2009, with Fu-Ren Catholic University as the host. The ***4th*** one in Hong Kong in 2011, with Baptist University as the host. And this one, the ***5th*** symposium, is now being held in Wuhan in 2013, with Central China Normal University as the host.

In the past over 10 years, the symposiums held in series have become an effective communication platform for professionals from translation and intercultural communication studies from the four regions. In fact, the symposiums have been gradually recognized as a brand name in the field and enjoyed reputations overseas. Serving as records of the activities of the events, the proceedings of the symposiums in the series, which have been generously published by a prestigious publisher in China—Shanghai Foreign Language Education Press—also have become important reference books for translation teachers, postgraduates, theoreticians and practitioners.

With our 10 years of experience, and especially with the "get-together" of all four presidents of the translation associations of the four regions, we are going to discuss establishing the Cross-Taiwan Straits Association for Translation and Intercultural Communication so as to more effectively serve the scholars from the four regions.

By the way, the next Cross-Taiwan Straits Symposium on Translation and International Communication will be held in Taiwan in 2015. All are welcome. Please pay attention to our further conference announcements.

II. A Few Words About the International Symposiums on Eco-translatology

Eco-translatology is an emerging Eco-translation paradigm of translation studies from ecological perspectives. With metaphorical analogies between the translational ecosystem and the natural ecosystem and conceptual borrowings as its methodology, Eco-translatology probes translational eco-environments, textual ecologies, and "translation community" ecologies as well as their interrelationships and interplay. Regarding the scene of translation as a holistic ecosystem, it describes and interprets translation activities in terms of ecological principles of eco-holism, Eastern eco-wisdom, and *Translation as Adaptation and Selection*. Within the Eco-translation paradigm, "Translation as Eco-balance", "Translation as Textual Transplants", and "Translation as Adaptation and Selection" are taken as its core concepts.

On the basis of the research and development of over 10 years, we would like to call for raising the "three consciousnesses", namely, the consciousness of continued innovation, the consciousness of taking a lead, and the consciousness of further development.

By the so-called consciousness of continued innovation, we mean to develop and coordinate innovative studies on the basis of the previous original study.

By the consciousness of taking a lead, we mean to look for the developing trends, work out strategic plans, and raise the consciousness and ability to work in the academic frontier.

By the consciousness of further development, we mean to conduct down-to-earth studies on Eco-translatology, which has been specifically elaborated in the monograph entitled *Eco-translatology: Construction and Interpretation* to be published next month by the Commercial Press in Beijing. In short, the "three consciousnesses" can be said to be only one consciousness, that is, the consciousness of "marching on".

Well, by the way, the next International Symposium on Eco-translatology will be held in Macao in 2015. The general theme will be Eco-translatology: Eastern

Eco-wisdoms and Western Eco-philosophies. All are welcome. Please pay attention to our further conference announcements.

Ladies and gentlemen,

The concurrent holding of the 5th Cross-Taiwan Straits Symposium on Translation and Intercultural Communication and the 4th International Symposium on Eco-translatology can be regarded as a great event among translation and intercultural communication scholars and scholars for Eco-translatological studies. I sincerely wish the two symposiums a complete success!

To do right things is more important than to do things right.

That's all. Thank you all.

5. East-West Communication, Eco-translatology itself Improvement, and Two Developmental Orientations—Hugs Opening Speech at the 5th International Symposium on Eco-translatology (Tainan, June 26, 2015)

Distinguished scholars, ladies and gentlemen,

Good morning!

Congratulations

The year of 2015 witnesses the 15th year of Eco-translatological studies and marks the 5th anniversary of the founding of the International Association for Eco-translatology Research (IAETR, *Eco-trans International*). First of all, on behalf of the IAETR *Eco-trans International*, let me extend our congratulations for the successful opening of the 5th International Symposium on Eco-translatology.

Eco-translatology is an emerging Eco-translation paradigm of translation studies from ecological perspectives. With metaphorical analogs between the natural ecosystem and translational ecosystem, conceptual transplants, and interdisciplinary studies as its methodology, Eco-translatology probes translational eco-environments, textual ecologies, and "translation community" ecologies, as well as their interrelationships and interplay. Regarding the scene of translation as a holistic ecosystem, it describes and interprets translation activities in terms of the ecological principles of eco-holism, Eastern traditional eco-wisdoms, and *Translation as Adaptation and Selection*. Within the Eco-translation paradigm, "Translation as Eco-balance", "Translation as Textual Transplantation", and "Translation as Adaptation and Selection" are taken as the core concepts.

The Conference Theme: East-West Communication

The theme of this symposium is Eastern wisdoms and Western philosophies. Regarding this aspect, Eco-translatology can be taken as an example of pursuing Eastern and Western integration. As an emerging Eco-translation paradigm of translation studies, Eco-translatology is characterized by its academic pursuit of integrating the present understanding with the past one, linking the humanities/social sciences with the natural sciences, and combing the Eastern philosophical basis and ways of representation with the Western ones.

As far as the pursuit of Eastern and Western integration in the research and development of Eco-translatology is concerned, many specific embodiments can be found, for instance, the integration from the perspective of philosophical concepts (namely, the *unity between heaven and man* in the East and *holism* in the West). One more example is the aspect of theoretical bases (namely, Hugs' strong approach to *Translation as Adaptation and Selection* in the East and Charles Darwin's principle of *natural selection* in the West).

Facts show that simultaneous considerations of both Eastern and Western philosophies, cultures, theories, norms, terms and ways of expression, etc. in the construction of theoretical discourse systems in the field of translation studies will help understand, exchange and accept ideas/concepts between scholars of East and West. It is pointed out that the above academic pursuit of integrating the Eastern philosophical basis and ways of representation with the Western ones is not only the researchers' ceaseless pursuit but also a telling feature of R&D for Eco-translatology, and along with the Chinese culture "going global", it may reflect a trend of academic pursuit in general in the years to follow.

Improvement of Eco-translatology

The law of epistemology tells us that humans' epistemological knowledge goes from less to more, from lower to higher, ceaselessly, and towards perfection. This is also true for R&D of Eco-translatology. Although the theoretical discourse system of Eco-translatology has been primarily instituted, there are still many aspects that need to be further modified, improved, and perfected. This has been set as one of the tasks to be fulfilled within the next 5 years to come.

Two Developmental Orientations

On the basis of Eco-translatological studies over a decade or so, and centering around the abovementioned Eco-translatology itself improvement, two orientations towards the R&D for Eco-translatology have been gradually shaped. Specifically, we should orient Eco-translation studies towards the following two main directions of longitude and latitude. Longitudinally, Eco-translation can be developed towards the global "green theme", to be incorporated into human activities of ecological society (including environmental protection and translating "green revolution"), while latitudinally, or horizontally, Eco-translation studies can well interplay with other adjacent disciplines such as eco-language studies, eco-literature studies, eco-cultural studies, eco-critical studies, etc., which are intrinsically interactive with translation studies but are previously ignored in our studies as yet. Upon the longitudinal and latitudinal studies, an integrative and interdisciplinary study on Eco-translation can be conducted in light of ecological philosophies and eco-reasons.

Ladies and gentlemen, I am confident that upon the improvement of the theoretical discourse system of Eco-translatology, and especially upon the completion of the tasks set in the above two developmental orientations, Eco-translatology will certainly take on a new look.

"**Grit is sticking with your future.**"

We'll expect to see it. We'll work hard at it.

That's all. Thank you all.

6. Eco-civilization, Eco-translation, and New Orientations of R&D for Eco-translatology—Hugs Opening Address at The 6th International Symposium on Eco-translatology and The 2nd Chinese Ph.D. Forum on Eco-translation Studies (Zhengzhou, Oct. 27, 2018)

Prof. Zhang Qianhong,

Prof. Yang Ping,

Prof. Christina Nord,

Prof. Michael Cronin,

Dear fellow workers in the field of Eco-translatology,

Ladies and gentlemen,

Good morning!

Following the five successful symposiums organized in the past 8 years, the 6th International Symposium on Eco-translatology is being held in the "Central Plain" of China—Zhengzhou, the capital of Henan Province.

Since the theme of this symposium is New Developments and New Orientations of Eco-translatology, I would like to talk briefly about eco-civilization, Eco-translation, and new orientations of R&D for Eco-translatology.

In the course of social development, human societies have undergone three historical periods, namely, "primitive civilization", "agricultural civilization" and "industrial civilization", and are now in the stage of "ecological civilization" (eco-civilization). Different from the previous three periods, eco-civilization features transcending nations, countries and social classes and has reached a higher stage of civilization.

Without any doubt, great contributions have been made through translations in the previous human civilizations. However, in the new period of eco-civilization, what opportunities have been brought to translation studies? What can translation studies do to meet the needs of this new historical period? In the new times, what are new points of departure to enrich translation studies? How can translation studies make new contributions to the development of human eco-civilization? All these questions need to be answered by us translators and researchers in the field of translation studies.

More than 17 years ago, that is, at the beginning of the present new century, we started our Eco-translation studies. The next year, 2019, will be the rite of passage into the "Adulthood" of Eco-translatology.

Instead of celebrating the 18-year-old "Adult Ceremony", we should pay much attention to what we are going to do at present.

We are going to improve, upgrade and optimize our theoretical discourse system construction of Eco-translatology to increase its practicality, explainability, and acceptability. Some of the deliberations and ideas will be presented, illustrated, exchanged, and discussed in our following two conferences.

We will set up a database through which we can offer services to scholars and students, both domestically and internationally, for their studies in the field of Eco-translatology. Suggestions and comments are welcomed to meet the end.

We are going to make the eco-paradigm of translation studies known by more and more people, in and out of the circles of translation studies, by way of paper publications, continuous conferences and seminars, and our international research centers, branches, and other set-ups.

Dear fellow workers in the field of Eco-translatology:

We will surely prefer the sense of participation and the sense of pride—"Be part of it, and be proud of it!"

We will certainly cherish the sense of responsibility and the sense of mission—"I grow up together with the new area!"

Let us work together and make our due contribution to the advancement of Eco-translatology in the spirit of "TEAM"— because

Together, ***E***veryone ***A***chieves ***M***ore!

Thank you all.

III. 3. Figures and Facts of R&D in Eco-translatology (Based on *EcoTrans NewsReel* published by the Secretariat of the International Association for Eco-translatology Research, No. 2, 2018)

- **Academic Papers**—Over 2000 articles published in and outside China (mainly in Chinese). Among them, over 600 are MA theses and Ph.D. dissertations inspired by Eco-translatology from over 150 universities and higher educational institutions.
- **Monographs and books**—Over 10 monographs and books published, including *An Approach to Translation as Adaptation and Selection* (HU, 2004); *Translator's Adaptation and Selection: Studies of Translation Process in International Media Communication* (LIU, 2010); *An Ecological Study on Translated Literature Systems in Contemporary China* (LU, 2012); *Eco-translatology: Construction and Interpretation* (HU, 2013); *Eco-translation of Public Signs* (YUE & YU, 2014); *Eco-translatology and Its Different Applications* (HAN & GUO, 2015); *System Construction of Eco-translation Criticism* (YUE & YU, 2016); *Eco-translatology and Literary Translation* (JIA & YU, 2017); and *Cultural*

Translation of Minority Nationalities for Global Communication: An Eco-translatological Perspective (ZHANG, 2018).

- **International symposiums**—6 International Symposiums on Eco-translatology organized since 2010, with over 500 scholars and presenters from different continents of the world.
- **Scholarly journals**—*Journal of Eco-translatology* started its first issue in 2010; *Studies on Eco-translatology* published with ISBN in 2014, and *Journal of Eco-translation Studies* (ISSN 2520-7911) in 2017.
- **Research projects**—Over 80 research projects were conducted in light of Eco-translatology at different levels (university, provincial and national).
- **The International Association and its branches and centers**—The International Association for Eco-translatology Research (IAETR) was founded in 2010 and headquartered in the Macao Special Administrative Region, People's Republic of China. Professor Hugs, HU Gengshen, is the founding President and Professor Michael CRONIN is Vice President of IAETR; Dr. Rindon KUNDU and Dr. Yara ISMAIL are Executive Directors of the South Asia Branch, North Africa and Arab Branch of IAETR. Prof. Shelly OCHS serves as Executive Director of the North America Committee of Eco-translation and Chinese Medical Cultural Communication, IAETR. In addition, with other centers and institutes established across China, Zhengzhou University Academy of Eco-translatology was set up recently in the central plains of China, where national and international Ph.D. students and postdoctoral researchers can be enrolled for advanced studies. In addition, the National Society for Doctoral Students in Eco-translatology is also founded and headquartered in Hangzhou, China.
- **International Seminars**—Since 2013, various seminars, lectures, workshops, and forums on Eco-translatology have been organized in different countries and regions, such as related events taking place at the Royal University of Technology, Melbourne, Australia (2014); University of Helsinki, Finland (2015); University College of London, U.K. (2016); and Kent State University, U.S.A. in 2018.
- **Communication channels**—In addition to the *Journal of Eco-translation Studies* and the International Symposiums on Eco-translatology in series, the IAETR has its website, www.eco-translatology.com; its newsletter, *EcoTrans NewsReel*; and its WeChat public identifier, *EcoTrans International*, to facilitate internal and external communications.

生态翻译学学刊
Journal of
Eco-Translatology
国际生态翻译学研究会会刊(中文版)
International Association for Eco-Translatology Research
(Chinese Edition)
No.1 (Serial No. 1) 2011

Events & Activities in Branch Associations
世界各地分会活动

Postscript

There is a saying that it is easier to criticize and complain than to act and practice in translation circles.

We cannot find fault with this claim if we see it from the perspective of the history of translation theory. The critics and complainers should be thanked because their remarks lead us to become aware of the overlooked dimensions in translation scholarship that deserve our attention and efforts in research. In this sense, critical comments turn out to be the impetus for scholarship in general and translation studies in particular.

In China, there have been many calls for further research on the theoretical study of translation, especially on filling in the blanks in pure theories, as well as appeals for a dialogue between China and the West on an equal platform and for new breakthroughs in translation studies. In this vein, what I have been doing can be regarded as a response to the appeal or as an explorer who made a discovery in the field of translation studies. Therefore, I am not ashamed of being called the initiator of Eco-translatology, but I would rather be seen as a practitioner to explore Eco-translatology.

This book offers a panoramic view of the emerging eco-paradigm of translation studies known as Eco-translatology and presents a systematic study of the theoretical discourse from ecological perspectives in the field of translation studies. Eco-translatology describes and interprets translation activities in terms of the ecological principles of eco-holism, traditional Eastern eco-wisdom, and "Translation as Adaptation and Selection." Furthermore, Eco-translatology approaches the phenomenon of translation as a broadly conceived ecosystem in which the ideas of "Translation as Adaptation and Selection," as well as translation as "textual transplantation" promoting "eco-balance," are integrated into an all-encompassing vision. Finally, Eco-translatology reinforces contextual uniqueness, emphasizing the deep embeddedness of texts, translations, and the human agents involved in their production and reception in their own habitus. It is particularly encouraging, in this increasingly globalized world, to see a new paradigm sourced from East Asian tradi-

(Hugs) G. S. Hu, *Eco-Translatology*, New Frontiers in Translation Studies,
https://doi.org/10.1007/978-981-15-2260-4

tions but with universal appeal and applications, which adds to the diversity and plurality of global translation studies.

In general, a monograph is based on the collection of separate papers that were published earlier, so the book has the advantage of presenting a more panoramic view than individual papers; that is, it may lead one from leaves to trees and to forests. More importantly, one's understanding of the world and pursuit of the truth must undergo a long process of gradual improvement and perfection. All my articles and arguments in more than 10 years have been condensed into this volume, *Eco-Translatology: Towards an Eco-paradigm of Translation Studies*, which offers a decisive definition and interpretation of the technical terms and concepts of Eco-translatology.

Upon the completion of this manuscript, the author would like to thank the following institutions and periodicals for their financial support and facilities: the Chinese National Social Science Foundation; Tsinghua University; Macao Polytechnic Institute; City University of Macao, Chinese Translators Journal; Shanghai Translators Journal; Jouranal of to Foreign Languages; Foreign Language Research; Foreign Language Teaching; Foreign Language Teaching and Research; Foreign Language and Literature; Foreign Language Research and Teaching; Journal of Eco-Translatology; Hubei Education Press, Shanghai Foreign Language Education Press; Foreign Language Teaching and Research Press; and Commercial Press. The author is also thankful to two old friends, Professor Mengzhi Fang and Professor Cay Dollerup, for their graceful prefaces to this book.

Although the author has completed the tasks set by the first phase, this does not mean an end to the research; in fact, it is rather a beginning of a new adventure in translation studies. The author believes that Eco-translatology will continue to advance and progress. Specifically, the tasks set by the second phase will be encountered on the way—the developments and breakthroughs including putting forward and elaborating the guiding principle of Neo-ecologism, letting the text to "trans-live and re-live" from the source-text ecology to the target-text ecology, advocating green translation, and (more culturally and socially) linking Eco-translation studies with eco-civilization constructions in some of which have been underway since 2013, and all of which will be turned out in the author's next book.

Similar to other disciplinary research, Eco-translatology is not without limitations and incompleteness. With confidence, however, the author hopes to work with those who share similar interests for the healthy development of the Eco-translatology in terms of "ABD": altitude, breadth, and depth.

A calm mind leads one to achieve ambition; a steady movement brings one far.

MIX
Papier aus verantwortungsvollen Quellen
Paper from responsible sources
FSC® C105338
FSC
www.fsc.org

If you have any concerns about our products,
you can contact us on
ProductSafety@springernature.com

In case Publisher is established outside the EU,
the EU authorized representative is:
Springer Nature Customer Service Center GmbH
Europaplatz 3, 69115 Heidelberg, Germany

Printed by Libri Plureos GmbH
in Hamburg, Germany